Adaptive Systems: Modeling and Applications

Edited by **Lucas Lincoln**

Willford Press

New York

Published by Willford Press,
118-35 Queens Blvd., Suite 400,
Forest Hills, NY 11375, USA
www.willfordpress.com

Adaptive Systems: Modeling and Applications
Edited by Lucas Lincoln

International Standard Book Number: 978-1-68285-204-0 (Hardback)

Printed in China.

Contents

Preface

The concept of adaptive system has been significantly researched upon in the recent past. This book compiles researches of renowned experts from across the globe exploring new applications of adaptive systems. Segregation mechanisms, performance modeling, applications of complex networks, agent-based models, etc. are some of the topics that have been thoroughly looked at. The aim of this text is to present topics that have transformed this discipline and aided in its advancement. As this field is emerging at a rapid pace, the contents of this book will help students and research scholars pursuing this field to better understand the modern concepts and applications of the subject.

After months of intensive research and writing, this book is the end result of all who devoted their time and efforts in the initiation and progress of this book. It will surely be a source of reference in enhancing the required knowledge of the new developments in the area. During the course of developing this book, certain measures such as accuracy, authenticity and research focused analytical studies were given preference in order to produce a comprehensive book in the area of study.

This book would not have been possible without the efforts of the authors and the publisher. I extend my sincere thanks to them. Secondly, I express my gratitude to my family and well-wishers. And most importantly, I thank my students for constantly expressing their willingness and curiosity in enhancing their knowledge in the field, which encourages me to take up further research projects for the advancement of the area.

Editor

1

Segregation mechanisms of tissue cells: from experimental data to models

Előd Méhes[1*] and Tamás Vicsek[1,2*]

* Correspondence: emehes@angel.elte.hu; vicsek@angel.elte.hu
[1]Department of Biological Physics, Eötvös University, Budapest, Hungary
[2]Statistical and Biological Physics Research Group of the Hungarian Academy of Sciences, Budapest, Hungary

Abstract

Considerable advance has been made in recent years in the research field of pattern formation by segregation of tissue cells. Research has become more quantitative partly due to more in-depth analysis of experimental data and the emergence modeling approaches. In this review we present experimental observations, including some of our new results, on various aspects of two and three dimensional segregation events and then summarize the computational modeling approaches.

Keywords: Segregation, Cell, Culture, Sorting, Motility, Differential adhesion, Development, Model, Simulation

Review

Introduction

Pattern formation by concerted motion of large populations of different cell types is a ubiquitous phenomenon in several biological events, such as embryonic development of various species as well as tumor formation and even with some impact in tissue engineering. Patterns can form as a response by cells to external guidance cues such as morphogens or chemotactic substances or as a process where instead of external cues the local cell-cell interactions and the inherent mechanical or motility characteristics of the cell types will give rise to various multicellular patterns through physical segregation (or sorting) of these cell types. In this review we collect a naturally incomplete list of experimental observations on various aspects of 2- and 3-dimensional segregation events, and then summarize, where available, some computational models on segregation and compare them with experimental findings.

Experimental data on segregation

Two-dimensional segregation: pattern formation on a substrate

Basic drives and mechanisms of pattern formation events taking place e.g. during embryonic development can be studied in simplified experimental systems where complexity is reduced and the events are more accessible for quantitative analysis.

In 2-dimensional co-cultures of adherent cells on a rigid substrate Méhes et al. (2012) studied the dynamics of segregation of two initially mixed cell populations into distinct clusters by cell migration in an environment lacking pre-defined external cues. They have found that segregation into large multicellular clusters is facilitated by collective effects in cell motion such as an increase in the directional persistence of constituent cells. The growth of such multicellular clusters by consecutive fusion of smaller clusters follows algebraic scaling law with characteristic exponents depending on the collective effects (Figure 1 also see Additional file 1). The growth exponent values measured in this cell culture system with self-propelled collective motion exceed the exponent values resulting from computer simulations with diffusively moving segregating units detailed in (Nakajima & Ishihara 2011).

Inducing directional migration in the form of migration down a cell density gradient during the segregation of initially mixed co-cultures had no considerable impact on the dynamics of segregation. Clusters of similar size formed with similar dynamics as compared to mixed co-cultures with no cell density gradient (Figure 2).

Segregation events in three dimensions: gastrulation and tissue organization

Three-dimensional segregation of cell populations is most prominent during gastrulation, the early phase of embryonic development resulting in the formation of main germ layers that later on give rise to all tissues. Gastrulation is a spectacular event under the microscope involving collective motion of large number of cells, but although gastrulation events have been known since early embryonic works at the beginning of the 20th century, the basic mechanisms that provide for both its accuracy and robustness are just being uncovered. Segregation of cell populations adopting different cell fates into distinct domains is governed by their mechanical properties and active motion, and constitutes an important driving mechanism of gastrulation and tissue organization. Segregation is also important in other embryonic processes, such as

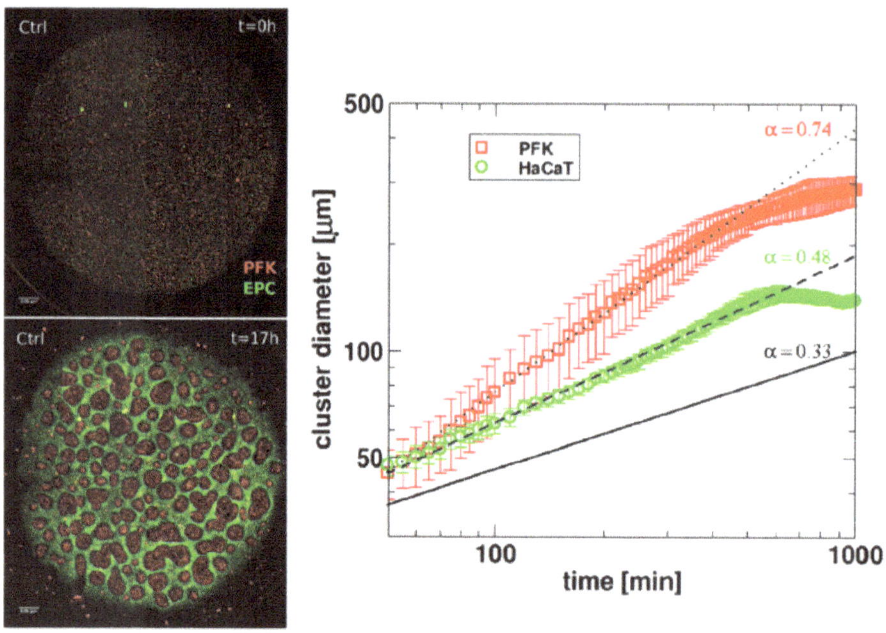

Figure 1 Dynamics of 2-dimensional segregation of keratocytes in culture. Left: Segregation in mixed co-cultures of primary goldfish keratocytes (PFK, red) and EPC fish keratocytes (EPC, green), consisting of >250 000 cells. Upper panel shows initial stage after cell attachment, lower panel shows final stage after 17 hours of cell migration. Scale bar is 100 μm. An additional movie file shows the dynamics of this in more detail [see Additional file 1]. Right: Average cluster diameter growth curves calculated from experiments with primary goldfish keratocytes (PFK) or human keratocytes (HaCaT). Exponent values obtained from fitting straight line segments to experimental curves are shown. Cluster growth curve of simulated segregation of cells without collective motion characterized by exponent value α = 0.33 is shown for reference (black solid line). From Mehes et al. (2012).

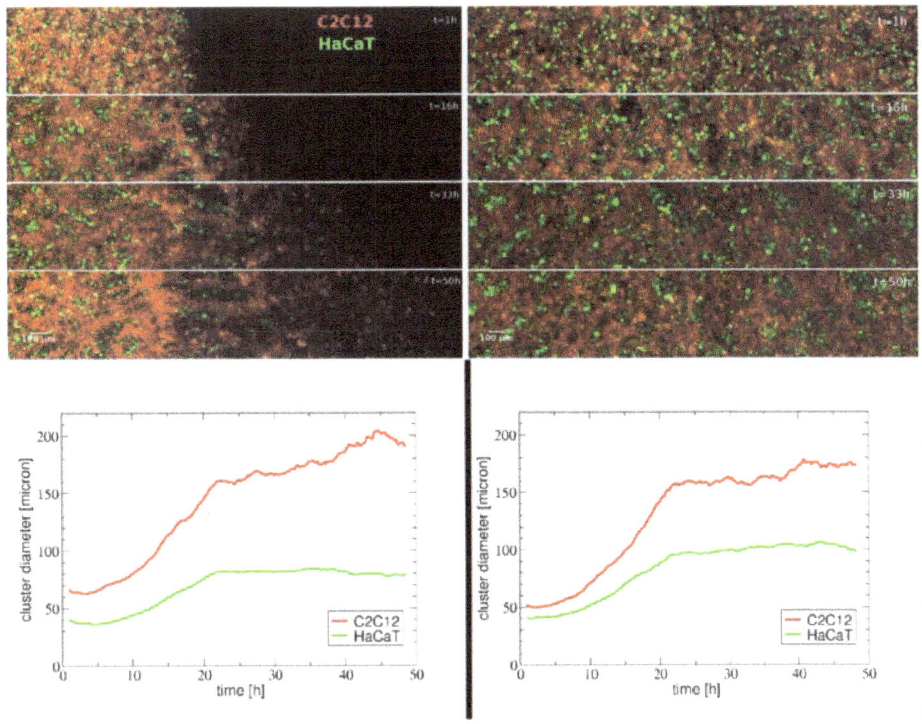

Figure 2 Impact of cell density gradient on segregation in co-cultures. Upper panel: Time-sequence images showing the process of segregation in mixed co-cultures of C2C12 mouse myoblasts (red) and HaCaT human keratocytes (green). Left: A co-culture with cell density gradient evoking directional migration from the dense monolayer toward the cell-free space to the right. Right: A similar co-culture with homogeneous cell density. Lower panel: Growth of cluster sizes calculated from two-point correlation measurement. The curves and line colors correspond to co-cultures shown in the upper panel.

blastocyst formation in mammals or the development of the initial blood vessel network in birds, where precursor cells differentiate in a spatially random pattern and they have to find their way to join and form an organ.

Cell segregation was first demonstrated by the experiments of Townes and Holtfreter in which presumptive neural and epidermal cells were isolated from amphibian gastrulas, subsequently they were mixed and they autonomously sorted into separate tissues (Townes & Holtfreter 1955). In similar early experiments, mixed cells isolated from the adult Hydra were shown to segregate and form separate tissues (Gierer et al. 1972).

Segregation of various cell types in 3-dimensions was studied is several works (Foty et al. 1996; Beysens et al. 2000; Foty & Steinberg 2005; Krieg et al. 2008; Schötz et al. 2008), aiming to explain the observed configurations of segregated domains, typically the envelopment of one cell type by the other, evolving from an initial mixture of cells. These in vitro segregating systems are considered to be analogous to non-mixing liquids and their segregation is shown to be driven by differences in tissue surface tension (TST) of the constituent cell types (Foty et al. 1996). Several works tested the contribution of cell-cell adhesion (Foty & Steinberg 2005; Schötz et al. 2008) and cell cortex tension (Krieg et al. 2008; Maître & Heisenberg 2011) to TST. The developmental aspects of cell sorting and TST are detailed in a recent review by Krens and Heisenberg (2011).

Experimental proof on the relative weights of adhesion and cell cortex tension in controlling cell-cell contact formation in zebrafish germ layer progenitors and determining the experimentally measurable separation force between cell pairs was provided by Maître et al. (2012). Cells are described as fluid objects with viscoelastic cortex under tension and adhesive bonds maintaining cell-cell contacts. Contact expansion is controlled by cell cortex tension at the contact, generated by myosin-2 activity, while adhesion by cadherins (membrane-spanning adhesion molecules) mechanically couples the adhering cells, and such coupling is limited by cadherin anchorage to the sub-membrane cortex. Cell segregation is mainly driven by contact formation by the active reduction of cell cortex tension at cell-cell interface, which leads to decrease in cell-cell interface tension, while cell cortex tension at the cell-medium interface is maintained, accounting for maintained TST. Adhesion is shown to have little direct function in contact expansion. Considering the typical cadherin density, the adhesion energy per unit area of the cell surface ($\sim 1 \times 10^{-7}$ N/m) is several orders of magnitude lower than typical TST measured in cell aggregates (being in the order of 1×10^{-3} N/m) (Amack & Manning 2012).

Emerging hypotheses

Two opposing hypotheses have been developed for explaining the origin of tissue surface tension. One is the differential adhesion hypothesis (DAH), developed by M. Steinberg (1962a); (1962b); Steinberg 1963; Steinberg 1996) postulating that tissue surface tension is proportional to the intensity of adhesive energy between point object cells. This hypothesis was elaborated in extensive modeling approaches by J. Glazier (1992). Experimental studies showed that TST is proportional to cadherin levels (Foty & Steinberg 2005).

The other hypothesis is the differential interfacial tension hypothesis (DITH), developed by Harris (1976), Brodland (2002; Brodland 2003) and Graner (1993), postulating that tissue surface tension arises from cortical tension of individual cells generated by actomyosin contractility, while a cell's mechanical energy changes with cell shape. This model was also supported by experimental data on cell cortex tension and TST (Krieg et al. 2008).

A model integrating cell-cell adhesion and contractility of cell interfaces in the generation of tissue surface tension, the driving force of cell segregation and tissue spreading, was provided by Manning et al. (2010). This model specifies an explicit relationship between surface tension and the ratio of effective adhesion (γ) to cortical tension (β). Effective adhesion, γ, is the net energetic contribution of contacting cell surfaces, which depends on the free energy of cadherin bonds and local changes in cell cortex tension. Surface tension exhibits a crossover at $\gamma/\beta \sim 2$ from adhesion-dominated behavior (DAH) in the regime of $\gamma/\beta < 2$ to a dependence on cortical tension and other mechanical effects in the regime of $\gamma/\beta > 2$.

A recent review emphasizes the role of boundary cells in TST as they can actively change their mechanical properties generating different cortical tensions along their internal and external interfaces. Such 'mechanical polarization' is suggested to exert the same net mechanical effect on the tissue as if extra adhesion was introduced among all cells and it is hypothesized to dominate TST instead of the mechanical energy of adhesive bonds (Amack & Manning 2012). Strong apical-basal actin polarization was indeed shown in surface cells in zebrafish embryonic explants (Krens et al. 2011). Considering the low adhesion energy of cadherins, the findings that TST is proportional to the number of surface cadherins (Foty & Steinberg 2005) can also be interpreted in a way that it is signaling through more cadherins leading to increased actomyosin contractility and resultant cell cortex tension which generates higher TST.

Experimental observations

The dynamics of growth of segregated domain size in 3-dimensions was studied by Beysens et al., (2000) using mixed cultures of embryonic pigmented epithelial cells and neural retinal cells, which segregated and formed enveloped structure over time in a configuration determined by surface tensions of the cell types. As a comparison, the segregation of gas and liquid phases was studied under microgravity resulting in similar segregated configuration determined by surface tension (Figure 3). Authors have found a linear increase in the size of segregated cell domains in time.

In a study quantifying the adhesive and mechanical properties of zebrafish germ line progenitor cell types Krieg et al. (2008) investigated the role of tensile forces in cell segregation. Using single-cell force spectroscopy they have measured the cell-cortex tension of these cell types (ectoderm, mesoderm and endoderm) while specifically interfering with actomyosin-dependent cell-cortex tension. Performing segregation experiments using cell types with altered myosin activity they have demonstrated that differential actomyosin-dependent cell-cortex tension is required and sufficient to direct the segregation of cell types and determine the final configuration of the segregated domains.

The dynamics of 3-dimensional segregation of mixed germ line progenitors of the zebrafish was studied by Klopper et al. (2010). As segregation proceeds in this system, the domain consisting of mesoderm cells gradually engulfs the ectoderm domain, which

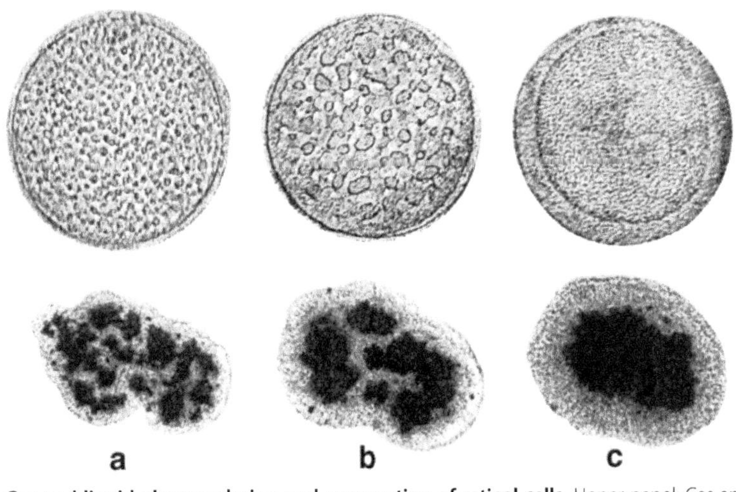

Figure 3 Gas and liquid phase ordering and segregation of retinal cells. Upper panel: Gas and liquid phase ordering in SF6 under reduced gravity, after a thermal quench of 0.7 mK below the critical point (45.564 C). Gas and liquid eventually order with the liquid phase wetting the container wall and surrounding the gas phase, corresponding to wall liquid interfacial tension $<$ wall-gas interfacial tension. **a**, **b** and **c** correspond to 120 s, 275 s and 3960 s after quench, respectively. Lower panel: Sorting out of chicken embryonic pigmented epithelial cells (dark) from chicken embryonic neural retinal cells (light). The average aggregate size is 200 μm. At the end of sorting, neural retinal cells preferentially wet the external tissue culture medium surrounding the aggregates. Medium-neural retina and medium-pigmented epithelium interfacial tensions are 1.6 dyne/cm and 12.6 dyne/cm, respectively. *a*, *b* and *c* correspond to 17 h, 42 h and 73 h after initiation of sorting, respectively. From Beysens al. (2000) with permission of Proc. Natl. Acad. Sci.

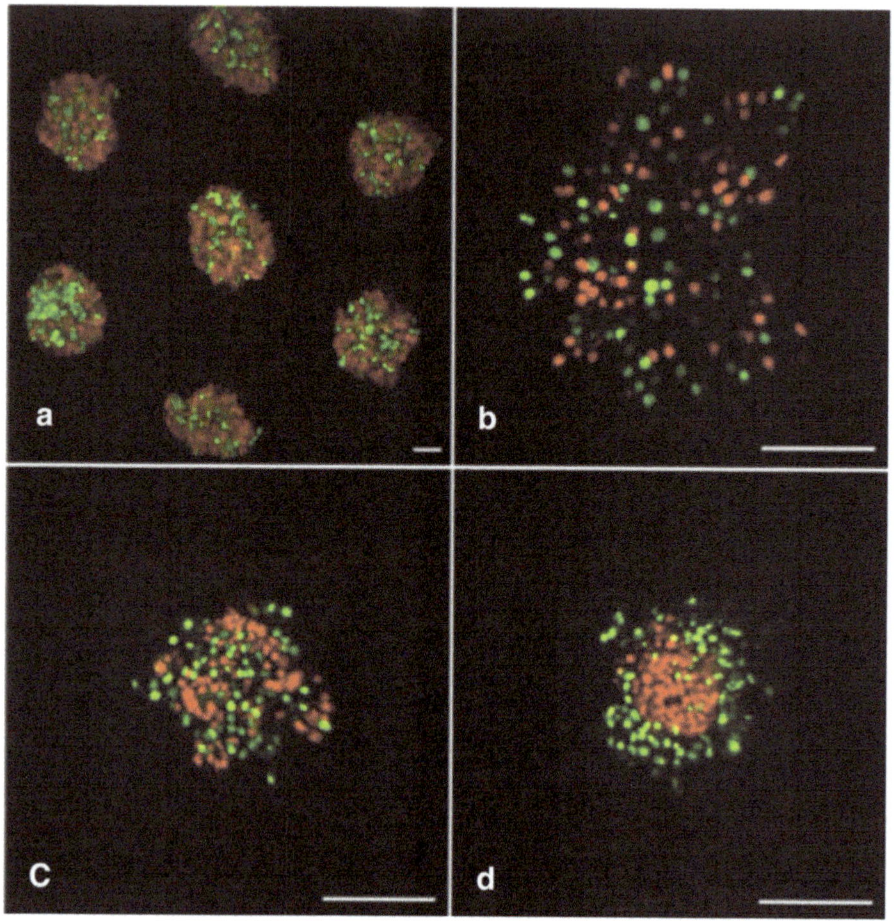

Figure 4 Imaging data for different time points in a segregation experiment with zebrafish ectoderm and mesoderm cells in culture. (**a**) Micro-molds are used to isolate small populations of ecto- and mesoderm cell mixtures labeled fluorescently with red and green nuclei, respectively. (**b**) Initial images show homogeneously mixed cells distributed throughout the mold. (**c**) Cells aggregate together on a time scale of roughly 100 minutes. (**d**) Imaging after sorting clearly shows the segregation of the two cell populations. Scale bar = 100 micron. From Klopper et al. (2010) with permission of Eur. Phys. J. E.

eventually takes the inner position (Figure 4). Authors have monitored the dependence of the local segregation order parameter on system size and found algebraic scaling and different characteristic exponent values for enveloping and engulfed cells.

In a similar in vitro system composed of two mixed epithelial cell types suspended in micromolds, Vicsek and coworkers have recently studied the dynamics of 3-dimensional segregation (manuscript in preparation). The two keratocyte types, primary goldfish keratocytes and EPC fish keratocyte cell line, were labeled with different fluorescent cell tracers and mixed in equal volume ratio. The mixed co-culture was transferred in agarose micromold wells and the ongoing segregation process was imaged by time-lapse videomicroscopy with optical sectioning. Consecutive optical sections were used for reconstructing the evolving 3-dimensional structure of segregated domains as well as analyzing the growth of clusters by two-point correlation method yielding characteristic cluster diameter values. Figure 5 shows snapshots of the evolution of segregation where the forming domains are adjacent and unlike zebrafish germline progenitors there is no engulfment of one domain by the other (see Additional file 2). It was found that the

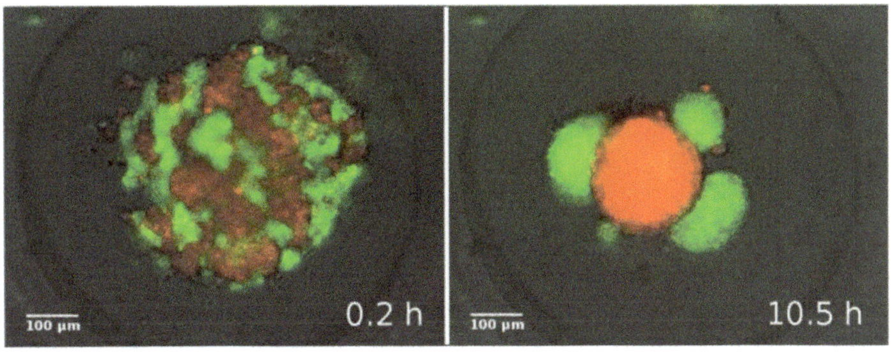

Figure 5 Snapshots from a segregation experiment with two keratocyte types in culture. Left: Initial mixture of primary goldfish keratocytes (red) and EPC fish keratocytes (green) in a micro-mold. Right: Homotypic cell clusters formed through segregation. An additional movie file shows this in more detail [see Additional file 2].

growth of segregated domain size follows algebraic scaling law and it is fast, typically completed within 6 hours. These observations are in contrast with earlier simulations (Belmonte et al. 2008) suggesting a much slower process (discussed in the section "Segregation by collective motion and adhesion").

In early embryonic development, during the blastocyst formation in mice, cell fate specification of epiblast and primitive endoderm cells occurs in a spatially mixed way, followed by their active segregation into distinct domains (Morris et al. 2010). Experimental studies of this process have revealed that segregation cannot be attributed to differential adhesion only but other processes such as active motion and apical-basolateral polarization of surface endoderm cells eventually surrounding the epiblast cells appear to contribute to it (Moore et al. 2009). It is also proposed and that repositioning of cells in turn influences their fate.

Pattern formation by segregation is a process that is not confined to embryonic development. In a recent publication Inaba et al. (2012) studied the formation of skin pigment patterns in the adult zebrafish. They have demonstrated that segregation of the two pigment cell types eventually forming the stripe pattern is governed by their short-range repulsive electric interactions that spatially orient their migration.

Models on segregation

Several computational models exist that attempt to explain and reproduce the experimentally observed segregation processes in various systems. While experimental works tend to focus on the role of tissue surface tension and adhesion in 3-dimensional cell segregation, modeling approaches also include autonomous cell motion and focus on 2 dimensions. A widely accepted model based on the Potts model and the idea of differential cell adhesion was developed by J Glazier and co-workers, later termed as Glazier-Graner-Hogeweg model or Cellular Potts Model (Graner & Glazier 1992). Variants of this model have been successfully employed in simulation works.

Impact of motility on segregation

Dynamic segregation in 2-dimensions was studied by Kabla (2012) using Cellular Potts Model simulations with self-propelled motile and non-motile cells characterized by

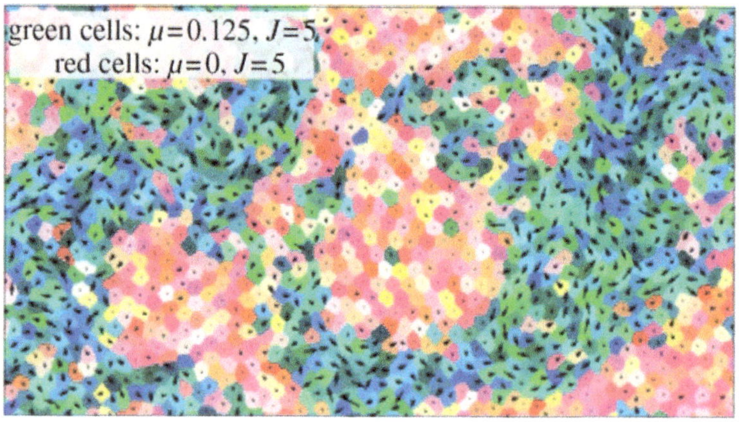

Figure 6 Simulated segregation of motile and non-motile cells. A snapshot of the simulated segregating tissue of motile and non-motile cells at $t = 10^6$ MCS (Monte-Carlo steps). Membrane tension, *J*, and motile force, *μ*, of cells are indicated. From Kabla (2012) with permission of J. R. Soc. Interface.

identical adhesive properties. Segregation efficiency has been found to depend on the motile forces controlling cell speed, and efficiency reaches maximum at motile forces close to the threshold required for streaming transition. It is also shown by these simulations that differences in motility are sufficient to drive the segregation of cell populations even without difference in adhesion and as a result motile cells will surround the islands of non-motile cells (Figure 6).

Using Brownian dynamics simulations McCandlish et al. (2011) studied dense mixtures of self-propelled and passive rod-like particles in 2-dimensions where only excluded volume interactions can occur. Adhesion properties do not play a role here, particles only differ in motility. Spontaneous segregation of the two particle species generates a rich array of dynamical domain structures with properties depending on particle shape and propulsion velocity or the combination of these two in the form of particles' Péclet number, a measure similar to the directional persistence of real cells.

Recently, Nakajima et al. (2011) used Cellular Potts Model (CPM) to study the dynamics of the segregation of mixtures of non-self-propelled cell types with diffusive motion. They have found that the increase in the size of segregated domains follows power law and the growth exponent is n = 1/3 for mixtures with 1:1 initial ratio of cell types where segregation proceeds via smoothing of the domain boundary. This is in contrast with previous works with CPM on smaller simulated systems displaying slower logarithmic growth for domain size (Graner & Glazier 1992; Mombach et al. 1995). CPM simulations with self-propelled cell types characterized by identical adhesive interactions as for the simulations in (Nakajima & Ishihara 2011) also yield domain growth exponent n = 1/3 (A. Czirók, personal communication).

Impact of adhesion on segregation

The role of adhesion in cell segregation was studied by Zhang et al. (2011) using Cellular Potts Model for simulations. In their model they consider variations in the distribution of adhesion molecules per cells. The speed of segregation is found to increase strongly with interfacial tension that depends on the maximum difference in the

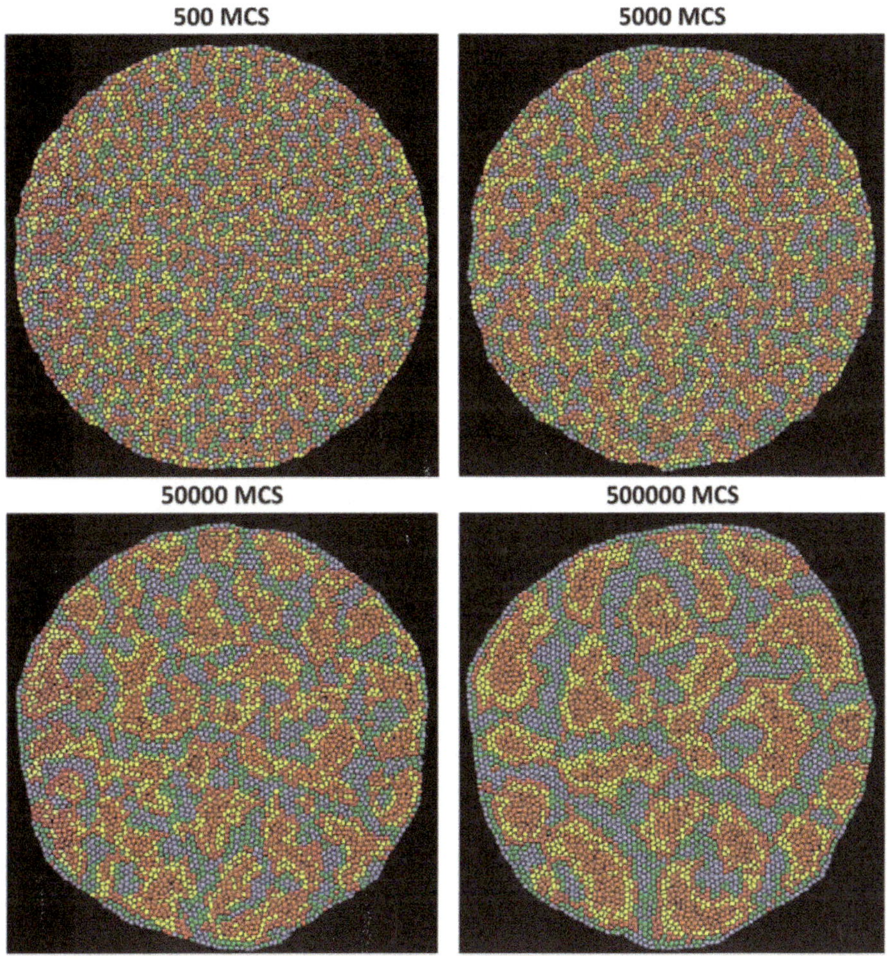

Figure 7 Clustering dynamics of cells with different adhesion characteristics. Snapshots taken from a 5000 cell aggregate simulation with five levels of cadherins showing the dynamics of cluster formation. Time points are denominated as Monte-Carlo steps (MCS). From Zhang et al. (2011).

number of cadherin adhesion molecules per cell and the reaction-kinetic models of cadherin binding (Figure 7).

Qualitative description of the dynamical features and the geometry of cell segregation depending on intercellular adhesion parameters is provided by Voss-Böhme et al. (2010) using a stochastic interacting particle model. In this model the hierarchy of segregation is determined by the strengths of adhesive interactions between cells and the boundary.

In a unique paper combining experimental data and modeling Krieg et al. (2008) studied the role of cell-cortex tension and adhesion in segregation. Carrying out simulations using Cellular Potts Model with cell adhesion and cell-cortex tension data derived from experiments they could reproduce the experimentally observed final configurations of segregating germ line progenitor cell types.

Segregation by collective motion and adhesion

To study cell sorting events Belmonte et al. (2008) developed a model combining the Vicsek Model of collective motion (Vicsek et al. 1995; Vicsek & Zafeiris 2012) with the differential adhesion hypothesis (DAH). In their model N particles move in 2-dimensional space with constant velocity υ_0. The velocity and the angle of orientation of particle n at time t is denoted by υ_n^t and θ_n^t, respectively. The new orientation θ_n^{t+1} of particle n is

$$\theta_n^{t+1} = \arg\left[\sum_m \left(\alpha_{nm}\frac{\vec{v}_m^{\,t}}{v_0} + \beta_{nm} f_{nm}^t \vec{e}_{nm}^{\,t}\right)\right] + \vec{u}_n^{\,t}$$

where $f_{nm}^t\ e_{nm}^t$ is the force exerted by particle m on particle n along the direction e_{nm}^t pointing from particle m to n. The noise is taken into account by u_n^t is a unit vector with random, uniformly distributed orientation. α_{nm} and β_{nm} are control parameters: α controls the relative weights of the alignment interaction and β shows the strength of the radial two-body forces f_{nm}, defined as

$$f_{nm} = \begin{cases} \infty & if \quad r_{nm} < r_c, \\ 1 - \dfrac{r_{nm}}{r_c} & if \quad r_c < r_{nm} < r_0, \\ 0 & if \quad r_{nm} > r_0 \end{cases}$$

that is for distances smaller than a core radius r_c it is a strong repulsive force, around the equilibrium radius r_e it is a harmonic-like interaction, whereas for distances larger than the interaction range r_0 it is set to zero. Having the classic experiments with Hydra cells in mind, authors defined two kinds of particles, "endodermic" and "ectodermic", denoted by 1 and 2, respectively. Accordingly, β_{11} and β_{22} stand for adhesion within the given cell type, whereas $\beta_{12} = \beta_{21}$ account for symmetric inter-cell-type interactions. Differential adhesion is described by different beta values for symmetric interactions between different cell types. The simulations were performed with cells on a square domain with linear size several magnitudes larger than cell size. Figure 8 shows snapshots from the evolution of the segregation process. Segregation is characterized by an index, γ, showing the average ratio of dissimilar cells around a cell, for either cell types. This index is decreasing as

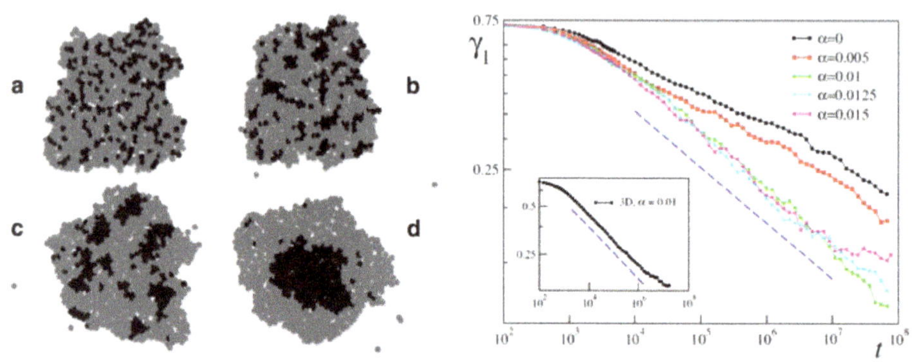

Figure 8 Segregation dynamics of simulated ectoderm and endoderm cells. Left: Cell sorting of 800 cells. The endodermic and ectodermic cells are represented by black and gray circles, respectively. (**a**) The initial cluster with mixed cell types. (**b**) The cluster after 3000 time steps and (**c**) is taken at $t = 3 \times 10^5$. Clusters of endodermic cells form and grow as time passes by. (**d**) At $t = 2 \times 10^6$ a single endodermic cluster is formed, but isolated cells remain within the ectoderm tissue, in agreement with experiments of Rieu et al. (1998). Right: Cell sorting in two dimensions from a random, roughly circular initial aggregate of N = 6400 cells in a proportion of 1:3 endodermic to ectodermic cells. Evolution of the segregation index, γ, for different α values. The dashed line has a slope $-\lambda = -0.18$. Inset: Same in three dimensions but with $\alpha = 0.01$ and $\beta_{11} = 8.3$. The dashed line has a slope -0.16. From Belmonte et al. (2008) with permission of Phys. Rev. Lett.

segregation proceeds and it is expected to approach zero in large systems. Authors have found that segregation is characterized by algebraic scaling laws and introducing even a moderate amount of local coherent motion will considerably speed up the segregation process (Figure 8). A variant of this computational model is published by Beatrici et al. (2011) investigating the segregation of self-propelled particles in 2 dimensions, driven by differences only in motility but not in adhesion. In this model, the faster cells envelope the slower cells forming islands as segregation proceeds.

Taking a different approach Honda et al. (2008) studied pattern formation and emergence of asymmetry in mammalian blastocyst formation by simulations using a 3D vertex dynamics model. Endowing the simulated cells with equivalent mechanical properties and assuming that due to basement membrane formation the cell-medium surface in the forming blastocyst cavity is stiffer than cell-cell surfaces, the simulation successfully produced blastocyst structure with variable axis. Enclosing the simulated aggregate in an ellipsoidal capsule corresponding to the zona pellucida, the structure surrounding the blastocyst in vivo, eventually oriented the axis. Authors have shown the autonomous emergence of asymmetry as well as the formation and segregation of distinct cell populations of different sizes in an initially equivalent cell population without preexisting intrinsic differences.

Conclusions

Our understanding of the events during embryonic development such as multicellular segregation and pattern formation is gradually shifting from a descriptive view towards a causative understanding of the mechanisms. Integrative biological attempts to promote this understanding have been successfully employing various approaches ranging from experimental embryology to statistical physics. Combination of experimental data with multi-disciplinary approaches including computational modeling and statistical physics can help us design more focused experimental tests or predict yet unseen outcomes and these can even further extend our understanding of the dynamic organization of multicellular biological systems.

Additional files

Additional file 1: mov. 2-dimensional segregation of mixed tissue cells. Segregation in a mixed co-culture of primary goldfish keratocytes (PFK, red) and EPC fish keratocytes (EPC, green) migrating on tissue culture plastic surface is imaged by time-lapse videomicroscopy. Fluorescent cell labels: red: cell tracker CMPTX, green: cell tracker CMFDA. Videomicroscopy duration: 20 hours, images were acquired every 10 minutes by a Zeiss Axio Observer system.

Additional file 2: mov. 3-dimensional segregation of mixed tissue cells. Segregation in a mixed co-culture of primary goldfish keratocytes (PFK, red) and EPC fish keratocytes (EPC, green) suspended in agarose micromold is imaged by time-lapse videomicroscopy. Fluorescent cell labels: red: cell tracker CMPTX, green: cell tracker CMFDA. Videomicroscopy duration: 20 hours, images were acquired every 10 minutes by a Zeiss Axio Observer system.

Abbreviations

DAH: Differential adhesion hypothesis; DITH: Differential interfacial tension hypothesis; TST: Tissue surface tension; CPM: Cellular potts model; PFK: Primary goldfish keratocytes; MCS: Monte-Carlo steps.

Competing interests

The authors declare that they have no competing interests.

Authors' contributions

TV conceived the study, participated in its design and contributed to drafting the manuscript. EM performed experiments and wrote the manuscript. Both authors read and approved the final manuscript.

Authors' information

TV is a statistical physicist and he is the principal investigator of the EU ERC COLLMOT project on collective motion. EM is a cell biologist specialized in cell motion research and he is a research fellow in the EU ERC COLLMOT project.

Acknowledgement

We thank Enys Mones (Dept. Biological Physics, Eötvös University) and András Czirók (Dept. Biological Physics, Eötvös University) for useful consultations. This work was funded by the European Union ERC COLLMOT Project.

References

Amack JD, Manning ML: **Knowing the boundaries: extending the differential adhesion hypothesis in embryonic cell sorting.** *Science* 2012, **338:**212–215.

Beatrici CP, Brunnet LG: **Cell sorting based on motility differences.** *Phys Rev E Stat Nonlin Soft Matter Phys* 2011, **84:**031927.

Belmonte JM, Thomas GL, Brunnet LG, de Almeida RM, Chaté H: **Self-propelled particle model for cell-sorting phenomena.** *Phys Rev Lett* 2008, **100:**248702.

Beysens DA, Forgacs G, Glazier JA: **Cell sorting is analogous to phase ordering in fluids.** *Proc Natl Acad Sci USA* 2000, **97:**9467–9471.

Brodland GW: **The Differential Interfacial Tension Hypothesis (DITH): a comprehensive theory for the self-rearrangement of embryonic cells and tissues.** *J Biomech Eng* 2002, **124:**188–197.

Brodland GW: **New information from cell aggregate compression tests and its implications for theories of cell sorting.** *Biorheology* 2003, **40:**273–277.

Foty RA, Steinberg MS: **The differential adhesion hypothesis: a direct evaluation.** *Dev Biol* 2005, **278:**255–263.

Foty R, Pfleger CM, Forgacs G, Steinberg MS: **Surface tensions of embryonic tissues predict their mutual envelopment behavior.** *Development* 1996, **122:**1611–1620.

Gierer A, Berking S, Bode H, David CN, Flick K, Hansmann G, Schaller H, Trenkner E: **Regeneration of hydra from reaggregated cells.** *Nat New Biol* 1972, **239:**98–101.

Graner F: **Can surface adhesion drive cell-rearrangement?** *J Theor Biol* 1993, **164:**455–476.

Graner F, Glazier JA: **Simulation of biological cell sorting using a two-dimensional extended Potts model.** *Phys Rev Lett* 1992, **69:**2013–2016.

Harris AK: **Is Cell sorting caused by differences in the work of intercellular adhesion? A critique of the Steinberg hypothesis.** *J Theor Biol* 1976, **61:**267–285.

Honda H, Motosugi N, Nagai T, Tanemura M, Hiiragi T: **Computer simulation of emerging asymmetry in the mouse blastocyst.** *Development* 2008, **135:**1407–1414.

Inaba M, Yamanaka H, Kondo S: **Pigment pattern formation by contact-dependent depolarization.** *Science* 2012, **335:**677.

Kabla AJ: **Collective cell migration: leadership, invasion and segregation.** *J R Soc Interface* 2012, **9:**3268–3278.

Klopper AV, Krens G, Grill SW, Heisenberg CP: **Finite-size corrections to scaling behavior in sorted cell aggregates.** *Eur Phys J E Soft Matter* 2010, **33:**99–103.

Krens SF, Heisenberg CP: **Cell sorting in development.** *Curr Top Dev Biol* 2011, **95:**189–213.

Krens SF, Möllmert S, Heisenberg CP: **Enveloping cell-layer differentiation at the surface of zebrafish germ-layer tissue explants.** *Proc Natl Acad Sci USA* 2011, **108:**E9–E10.

Krieg M, Arboleda-Estudillo Y, Puech PH, Käfer J, Graner F, Müller DJ, Heisenberg CP: **Tensile forces govern germ-layer organization in zebrafish.** *Nat Cell Biol* 2008, **10:**429–436.

Maître JL, Heisenberg CP: **The role of adhesion energy in controlling cell-cell contacts.** *Curr Opin Cell Biol* 2011, **23:**508–514.

Maître JL, Berthoumieux H, Krens SF, Salbreux G, Jülicher F, Paluch E, Heisenberg CP: **Adhesion functions in cell sorting by mechanically coupling the cortices of adhering cells.** *Science* 2012, **338:**253–256.

Manning ML, Foty RA, Steinberg MS, Schoetz EM: **Coaction of intercellular adhesion and cortical tension specifies tissue surface tension.** *Proc Natl Acad Sci U S A* 2010, **107:**12517–12522.

McCandlish SR, Baskaran A, Hagan MF: **Spontaneous segregation of self-propelled particles with different motilities.** *Soft Matter* 2011, **8:**2527–2534.

Méhes E, Mones E, Németh V, Vicsek T: **Collective motion of cells mediates segregation and pattern formation in co-cultures.** *PLoS One* 2012, **7:**e31711.

Mombach JC, Glazier JA, Raphael RC, Zajac M: **Quantitative comparison between differential adhesion models and cell sorting in the presence and absence of fluctuations.** *Phys Rev Lett* 1995, **75:**2244–2247.

Moore R, Cai KQ, Escudero DO, Xu XX: **Cell adhesive affinity does not dictate primitive endoderm segregation and positioning during murine embryoid body formation.** *Genesis* 2009, **47:**579–589.

Morris SA, Teo RT, Li H, Robson P, Glover DM, Zernicka-Goetz M: **Origin and formation of the first two distinct cell types of the inner cell mass in the mouse embryo.** *Proc Natl Acad Sci U S A* 2010, **107:**6364–6369.

Nakajima A, Ishihara S: **Kinetics of the cellular Potts model revisited.** *New J Phys* 2011, **13:**033035.

Rieu JP, Kataoka N, Sawada Y: **Quantitative analysis of cell motion during sorting in two-dimensional aggregates of dissociated hydra cells.** *Phys Rev E* 1998, **57:**924–931.

Schötz E, Burdine RD, Jülicher F, Steinberg MS, Heisenberg CP, Foty RA: **Quantitative differences in tissue surface tension influence zebrafish germ layer positioning.** *HFSP J* 2008, **2:**42–56.

Steinberg MS: **On the mechanism of tissue reconstruction by dissociated cells. I. Population kinetics, differential adhesiveness and the absence of directed migration.** *Proc Natl Acad Sci USA* 1962a, **48:**1577–1582.

Steinberg MS: **Mechanism of tissue reconstruction by dissociated cells. II. Time-course of events.** *Science* 1962b, **137:**762–763.

Steinberg MS: **Reconstruction of tissues by dissociated cells. Some morphogenetic tissue movements and the sorting out of embryonic cells may have a common explanation.** *Science* 1963, **141**:401–408.
Steinberg MS: **Adhesion in development: an historical overview.** *Dev Biol* 1996, **180**:377–388.
Townes PL, Holtfreter J: **Directed movements and selective adhesion of embryonic amphibian cells.** *J Exp Zool* 1955, **128**:53–120.
Vicsek T, Zafeiris A: **Collective motion.** *Physics Reports* 2012, **517**:71–140.
Vicsek T, Czirók A, Ben-Jacob E, Cohen II, Shochet O: **Novel type of phase transition in a system of self-driven particles.** *Phys Rev Lett* 1995, **75**:1226–1229.
Voss-Böhme A, Deutsch A: **The cellular basis of cell sorting kinetics.** *J Theor Biol* 2010, **263**:419–436.
Zhang Y, Thomas GL, Swat M, Shirinifard A, Glazier JA: **Computer simulations of cell sorting due to differential adhesion.** *PLoS One* 2011, **6**:e24999.

2

Replication and replacement in dynamic delivery networks

Anita Sobe[1,2*] and Wilfried Elmenreich[1,3]

*Correspondence:
anita.sobe@unine.ch
[1]Institute of Networked and Embedded Systems/Lakeside Labs, Alpen-Adria-Universität Klagenfurt, Klagenfurt, Austria
[2]Computer Science Department, University of Neuchâtel, Neuchâtel, Switzerland
Full list of author information is available at the end of the article

Abstract

Purpose: Content delivery in dynamic networks is a challenging task, because paths may change during delivery and content might get lost. Replication is a typical measure to increase robustness and performance.

Method: In previous work we proposed a hormone-based algorithm that delivers content, and optimizes the distribution of replicas. Clients express demands by creating hormones that will be released to the network. The corresponding resources are attracted by this hormone and travel towards a higher hormone concentration. This leads to a placement of content close to their most frequent requesters. In addition to that the hormone-based delivery requires an appropriate replication and clean-up strategy to balance the replicas throughout the network without exceeding the nodes' storage limits or the networks communication capacity.

Results: We examine different combinations of replication and replacement strategies and evaluate them in realistic scenarios involving node failure and networks of different size and structure.

Conclusion: Results show that it is necessary to match the replication mechanisms with the clean-up mechanism and that the local hormone information can be used to improve the clean-up decision.

Keywords: Self-organization, Replication, Storage balancing, Artificial hormone system, Multimedia distribution

Background

Introduction

In peer-to-peer networks each node can act as a client or server for the other nodes in the network, allowing shared access to various resources such as files, peripherals, and sensors without the need for a central server. We consider peer-to-peer networks, where content is dynamically stored at different nodes and delivered to the client on request. Depending on the request patterns in the network, stored content is migrated from one node to another, replicated or dissolved. Such a dynamic delivery system needs to implement two functions: the primary function is to provide the clients with the requested content with short delay. The secondary function is to optimize the placement of replicas in the network to minimize network delay for future requests.

When considering limited communication capabilities, the fulfillment of the two functions mutually influence each other. The placement optimization will increase the network load for some time and therefore worsen the system's responsiveness to client requests. Over time, the re-arrangement and replication of content, however, will improve the delivery performance and reduce network load. In applications with smart sensors or mobile devices, the system operation is also significantly affected by the limited storage of embedded devices. Thus, it is necessary to clean-up unused replicas or to balance the storage of content according to the available storage on the nodes. Moreover, we assume that the network is dynamic and therefore it is impossible to build global knowledge about all nodes and the stored content. The dynamics of the network stem from new generation of content/deletion of content, devices joining/leaving the network, and changing locations of a device due to user mobility. In contrast to a centralized solution, this makes it hard to ensure that for each content unit at least one copy remains in the system at all time. Since lost content can usually not easily be recovered this is an important requirement for most content storage networks. In order to solve the distributed multi-optimization problem for storage balancing, request delay, and network load (within the network) we propose a distributed self-organizing and self-adapting replication and replacement strategy.

The hormone-based delivery algorithm (Sobe et al. 2010) is a promising candidate for delivering and distributing content in large networks. It assumes only local knowledge at each node. The central idea of this algorithm is to use artificial hormones to request transport and replication of content. The hormone-based algorithm is managed without global knowledge and consists of two parts: (1) Requests for content are represented by hormones. Hormones are created on an arriving request, are diffused to neighbors and evaporate over time; (2) hormones trigger the transport of corresponding content, which is done hop-by-hop. A further important mechanism is replication, which is not only a measure of robustness; the algorithm also exploits replication by placing replicas on the transport path. This reduces the search space.

In contrast to ant algorithms (Dorigo et al. 2006), which can be used for robust routing in dynamic networks, the hormone approach provides a means for optimizing the content placement in the network. In the endocrine system of higher mammals, hormones are created and travel around the body where they lead to specific actions in target cells. The same hormone might lead to different actions, depending on the type of the target cells (Xu et al. 2011). This is also the main principle of the hormone-based delivery system. Nodes in the network represent cells that are able to create, consume and forward hormones. As a reaction to hormones a node can consume, replicate, forward content or even do nothing. A further difference is that we do not assume static content location, such as the food source in ant-based systems. In our system content can replicate and move towards the receiver.

It is the purpose of this paper to examine different replica replacement strategies for large distributed systems to balance the storage of nodes. The results of our evaluation show that the selection of the appropriate clean-up mechanism, i.e., the particular strategy for removing or migrating units to ensure sufficient storage capability at a node to fulfill future requests has a crucial effect on the network performance. An initial summary of our results have been made available as a technical report (Sobe et al. 2011b) to enable discussion.

The paper is structured as follows: In the following section we introduce related work on replication strategies for bio-inspired distribution networks. To emphasize the relevance of our problem statement, we elaborate the context of the targeted problem followed by a detailed description of the hormone-based delivery and replication approach. A special focus is given on the different replacement mechanisms. We further consider the constraint that a clean-up action must never delete the last instance of a unit. The described strategies are implemented in a simulation model of a network of multimedia servers with limited storage and consumers requesting content according to a preference model. The simulation results allow for a quantitative comparison of the efficiency of the different variants. We evaluated realistic scenarios involving node churn and networks of different size and structure. The conclusion summarizes our recommendation for effective replication and replacement strategies in dynamic content delivery networks.

Related work

Typically, the popularity of content is dynamic and replicas of content with decreasing popularity have to be handled to balance the storage. This can be done by introducing clean-up mechanisms; however, the challenge is to identify unnecessary replicas.

Lee et al. describe in (Lee et al. 2008a) a utility based replication scheme, named SmartPin for pre-recorded IPTV content in peer-to-peer streaming systems. The IPTV content is divided into variable length segments. Depending on the utility, content and/or its metadata are actively replicated to other peers. E.g., for intermediate utility content, only the metadata containing the original place of the content is replicated. The utility depends on the popularity of the content, its benefit and its costs. As benefit the authors define the perceived quality and the required bandwidth, whereas the costs regard the delivery and storage costs (Lee et al. 2008b). To increase the content availability the minimum number of needed replicas is calculated centrally.

A utility function measuring the importance of content is a first step to handle the dynamics of user preferences. Nonetheless, the utility depends on the popularity, which is not easy to derive in a system without global knowledge. The authors do not mention what happens to the replicas if the popularity of the content changes.

The author of (Rong 2008) proposes another replication scheme for peer-to-peer video systems that adapts the number of replicas according to their utilization rate. Temporary popular videos are replicated more often and the replicas are destroyed when the number of requests for the video decreases. The number of replicas is handled by a central coordinator. In a system with local knowledge only a more dynamic solution has to be provided.

In (Herrmann 2007) an example for service replica placement in ad-hoc grids is shown, where the cost of a request is measured by the number of message transmissions per time unit to serve it. The goal is to minimize those costs without global coordination. The author defines a parameter ρ which describes the coverage radius of clients. If the number of requests for this service is below a given threshold, the service replica removes itself from the system. Service and content replication is comparable, however, in this work services are independent from each other. If a composition of services is provided the cost function would have to be adapted to a more flexible system.

Forestiero et al. proposed a descriptor sorting and replication algorithm for peer-to-peer grids called *Antares* in (Forestiero et al. 2007). This approach is not only ant inspired, but can also be categorized as *brood sorting* such as described in (Mamei et al. 2006). The descriptors are bit strings encoded by locality sensitive hash algorithms (see Kulis and Grauman 2009). In contrast to standard hash algorithms, similar content results in similar hash values. This helps to sort similar content based on its hash value. The sorting is done by ant-inspired agents that travel the grid and pick and drop descriptors. Agents operate in two modes, copy and move. In copy mode the agent creates a replica of the descriptor, whereas in move mode not. The transition between these states is pheromone based. In the beginning, each agent starts in the copy mode. At some point, when the resources are better sorted, the activity of the agent decreases. Then, the agent starts to increase its pheromone level until a given threshold and subsequently switches to the move mode.

It is interesting that Antares has also been adapted to support QoS-based picks and drops (Forestiero et al. 2008). Descriptors with better QoS are more likely to be replicated. The QoS is described as a non-negative real value, where higher values describe higher QoS. Analogous to the pick probability the drop probability increases if a descriptor with high QoS is carried by an agent. Although the placement of replicas is handled by Antares, the number of replicas is never regulated. An extension of the model by introducing a *delete agent* could be done by using one of the mechanisms described in the current paper.

In (Sobe 2012) we give a more detailed overview and some more related work on these topics.

Context

Our evaluation is based on a scenario of a social event like a triathlon. Visitors follow such an event along a very large area. To see what is happening around the track they have to rely on videos provided by the organizers on video walls. The possibility of querying content such as *"I want an overview of interesting parts of the last 30 minutes"* or *"I want to see more of the athlete with no. 1234"* is missing. Visitors produce masses of multimedia data, but there is no specific possibility to exploit this data for live sharing with other visitors. Visitors should be able to create their individual presentations depending on their interests (Böszörmenyi et al. 2011).

We define the term multimedia unit, which can be either a photo or a short video sequence, e.g., generated at a live sports event by a visitor or a picture. We further assume dynamic access patterns. Users can "compose" their presentations consisting of a number of different units in their preferred order. In (Sobe et al. 2010) we also support parallel presentations, such as overviews in a split screen. Such access patterns allow flexibility, but also introduce complexity, because the typical pre-defined video sequence as known from movies is not existing anymore.

The triathlon scenario further shows that content gets more and more important both on the consumer and on the producer side. In this scenario visitors produce multimedia content all around the area, and should also be able to consume anything (multimedia content), anytime and anywhere they want. However, the transport of content is still limited to traditional, mostly static, delivery methods. Future Internet discussions such as described by Hausheer et al. in (Hausheer et al. 2009) show that flexible solutions for delivery are needed.

The used delivery algorithm is capable of handling this complexity relying on simple local information. The algorithm is inspired by two existing bio-inspired approaches. A specific ant-based application for search in peer-to-peer networks was introduced as SemAnt (Michlmayr 2007). The second approach is an artificial hormone-based agreement for task allocation introduced by Brinkschulte et al. in (Brinkschulte et al. 2007).

We adopted the keyword search from SemAnt and introduce one type of hormone by keyword. The hormone value expresses the current demand (the *goodness*, as adopted from Brinkschulte et al.) for the corresponding keyword. SemAnt assumes static content locations, which keeps the search space very large. To reduce the search space we replicate content and exploit the unstructured overlay by letting the content travel towards higher levels of its corresponding hormone. This allows intermediate nodes to decide if the traveling unit is needed for future requests, thus predicting places to reduce search space and delivery costs.

This work relies on a bio-inspired delivery algorithm introduced in (Sobe et al. 2010). This basic version of the algorithm only considers the hormone creation, diffusion and transport of content based on hormones. Different replication mechanisms have been evaluated in (Sobe et al. 2011a).

Since peers are not likely to provide unlimited storage for other peers and in a dynamic system the popularity of units changes, in this paper we investigate different measures to efficiently balance the storage of the peers. We periodically apply different strategies if a certain storage level is reached. We compare LRU (Least Recently Used), LFU (Least Frequently Used) and a hormone-based clean-up. We show that the chosen clean-up mechanism has an impact on the delivery performance and that the system is still robust against peer failure.

Method

Hormone-based delivery and replication

The hormone-based delivery approach introduced in (Sobe et al. 2010) involves three components: hormone creation, hormone diffusion, and the behavior of units in presence of a corresponding hormone. The distributed, self-organizing nature of the approach assists with handling the complexity of requests and the search for units in the network with comparably simple decision algorithms based on local knowledge.

Upon request of a particular unit, a corresponding hormone is issued at the requesting node. The hormone is represented as a real value – the current value of a hormone is the *hormone level*. The hormone is diffused via the neighboring nodes to the network, creating a hormone gradient towards the requesting node.

The diffusion of hormones depends on the network structure – a node only a few hops away from the requesting node will get more hormones than a more distant node.

Figure 1 shows an example of the diffusion of hormones from a requesting node P_1. Assume a network of seven nodes P_1 to P_7. Links have different quality, which is indicated by line thickness. Node P_1 creates hormones (`total`) for the requested unit. To create a hormone gradient P_1 splits the available hormones and reserves a part for diffusion to neighbors (`to_diffuse`). How much each neighbor gets from `to_diffuse` depends on the quality of the link to it. Thus, P_2 gets the highest rate and P_4 the lowest rate. When the neighbors get the hormones, they forward again a part of it. The forwarding is repeated until a node is reached where the requested content is located (here, P_7). The unit can now move hop-by-hop towards the highest hormone concentration, which is found at the requester.

Beyond this simple example we further implemented continuous hormone creation. As long as a request is unfulfilled, the hormone level is raised periodically, thus increasing

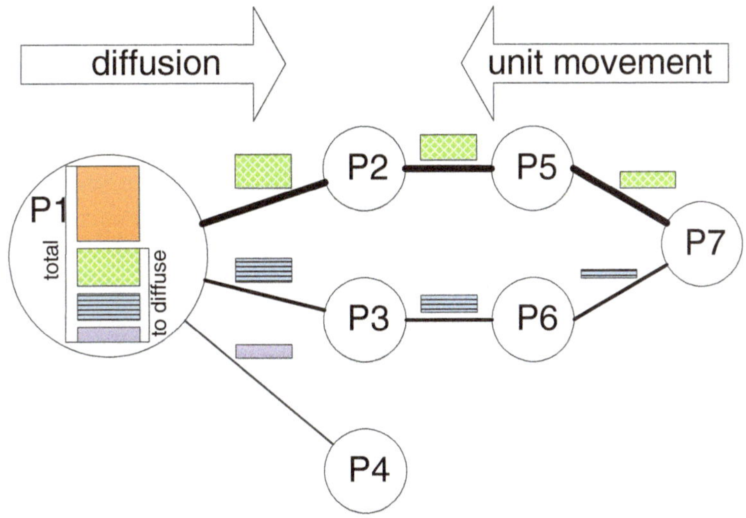

Figure 1 Example of the artificial hormone diffusion. Interplay of diffusion and unit movement in hormone-based delivery.

the concentration of hormones at the sender end, encouraging diffusion of hormones towards the receiver, until the request is fulfilled.

A unit that corresponds to the given hormone will move towards the nodes with the most of that hormone and relies on the diffusion mechanism to choose currently the best path. In order to reduce attracting multiple copies of a unit, the diffusion of a hormone is stopped if the respective unit already resides on the node. The hormone-based delivery creates a feedback loop between network conditions. The stability of this feedback loop depends on the parameter settings, which are discussed in the following section. Multiple requests for different units lead to a different set of hormones being handled in parallel by the network. Requests for the same unit result in a superimposed hormone landscape for that unit. In this case, a unit might be attracted by two hormone trails. Without replication, the unit must move to different requesters in order. The requester that receives the unit first is determined by the strength of hormone reaching the unit (from the requester). In order to avoid such detours, an intelligent replication mechanism has to take care of this issue.

Parameter settings

The proposed algorithm is self-organizing and depends on a number of configuration parameters, which are listed in Table 1. The most important ones are η_0 and η, α and ϵ, because they are defining how many hormones are created, diffused and evaporated per time step. η_0 controls the hormones created at the issuing of the request, η defines the hormone per time unit that are added at the requesting node until the request is fulfilled. α is the percentage of hormones that can be diffused to neighbors (the result of $\alpha * total$ corresponds to `to_diffuse` in Figure 1). The evaporation value ϵ will be subtracted for reducing the hormones on alternative paths. These parameters need to be tuned dependent on each other. E.g., if η_0 and η are low and the evaporation value ϵ is high, the movement of units is limited to a fewer number of hops. The greater the amount of hormones created, the further the hormones can travel and thus the search

Table 1 Parameters to configure at system startup

ID	Explanation
η_0	Hormone strength of a unit at new request
η	Increase of hormone after each time step by the requester
α	Percentage of hormones to be forwarded to the neighbors
ϵ	Hormone evaporation value
t	Significance threshold for hormones
m	Minimum hormone strength difference to move unit

space increases. Further parameters regard the minimum hormone strength difference to move a unit m, which controls the mobility of units. If m is high, the units need a higher hormone concentration to move to a neighbor, leading to a longer waiting time for the requester. t is the minimum hormone strength, if a hormone value is below this threshold it is considered as insignificant and can deleted. In general, the parameter settings are essential for the algorithm to work and their inter-dependencies and combinations make it hard to tweak them manually. We therefore optimize them using an evolutionary algorithm as described in (Elmenreich and Klingler 2007).

Initially, the algorithm creates a random population of parameters. It then uses elite selection for creating the next generation. The candidates are sorted according to their fitness and the best x candidates are chosen. These candidates propagate to the next generation. To reach the same population size as the last generation, the rest of the slots are reserved for mutation, crossover and new candidates. For mutation and crossover random elite candidates are chosen. Finally, random new candidates are added to the population.

Replication mechanisms

We target content delivery systems that set up on top of unstructured peer-to-peer overlays and we investigated in (Sobe et al. 2010) that efficient replication mechanisms are necessary, thus we compared existing and proposed replication mechanisms in (Sobe et al. 2011a). The following categorization is based on (Androutsellis-Theotokis and Spinellis 2004) and (Yamamoto et al. 2005).

Existing replication mechanisms

Owner replication

The content is replicated at the requester's node (Lv et al. 2002) and is also called passive replication. Typically, this replication technique is used in file sharing systems based on BitTorrent (Cohen 2003). BitTorrent supports direct download, if a resource is found it is copied to the requester. Only nodes that are interested get the resource.

Path replication

In a multi-hop network where content is not transported directly such as in Freenet (Clarke et al. 2001), it is possible to cache one replica of the content at each intermediate node. Since the intermediate nodes are acting as caches, path replication is also called cache-based replication. It is assumed that intermediate nodes provide storage space for replicas even if they are not interested in the content. Path replication leads to a high number of unused replicas.

Therefore, an improved approach replicates the content on an intermediate node according to a fixed replication rate (*path random replication* (Yamamoto et al. 2005)). The advantage of this approach is a compromise between a higher replica usage and limited hop distance to other replicas. The difficulty of this approach is to specify a suitable replication rate for each file in advance, which is hard if the files are not known at system startup.

An alternative is to specify a node specific replication probability, where nodes decide ad-hoc if a file is replicated or not. The replication probability is dependent on the peer's resource status and optionally refers to the replication rate, too. The authors in (Yamamoto et al. 2005) refer this strategy to as *path adaptive replication.*

Active replication

The goal is to place the right number of replicas at the right locations *before* they are requested. Researchers investigated the optimal number of replicas in the context of robustness. In (Cohen and Shenker 2002) and (Lv et al. 2002) the authors investigate *random, proportional, and square root replication.* When applying random replication a uniform number of replicas for each object are created. Proportional replication creates replicas proportional to their query rate. The authors showed that square root replication determines the optimal replication rate r_i for object i, which is calculated as $r_i = \lambda\sqrt{q_i}$, with $\lambda = R/(\sum_i \sqrt{q_i})$, where q is the query rate and R is the number of object replicas in the system. Square-root replication does not consider the location of replicas. All strategies require global knowledge on the number of currently existing replicas and the current query rate for each of the replicas.

To reach square-root replication with limited knowledge researchers proposed *Pull-then-Push Replication* introduced in (Leontiadis et al. 2006). The first phase of this method regards the search of the content, with any existing algorithm. The second phase regards the replication of content to the neighbor nodes. To reach square root replication, the authors suggest that for the pull and push phase the same algorithms are used, because the number of replicas should be equal to the number of nodes visited during search. The authors evaluated typical algorithms, such as flooding and random walks. Their focus is on robustness even on update situations. As multimedia content is usually not updated after creation and this algorithm only considers the number and not the location of content, this approach is out of scope of this work.

Local knowledge replication mechanisms

Although a unit is delivered hop-by-hop the basis for the following replication mechanisms is owner replication, since the units are consumed for some time at the requester and therefore need to be replicated to be further usable by other nodes. Units replicated at the requester can only be supportive for the immediate neighborhood. To serve future requests, replicas should also be created on the delivery path. If the hormone concentration of a neighbor attracts a stored unit, the peer has to decide whether to move or to copy the unit. The simplest solution would be to apply path replication, but then the utilization of replicas would drop and the storage space is not used efficiently. Therefore, the goal is to find a replication mechanism that balances replica utilization and delay without the need of global information.

Simple hormone

If a unit is requested by peers from different parts of the network, the unit has to move first to one requester and afterwards to the other requester. This can lead to long traveling paths, which can be avoided by replicating a unit if more than one neighbor holds hormones for it. Note that it is not possible to differentiate if hormones on the neighbor are created by different peers. Thus, it is possible that unnecessary replications are made.

Local popularity

Each node uses the local request history of the corresponding content to decide if it is likely to be requested again in the future. If the rank of the content is among the best 30% the corresponding unit is replicated, so popular units are more likely to be replicated, but popularity information from neighbors is ignored. With this method the communication effort is minimized.

Neighbor popularity ranking

After collecting the popularity ranks for the content from the neighbors, the peer decides if it is worth to replicate the corresponding unit. The ranks are aggregated to a region rank (see (Sobe et al. 2009)), which is calculated as follows:

$$R = \frac{1}{n}\sum_{i=1}^{n} \ln(r_i)$$

n represents the number of neighbors and r is the rank of the specific unit at a neighbor i, where 0 is the best rank. To reduce the impact of peak ranks (e.g., one unit is best ranked at two nodes, but worst ranked on the third node) the logarithm is used. If the region rank R is lower than a given threshold (e.g., the best 30% at all neighbors) the unit is replicated.

Neighbor hormone ranking

Analogue to the popularity ranking the units can also be ranked by their hormone values at the neighbors. The higher the hormone value for a unit on a neighbor, the better is the unit's rank. The collected ranks can be aggregated as before and if the region rank is lower than a given threshold (e.g., the best 30%), the unit is replicated.

Clean-up/Replacement strategies

We concluded in (Sobe et al. 2011a) that efficient replacement or clean-up has to be done in order to avoid blocking the transport of units. A clean-up is triggered if a certain storage level is reached, which leads to a balanced storage load of the system.

However, if we want to have at least one instance of the unit available in the system, the choice of which unit to delete becomes more complicated. We solve this issue by applying a simple mechanism. Before deleting a unit, the node checks if at least one copy of it is on one of its neighbor nodes. We assume that the nodes follow a simple coordination protocol to avoid concurrent deletion of the last two units in a system.

The goal is to find an efficient strategy that does not increase the delay, but increases the utilization of replicas. We compare three mechanisms: least recently used (LRU), least frequently used (LFU) and hormone-based clean-up. LRU takes care of popularity changes of units, thus removes those units for which the most time steps passed since their last presentation. LFU targets units of low popularity, i.e., the least frequently presented units are removed. Hormone clean-up exploits local knowledge about hormone concentration.

A unit is deleted if there are no hormones for it on the neighbors, thus there is no current demand for it. Units currently in delivery are not deleted.

Simulation setup

We implemented a simulator[a]. At start-up, the simulator generates the network topology, initializes the nodes with initial storage, configures the client settings.

A simulation run is a loop of actions performed at each time step. First, the clients generate new requests or consume content presentations, after that the peers are diffusing hormones and moving content. For our evaluations we average the results of 10 simulation runs. One simulation run lasts for 500s, which is sufficient to show that the content placement stabilizes.

Our simulation evaluates the capability of nodes to adapt to a high number of new content units, therefore we randomly create content at the beginning of the simulation. Clients request content and consume content (watching units). To stress the system, clients can submit a new request immediately after the consumption of the former content.

We consider two scenarios. One where we assume a social event with 50 attendees (e.g., a fair for selected consumer groups), and a larger social event (e.g., a local triathlon) with 1,000 attendees. We assume that each person is represented by one node. For both scenarios different settings apply, which we describe in the following.

Network topology

We assume for small overlay networks of 50 nodes a connected Erdős-Rényi random graph with a diameter of 6. For larger networks, e.g., with 1,000 nodes, we assume a scale-free network topology. To generate such a network the Eppstein Power Law Algorithm (Eppstein and Wang 2002) is used. The algorithm gets as input a random graph and by repetitively removing and adding edges a power law distribution is reached. The network diameter of the scale-free graph is 13. The bandwidth was set to 100 Mbit/s. In a different scenario, we considered a lower bandwidth with comparable delay results (Szkaliczki et al. 2012).

Initial storage

At the beginning of a simulation run each node creates units until 30% (50 nodes) respectively 3% (1,000 nodes) of its storage of 900 MB is filled. This results in approx. 5,000 units for the 50 nodes scenario and 15,000 units for the 1,000 peers scenario. The units do not have any specific placement.

We expect that in a scenario with 50 motivated persons, each person will contribute with equal probability. In a scenario with 1,000 visitors we expect that there will be few highly motivated persons and a high number of less motivated persons. Thus, nodes with a higher degree will provide more storage space than others.

The average size of a unit is 2.6 MB, the maximum size is 16 MB and the minimum size is 190 KB, with a playback bit rate of 1 Mbit/s. These unit sizes are the result of a project use case "The long night of research"[b]. This use case was part of a university-local event where all research groups presented their work to the public, similar to a fair. We encouraged the visitors to upload their photos and videos to one of our servers. The visitors could browse and tag the uploaded content on a web-page, but were not able to share their content directly with other visitors. We analyzed the keywords for the uploaded content and additionally argue that one can compare the tag popularity to content popularity of user-generated content (Cha et al. 2007). Thus, we assume that the tags follow a Zipf-like distribution. On content creation, keywords are mapped to units, whereas one keyword might be used for several units.

Request generation

One request consists of 1–10 units identified by keywords. Thus, a client is requesting content types, not specific files. The request is fulfilled if for each keyword at least one unit is stored on the node. We further implemented a taste change, i.e., if a user likes the content just watched, her taste for future requests is with 10% probability similar to the currently watched unit.

In this paper we do not consider any order of the units, thus, if a requested unit arrives, it is presented to the user. Sequential and parallel dependencies have been handled for the delivery in (Sobe et al. 2010).

We further introduce a deadline for each unit, until which it has to be delivered. The deadline is dependent on the size, the link bandwidth and the maximum number of hops a unit can travel `maxhops` (see Equation 1).

$$deadline_u = t + maxhops \cdot \frac{size_u}{datarate} \tag{1}$$

If a deadline is missed, no further hormones for that unit are created to stop attracting content. A request is considered as failed if none of the requested units could fulfill their deadline.

A user can only submit one request at a time. If this request is fulfilled or failed, a new request will be generated. This leads to around 10,000 requests during a simulation run in the 50 nodes scenario and around 400,000 requests for the 1,000 nodes scenario. A failed request usually leads faster to the creation of a new request than the presentation of units, thus the number of requests sent during a simulation might differ. Therefore, the metrics will consider the ratio of successful and submitted requests.

Simulation parameters

We used the evolutionary algorithm as described before with a fitness function targeting client satisfaction by optimizing the number of successful requests, ($f = \frac{requests_{fulfilled}}{requests_{submitted}}$). The evolutionary algorithm is part of our open source simulator that also runs the artificial hormone-system. For evaluating the fitness of a parameter set, the simulation is started with this parameter set for a number of runs and the results are averaged (we used hormone ranking replication with hormone clean-up). The parameter sets of one population are compared according to their fitness and the result of one generation is the parameter set with the highest fitness. The higher the number of generations the higher is the fitness of the resulting parameter set. The resulting parameter set can be used for all simulations and real implementations of the algorithm for which the system's configuration (e.g., number of nodes, replication type, etc.) is similar to the input of the genetic algorithm.

In Table 2, the resulting parameters are shown. The hormone creation values for a unit, η_0 and η, are high and hence lead to a wider travel range. Additionally, the diffusion of 45% (α) of the hormones supports longer travel distances. Compared to the creation amount η,

Table 2 Parameter settings

η_0	3.95
η	4.39
α	0.45
ϵ	0.16
t	0.23
m	0.23
c	60%
maxhops	10

Parameter configuration, optimized by a genetic algorithm.

the evaporation rate ϵ is low, meaning the hormones last for some time. The evaporation of hormones is a fixed value subtracted from the current hormone level. The migration threshold m describes the minimum hormone difference between two nodes to make a unit move. In this case the difference is very low in comparison to the creation amount of hormones. This leads to a very dynamic behavior of the units. The clean-up (c) is triggered by a node if its current storage level exceeds 60% (50 nodes) resp. 30% (1,000 nodes).

Metrics

We aim at evaluating both the request fulfillment and the utilization of replicas. The fulfillment of a request is quantified by the *delay*. The delay of a unit is shown in equation 2 is measured from the request time of a unit until the arrival of that unit on the node. A delay of 0 s means that the unit was already on the node.

$$\delta = t_{arrival} - t_{req} \tag{2}$$

For better readability the delay is presented as cumulative distribution function (CDF) (Gubner 2006). It shows the likelihood of a delay value to be less than or equal to a certain value at a given point during the simulation.

The *deadline miss rate* represents the number of units out of all units U (not requests), for which the deadline is missed (see Equation 3).

$$M := \{u_i | u_i \in U : \delta_{u_i} > t_{deadline} - t_{req}\}$$

$$\text{deadline miss rate} = \frac{|M|}{|U|} \tag{3}$$

If a unit missed its deadline, the delay is calculated as deadline minus request time (max. delay) as shown in equation 4.

$$\delta_{max} = t_{deadline} - t_{req} \tag{4}$$

The *request failed rate* indicates requests R_i out of submitted requests R, from which all units u_i missed their deadline as shown in equation 5.

$$F := \{u_i | u_i \in R_i, R_i \subseteq R : \delta_{u_i} \geq \delta_{max}\}$$

$$\text{request failed rate} = \frac{|F|}{|R|} \tag{5}$$

A unit is presented for some time, and as *unit utilization* we measure the rate of units (u_i) that are currently being presented (are part of P). The more unit presentations started in comparison to the number of their replicas U_i, the better the *unit utilization*.

$$P := \{u_i | u_i \text{ currently being presented}\}$$
$$C_i := \{r_i | \exists u_i \in P : r_i \text{ is replica of } u_i\}$$

$$\text{utilization} = \sum_{i=0}^{|P|} \frac{1}{|C_i|} \tag{6}$$

The utilization and the request failed rate will be depicted as box plot with 1.5 interquartile range whiskers.

Results and discussion

As explained in the former section we apply two scenarios, a 50 nodes scenario and a 1,000 nodes scenario. We investigate combinations of replication mechanisms and clean-up mechanisms and their impact on delay, utilization and request failure rate. We further analyze the impact of node failure to the replication and clean-up mechanisms.

In the following we list the used replication techniques and their names used in the figures alphabetically:

- `hormone` replicates if for the given unit hormones exist on the neighbors
- `hranking` ranks the hormones of neighbors to decide the need of replication
- `owner` only replicates on the requesting node
- `path` replicates always
- `path_adapt` replicates with a given probability on the current node
- `pop` replicates if the popularity of this unit is high enough on the current node
- `pranking` ranks the unit's popularity in the neighborhood to decide the need of replication

50 Nodes random network

In this scenario we start with a wrap-up of the base replication scenarios without any clean-up, then we show the differences, if clean-up is applied.

In Figure 2 the base replication cases are shown as CDF. It depicts the likelihood of delays to be below or equal a specific value. Thus, the more the curve is located on the left side, the lower is the delay. In the legend we order the replication names according to their resulting delay. `owner` replication leads to the highest delay, followed by `hormone` and `pranking`, then `path_adapt` and `pop`. It can be seen that `hranking` even outperforms the `path` replication mechanism. The reason for this is that `path` replication creates too many copies and therefore some nodes do not have any storage space left for transport[c]. The storage space balancing is not only a problem of `path` replication, the other replication mechanisms also fill a number of nodes that cause the blocking of transport. In the following we compare the delay, the request failure rate and the utilization of the three introduced clean-up mechanisms hormone, LRU and LFU.

In Figure 3 the delay CDF for hormone clean-up is shown. In general, the order of delay curves remains the same except for `path` and `hranking`, which switch their position.

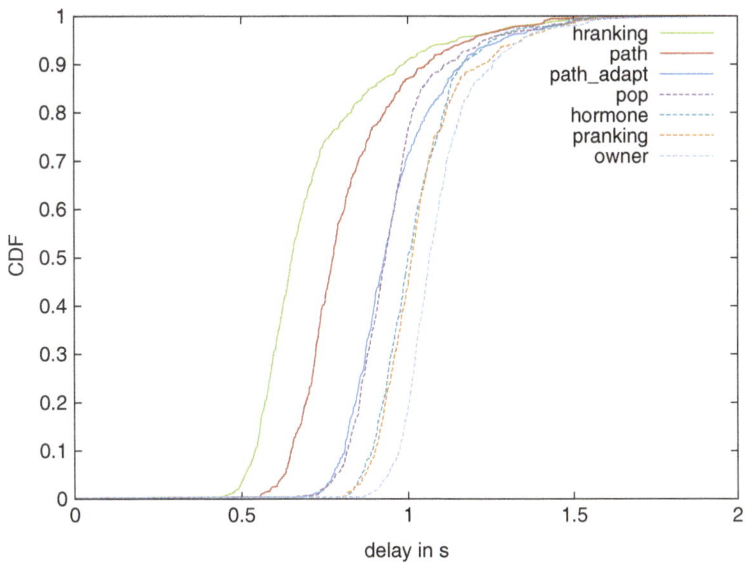

Figure 2 No clean-up. Delay of replication mechanisms without clean-up in a 50 nodes random network.

However, the clean-up has a negative impact on all replication mechanisms as their delay curves are located closer to 1s. In comparison to the base case, the most conspicuous example is the combination with local popularity replication (`pop`), where a high delay jitter results in a gentle curve. The following `hormone` delay curve shows a similar, although less extreme delay jitter increase.

The LRU clean-up is depicted in Figure 4. With this clean-up mechanism `path` replication is also affected, which results in a similar delay of `path`, `hranking` and `path_adapt`. LRU further increases the delay jitter of `path` as shown by a more gentle

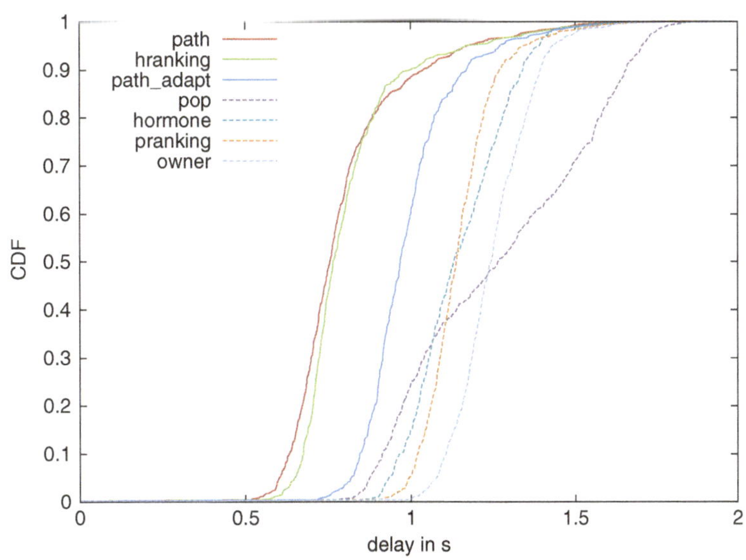

Figure 3 Hormone clean-up. Delay of replication mechanisms in combination with hormone clean-up in a 50 nodes random network.

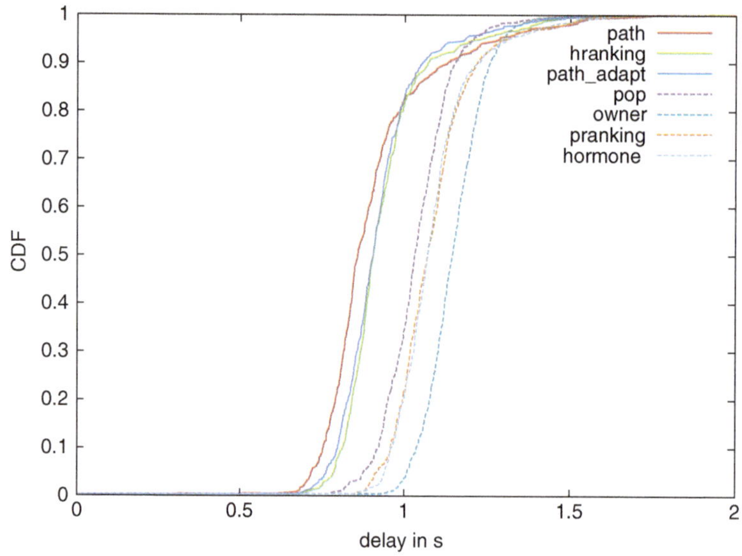

Figure 4 LRU clean-up. Delay of replication mechanisms in combination with LRU clean-up in a 50 nodes random network.

curve than `hranking`. In this graph the delay of `hormone` is even higher than `owner` replication.

LFU, as depicted in Figure 5, results in high delay increases for all replication mechanisms. In particular the `hormone` replication mechanism and `owner` replication suffer from wrong decisions and result in gentle curves. `path` replication results in doubled delay, which leads to a similar delay as `pop`. `hranking` results in a similar delay as `path_adapt`, but with a lower delay jitter its curve is steeper.

When comparing the different clean-up mechanisms, one can see that all of them have a negative impact on the delay. In general, hormone clean-up has the lowest impact if

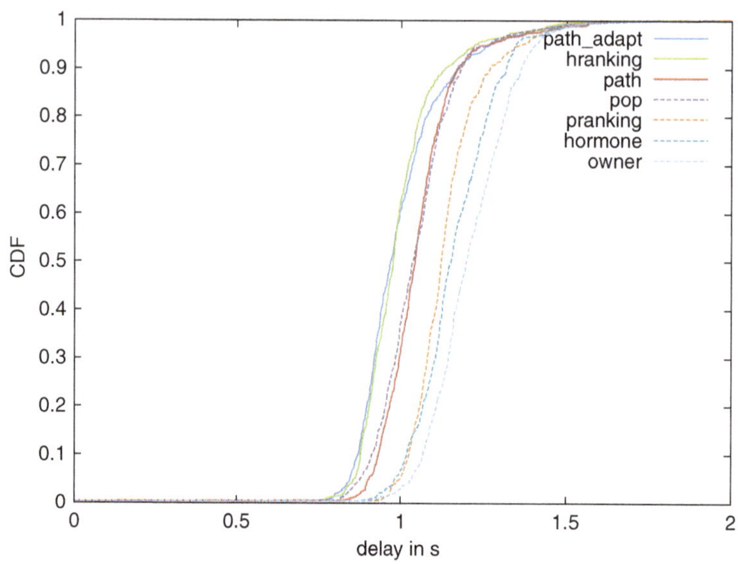

Figure 5 LFU clean-up. Delay of replication mechanisms in combination with LFU clean-up in a 50 nodes random network.

considering the delay of the best combination (here, path replication). However, it has a very bad impact on pop and hormone replication and further results in the highest delay for owner. LRU has the lowest impact on path_adapt in comparison to the base case. Although the delay of the replication mechanisms increases it leads to steeper curves than hormone clean-up and LFU. LFU has the highest impact if considering the delay increase of the best combination (here, path_adapt replication).

hormone replication seems to suffer from any clean-up. The reason is the low number of replicas created by this mechanism. Additionally, the placement is not ideal such that any removal is crucial. pranking replication does not create enough replicas and therefore performs similar to hormone replication. The combination of pop and hormone clean-up does not fit, because the replication mechanism considers long-term popularity (based on the history of the requests) and the clean-up mechanism considers short-term popularity. hranking replication creates many replicas on strategic positions and therefore the transport might be blocked.

Although the clean-up mechanisms have a negative impact on the delay, it is interesting to see whether the utilization of the units has been improved.

The utilization measures the improvement in storage efficiency of the single replication mechanisms. In Figure 6 we show the box plot of all combinations, including the base case. We group the results per replication mechanism.

In comparison to the base case, hormone replication does not take advantage of any clean-up mechanism, because it already creates only a small amount of replicas. The utilization of path and pop replication lowers if clean-up is applied. path_adapt and owner lead in any combination to a stable utilization. The highest utilization improvement, in comparison to the base case, is shown by hranking in combination with hormone clean-up. Also LFU would make a good fit, but results in a higher delay.

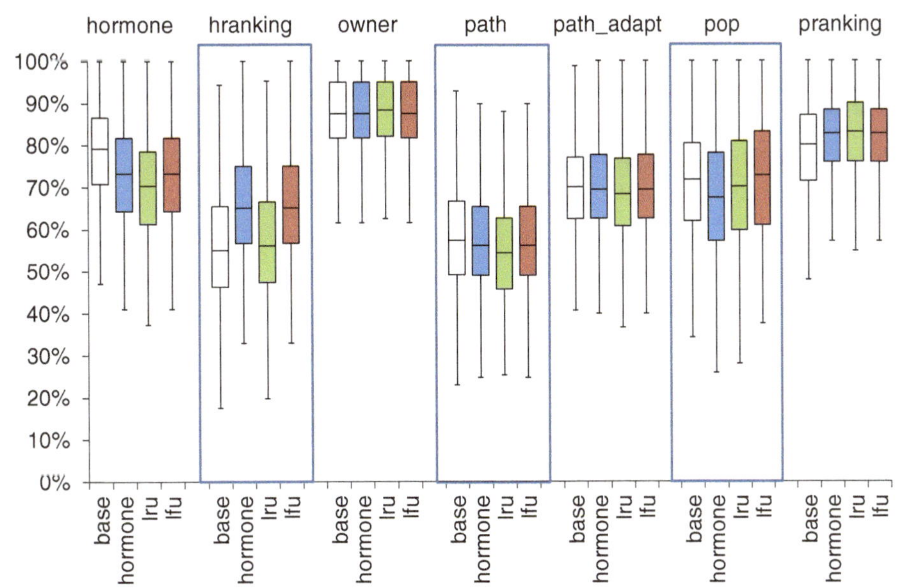

Figure 6 Utilization. Utilization comparison of hormone clean-up, LRU and LFU in a 50 nodes random network.

path replication leads to the lowest utilization if combined with LRU. As already seen in the delay comparison, pop replication with hormone clean-up do not fit. hranking has the worst utilization if combined with LRU. The highest utilization is reached by owner replication, because it creates the lowest number of replicas. However, has also the highest delay.

The utilization shows the effects of the clean-up, however the quality of the approaches are further evaluated by the capability to fulfill user requests. We compare the failed request rate in the following paragraphs.

Figure 7 groups the box plots of each replication and clean-up combination including the base case. One can see that the average failed rate of all combinations is around 10 to 20%. The lowest failed rate is reached by hranking and path replication. The highest failed rate is reached by owner replication. This indicates that replicas on the transport path improve the service quality. As already shown by the delay and the utilization, pop replication with hormone clean-up does not fit. The outliers marked by (*) can be explained by high failure rates at the beginning (within the first 100s) of the simulation. owner and hormone replication show the highest variance of failure rates.

In general, the failure rate increases in comparison to the base case, because the number of replicas is reduced and not all future requests can be predicted. The lower the increase of the failure rate the better the combination fits. However, owner replication with LRU, path_adapt with LRU and path with hormone clean-up manage to lower the failure rate in comparison to the base case. This can be explained by a better balancing of the placement. LRU leads to the lowest increase of the failure rate with most of the approaches, except the combination of hranking and path with hormone clean-up. LFU leads to the highest increase of the failure rate.

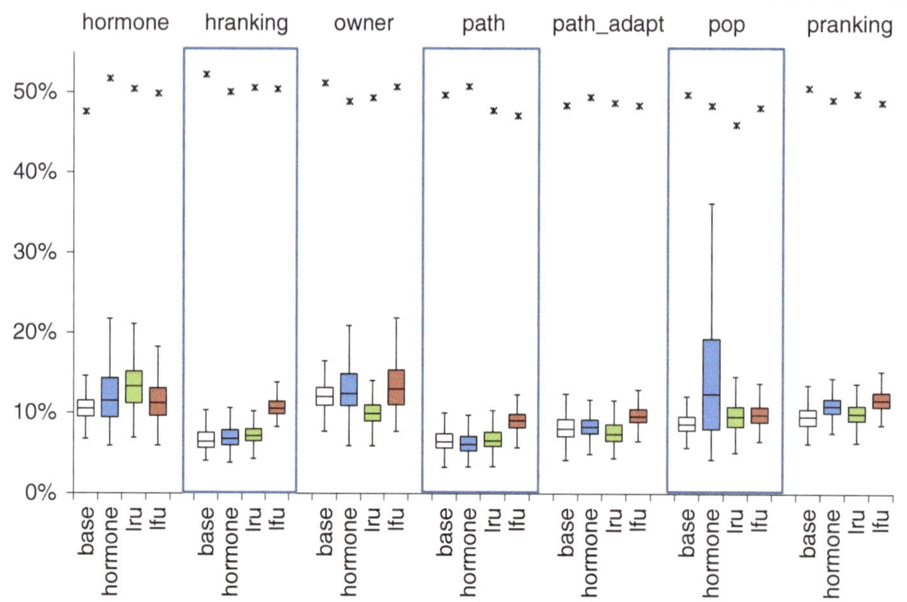

Figure 7 Failed requests. Hormone, LRU and LFU request failed rate comparison in a 50 nodes random network.

Considering delay, utilization and failure rate we select two of the best approaches and continue our discussion only with them. Although path replication has the lowest delay it creates too many replicas and leads to the lowest utilization and therefore we do not further consider this approach. We further omit all cases with LFU clean-up, because it leads to the highest delay, the lowest utilization and the highest failure rates.

There is no replication mechanism that is best in all cases. The second best candidates regarding delay are hranking and path_adapt. hranking has the lowest delay in combination with hormone clean-up. This combination leads to the best improvement of utilization in comparison to the base case and only results in a slight increase of the failure rate. The lowest impact on the delay has the combination of path_adapt and LRU, which is further leading to a lower failure rate, however also a slightly lower utilization as the combination with hormone clean-up. We decide for combining path_adapt with LRU because of higher chances to fulfill user requests.

A discussion of these two approaches is further interesting, because hranking is based on neighborhood knowledge and path_adapt represents an uninformed (traditional) approach.

Impact of node failure

We simulate node failure as randomly chosen nodes being removed periodically one-by-one. We do not handle isolated nodes after peer deletion. Thus, there is a performance gain potential if an overlay algorithm takes care of reconnecting such nodes. In other scenarios we also add new nodes (Szkaliczki et al. 2012), however, the results differ only marginally. We chose 5, 10 and 20 nodes to be removed. If nodes fail it cannot be guaranteed anymore that one copy of a unit exists, however, since a request consists of keywords and not unit ids, it is enough to have at least one unit per keyword. The scenario should further show how good the algorithms adapt the unit placement and change the transport paths.

Figure 8 shows the delay distribution of the hranking algorithm with hormone clean-up. One can see that it is capable of handling loss. Overall, the delay increases slightly, but interestingly the delay of 5, 10 and 20 removed nodes is very similar. This shows that the transport paths adapt fast to the new conditions.

The path_adapt replication with LRU leads to interesting results as shown in Figure 9. Here, the delay of the peer failure scenarios is lower than in the original case. This anomaly can be explained by referencing the clean-up failures of the original scenario. A clean-up fails if on the current node all units are currently in use or there is no copy of the current unit on a neighbor. A disadvantageous replica distribution might be that at every second hop a replica is placed, which means that a high number of replicas exist in the system, but because the nodes only see their neighbors, the units cannot be deleted. In the peer failure scenario, it is likely that a node that blocks the transport is removed and therefore opens the possibility of alternative paths. Hence, to reduce the delay of the original case, a clean-up mechanism would have to check the neighborhood with a larger hop-distance.

In Figure 10 we compare the request failure rates in case of node failure. One can see that the failure rates only increase slightly, which shows that both path_adapt and

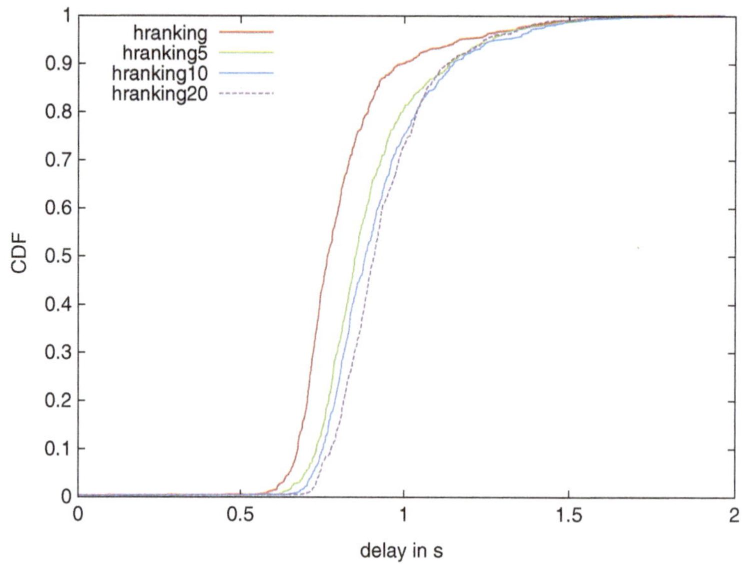

Figure 8 Node failure - hormone ranking. Delay distribution of hormone ranking with hormone clean-up if 5, 10, 20 nodes fail.

`hranking` are robust. However, if the number of failed nodes gets higher than 20, we expect a performance drop.

Both approaches are performing similarly for a small network, however, with a lower delay of `hranking`, but a more stable failure rate is reached by `path_adapt` if nodes fail. `hranking` places the units more efficiently, its utilization can be increased by a clean-up mechanism.

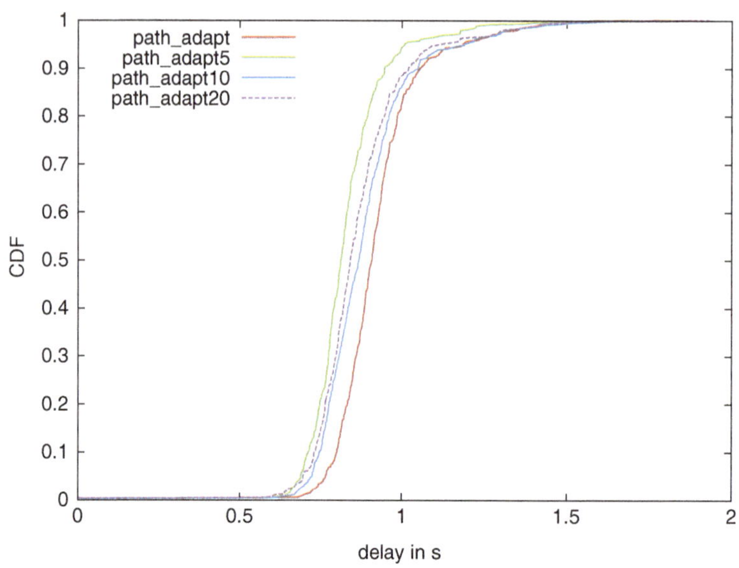

Figure 9 Node failure - path adaptive. Delay distribution of path adaptive with LRU if 5, 10, 20 nodes fail.

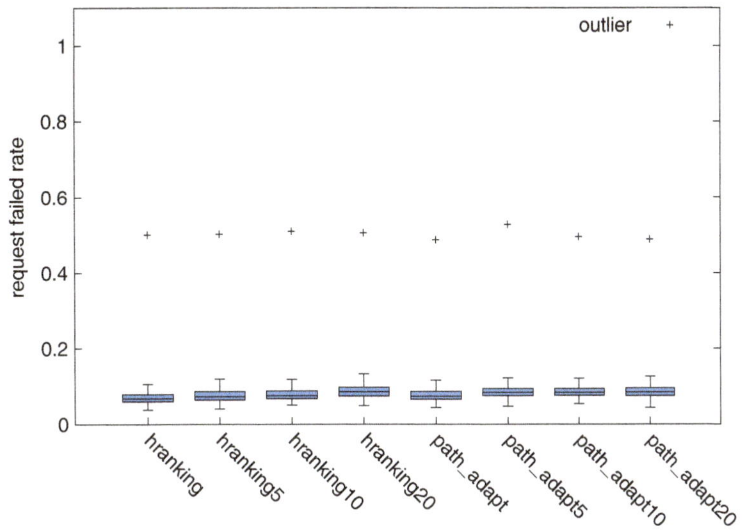

Figure 10 Failed requests and node failure. Failed request rate comparison in case of node failures in a 50 nodes network.

1,000 nodes scale-free network

In this section we evaluate the applicability of our delivery algorithm for scale-free networks. It is shown that the parameters for the 50 node network also work for the 1,000 node network. Specific optimization using the genetic algorithm could lead to even better results. For the delay comparison we immediately include the delay of failed nodes.

Figure 11 depicts the delay of `hranking` with hormone clean-up. In comparison to the small network scenario the delay is increased by around 500 ms and the curve is less steep. The paths in a scale-free network scenario are not as long as

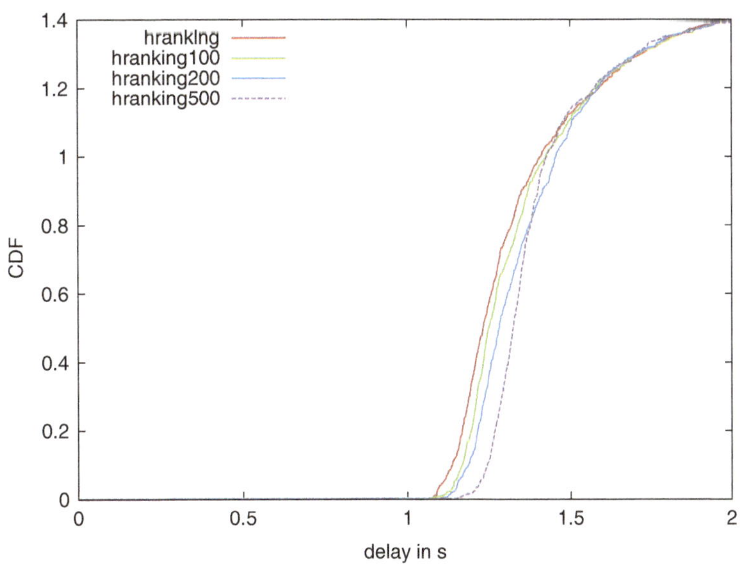

Figure 11 1000 nodes - hormone ranking. Delay distribution of hormone ranking with hormone clean-up if 100, 200, 500 nodes fail.

in a random network. The best performance is reached if the units are placed on high-degree nodes.

If 100, 200 and even 500 nodes fail the delay does not increase considerably, even less than in the small network scenario. Note that also high degree nodes may fail, because the nodes leaving the network are chosen randomly. Thus, only replicating on high-degree nodes would cause service interruptions in case of node failures. The replication and clean-up mechanisms find the right balance of placement, which can be explained by low delay increases.

Figure 12 shows that for `path_adapt` the problem with clean-up failures does not apply such as in the 50 nodes network. The reason is the network structure and its rather low diameter. Therefore, the delay of `path_adapt` replication is only a bit higher to the delay of `hranking`.

Both algorithms show slight increases of request failures also in the presence of node failure (see Figure 13). Overall, the request failed rate is 20% in comparison to around 10% in the small network case. Also the variance of the values is higher. The `hranking` replication is more robust than `path_adapt`.

Discussion

Although the results of the replication mechanisms are promising, the clean-up has a negative impact on the delay, which means that the goals of the clean-up are not reached. Ideally, a combination of replication and clean-up should lead to low delay, low failed rate and high utilization. If the settings regarding unit deletion are as strict as in this paper, `path` replication with hormone clean-up, although inefficient, performs best regarding delay. Alternatives could be `path_adapt` replication with LRU and `hranking` replication with hormone clean-up. The node failure scenarios showed that there are still nodes, which block the transport of units. To solve this issue the settings could be less strict regarding the deletion policy. Instead of deleting a unit only if there is a copy of it in the

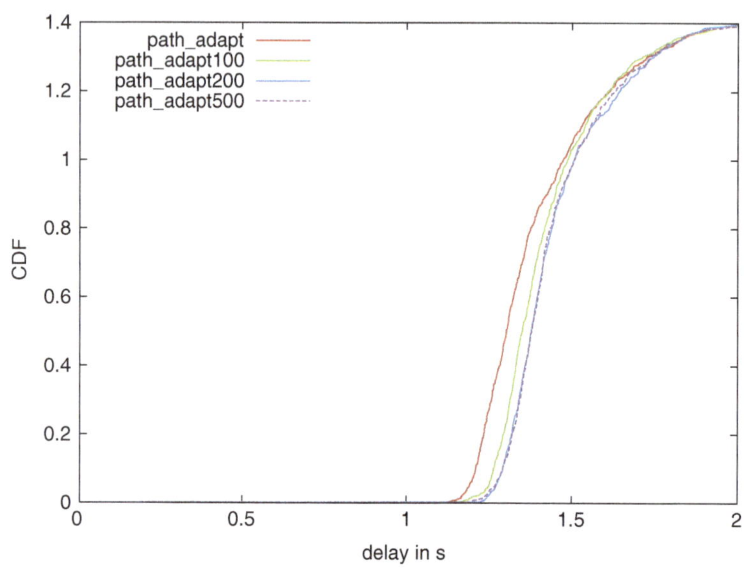

Figure 12 1000 nodes - path adaptive. Delay distribution of path adaptive with LRU if 100, 200, 500 nodes fail.

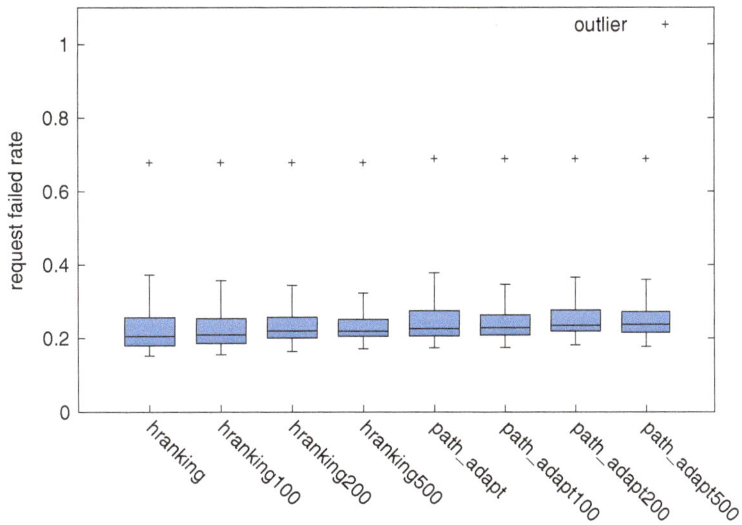

Figure 13 1000 nodes - failed requests. Failed request rate comparison in case of node failures in a 1000 nodes network.

neighborhood, it could be weakened to delete a unit if another unit covering the same keyword is in the neighborhood.

Conclusion

Replication is a typical measure to increase robustness, but usually nodes in a network do not provide unlimited storage space. A clean-up or replacement mechanism is necessary to balance the available storage space. If a network of nodes is fully connected, and a central server knows how many copies exist in the system, the decision to delete one of the copies is easy (although it is not easy to choose the optimal number of replicas in the system). In self-organized networked systems, where the knowledge of a node is limited to its neighborhood, storage balancing is a challenge. In this paper we investigated different replication mechanisms in combination with three clean-up mechanisms, under the constraint that content must not vanish from the system. The basis is a hormone-based delivery algorithm that allows for quality-aware content distribution based on local knowledge. We evaluated the delay impact of the clean-up mechanisms, as well as the impact on the utilization of replicas. Although reducing the number of replicas, the robustness of the system is still given. Similar results are reached if the system is applied to a scale-free network of 1,000 nodes.

The evaluated clean-up mechanisms are either uninformed, such as least recently used (LRU) and least frequently used (LFU), or use the hormone information of the neighborhood. We showed that it is necessary to match the replication mechanisms with the clean-up mechanisms, otherwise a major delay drop can be experienced. We also found out that uninformed replication mechanisms like path adaptive replication (random replication on the delivery path) in combination with LRU might be an alternative for dynamic delivery that does not rely on any indicator of interest such as hormones or pheromones. The hormone-based clean-up mechanism might be mapped to pheromone-based delivery systems, where specific ants could check the neighborhood for interest in a given content unit. As we have seen that any of the clean-up mechanisms influences the delay, future work should regard the replication mechanisms, i.e., to avoid unnecessary replicas in the system. Furthermore, an alternative deletion policy might be introduced that is less restrictive than the one chosen in this paper.

Endnotes

[a] The simulator used in this section is published under GNU GPL at http://code.google.com/p/videonetwork/

[b] SOMA Web-page http://soma.lakeside-labs.com/?p=253

[c] For a more detailed discussion of the base case please refer to (Sobe et al. 2011a)

Competing interests

Both authors declare that they have no competing interests.

Authors' contributions

AS conducted the simulation scenarios, implemented the artificial hormone algorithm and drafted the manuscript. WE implemented the genetic algorithm, helped with the concept of the artificial algorithm and contributed with text to the manuscript. All authors read and approved the final manuscript.

Acknowledgements

This work was supported by the Lakeside Labs GmbH, Klagenfurt, Austria, and funding from the European Regional Development Fund and the Carinthian Economic Promotion Fund (KWF) under grant KWF 20214|21532|32604 (MESON), and KWF 20214|22935|34445 (SMART MICROGRID). We would like to thank Lizzie Dawes for proofreading the paper.

Author details

[1]Institute of Networked and Embedded Systems/Lakeside Labs, Alpen-Adria-Universität Klagenfurt, Klagenfurt, Austria. [2]Computer Science Department, University of Neuchâtel, Neuchâtel, Switzerland. [3]Complex Systems Engineering, Universität Passau, Passau, Germany.

References

Androutsellis-Theotokis S, Spinellis D: **A survey of peer-to-peer content distribution technologies.** *ACM Comput Surv* 2004, **36**(4):335–371.

Böszörmenyi L, del Fabro M, Kogler M, Lux M, Marques O, Sobe A: **Innovative directions in self-organized distributed multimedia systems.** *Multimedia Tools Appl, Springer* 2011, **51**(2):525–553.

Brinkschulte U, Pacher M, Renteln A: **Towards an artificial hormone system for self-organizing real-time task allocation.** In *Software Technologies for Embedded and Ubiquitous Systems, Volume 4761 of Lecture Notes in Computer Science*. Edited by Obermaisser, R, Nah Y, Puschner P, Rammig F. Berlin Heidelberg: Springer; 2007:339–347.

Cha M, Kwak H, Rodriguez P, Ahn Y, Moon S: **I tube, you tube, everybody tubes: analyzing the world's largest user generated content video system.** In *Proceedings of the 7th ACM SIGCOMM Conference on Internet Measurement*. San Diego: ACM; 2007:1–14.

Clarke I, Sandberg O, Wiley B, Hong T: **Freenet: A distributed anonymous information storage and retrieval system.** In *Designing Privacy Enhancing Technologies Volume 2009 of Lecture Notes in Computer Science*. Edited by Federrath H. Berlin Heidelberg: Springer; 2001:46–66.

Cohen B: **BitTorrent Protocol Specification.** 2003. [http://www.bittorrent.org/beps/bep_0003.html]

Cohen E, Shenker S: **Replication strategies in unstructured peer-to-peer networks.** In *Proceedings of the 2002 Conference on Applications, Technologies, Architectures, and Protocols for Computer Communications*. SIGCOMM '02. New York: ACM; 2002:177–190.

Dorigo M, Birattari M, Stutzle T: **Ant colony optimization.** *IEEE Comput Int Mag* 2006, **1**(4):28–39.

Elmenreich W, Klingler G: **Genetic evolution of a neural network for the autonomous control of a four-wheeled robot.** In *Sixth Mexican International Conference on Artificial Intelligence, Special Session*. Aguascalientes: IEEE; 2007:396–406.

Eppstein D, Wang J: **A steady state model for graph power laws.** In *International Workshop on Web Dynamics, WebDyn 2002*. Honolulu: arxiv e-print, arxiv:cs/0204001; 2002.

Forestiero, A, Mastroianni C, Spezzano G: **Building a peer-to-peer information system in grids via self-organizing agents.** *2007 2nd Bio-Inspired Models Netw, Inf Comput Syst* 2007, **6**(2):125–140.

Forestiero, A, Mastroianni, C, Spezzano, G: **QoS-based dissemination of content in grids.** *Future Gen Comput Syst* 2008, **24**(3):235–244.

Gubner, J: *Probability and Random Processes for Electrical and Computer Engineers*. New York: Cambridge University Press; 2006.

Hausheer, D, Nikander P, Fogliati V, Wünstel K, Callejo M, Jorba S, Spirou S, Ladid L, Kleinwächter W, Stiller B, Behrmann M, Boniface M, Courcoubetis C, Man-Sze L: **Future internet socio-economics - challenges and perspectives.** In *Towards the Future Internet*. Netherlands: IOS Press Amsterdam; 2009:1–11.

Herrmann K: **Self-organizing replica placement-a case study on emergence.** In *Self-Adaptive and Self-Organizing Systems, 2007. SASO'07*. Cambridge: IEEE; 2007:13–22.

Kulis B, Grauman K: **Kernelized locality-sensitive hashing for scalable image search.** In *IEEE 12th International Conference on Computer Vision, ICCV2009*. Kyoto: IEEE; 2009:2130–2137.

Lee S B, Muntean GM, Smeaton AF: **User-centric utility-based data replication in heterogeneous networks.** In *IEEE International Conference on Communications Workshops, ICC Workshops 2008*. Beijing: IEEE; 2008a:290–294.
Lee S B, Muntean GM, Smeaton AF: **Smart PIN: Utility-based replication and delivery of multimedia content to mobile users in wireless networks.** In *2008 IEEE International Symposium on Broadband Multimedia Systems and Broadcasting*. Las Vegas: IEEE; 2008b:1–8.
Leontiadis E, Dimakopoulos V, Pitoura E: **Creating and maintaining replicas in unstructured peer-to-peer systems.** In *Euro-Par 2006 Parallel Processing Volume 4128 of Lecture Notes in Computer Science*. Edited by Nagel W, Walter W, Lehner W. Berlin Heidelberg: Springer; 2006:1015–1025.
Lv Q, Cao P, Cohen E, Li K, Shenker S: **Search and replication in unstructured peer-to-peer networks.** In *Proceedings of the 16th International Conference on Supercomputing*. New York: ACM; 2002:84–95.
Mamei M, Menezes R, Tolksdorf R, Zambonelli F: **Case studies for self-organization in computer science.** *J Syst Arch* 2006, **52**(8–9):443–460.
Michlmayr E: *Ant algorithms for self-organization in social networks*. Austria: Technical University Vienna; 2007.
Rong L: **Multimedia resource replication strategy for a pervasive peer-to-peer environment.** *J Comput* 2008, **3**(4):9–15.
Sobe A: *Self-organizing multimedia delivery*. Klagenfurt: Alpen-Adria Universität; 2012.
Sobe A, Böszörmenyi L: **Non-sequential Multimedia Caching.** In *2009 First International Conference on Advances in Multimedia (MMedia 2009)*. Colmar: IARIA/IEEE; 2009:158–161.
Sobe A, Elmenreich W, Böszörmenyi L: **Towards a self-organizing replication model for non-sequential media access.** In *ACM MM Workshop on Social, Adaptive and Personalized Multimedia Interaction and Access (SAPMIA)*. Florence: ACM; 2010:3–8.
Sobe A, Elmenreich W, Böszörmenyi L: **Replication for bio-inspired delivery in unstructured peer-to-peer networks.** In *Ninth Workshop on Intellingent Solutions for Embedded Systems (WISES 2011)*. Regensburg: IEEE; 2011a:6.
Sobe A, Elmenreich W, Böszörmenyi L: **Storage balancing in self-organizing multimedia delivery systems.** *Technical Reports, Institute of Information Technology, Alpen-Adria Universität Klagenfurt, Austria, arxiv e-print, arxiv:1111.0242* 2011b, **TR/ITEC/01/2.13:**16.
Szkaliczki T, Sobe A, Böszörmenyi L: **Discovering bounds of performance of self-organizing content delivery systems.** In *Self-Adaptive and Self-Organizing Systems Workshops (SASOW), 2012 IEEE Sixth International Conference on*; 2012:97-104. doi:10.1109/SASOW.2012.26.
Xu Qz, Wang L: **Recent advances in the artificial endocrine system.** *J Zhejiang Univ SCI C* 2011, **12**(3):171–183.
Yamamoto H, Maruta D, Oie Y: **Replication methods for load balancing on distributed storages in P2P networks.** In *The 2005 Symposium on Applications and the Internet, SAINT 2005*. Trento: IEEE; 2005:264–271.

3

Modeling population growth in online social networks

Konglin Zhu[1*], Wenzhong Li[1,2] and Xiaoming Fu[1]

*Correspondence: zhu@cs.uni-goettingen.de
[1]Institute of Computer Science, University of Goettingen, Goettingen, Germany
Full list of author information is available at the end of the article

Abstract

Purpose: Online social networks (OSNs) are now among the most popular applications on the web offering platforms for people to interact, communicate and collaborate with others. The rapid development of OSNs provides opportunities for people's daily communication, but also brings problems such as burst network traffic and overload of servers. Studying the population growth pattern in online social networks helps service providers to understand the people communication manners in OSNs and facilitate the management of network resources. In this paper, we propose a population growth model for OSNs based on the study of population distribution and growth in spatiotemporal scale-space.

Methods: We investigate the population growth in three data sets which are randomly sampled from the popular OSN web sites including Renren, Twitter and Gowalla. We find out that the number of population follows the power-law distribution over different geographic locations, and the population growth of a location fits a power function of time. An aggregated population growth model is conducted by integrating the population growth over geographic locations and time.

Results: We use the data sets to validate our population growth model. Extensive experiments also show that the proposed model fits the population growth of Facebook and Sina Weibo well. As an application, we use the model to predict the monthly population in three data sets. By comparing the predicted population with ground-truth values, the results show that our model can achieve a prediction accuracy between 86.14% and 99.89%.

Conclusions: With our proposed population growth model, people can estimate the population size of an online social network in a certain time period and it can also be used for population prediction for a future time.

Keywords: Spatiotemporal scale-space; Population distribution; Population growth; Online social networks

Background

Nowadays online social networks (OSNs) are considered as the most popular applications on the web, which offer platforms for people to interact, communicate and collaborate with others. The user population of online social networks is growing expeditiously. It is reported that Facebook (2013) has reached 900 million users in April 2012. Meanwhile, Twitter (2013) has also surpassed 500 million users in July 2012. The rapid development of OSNs facilitates people's daily communications. However, the growth of user population also causes problems to service providers, such as overload of servers. One example is the "fail whale" phenomenon in Twitter, where the requested page returns a "fail whale" image when too many burst requests occur.

The issues and patterns of population growth in OSNs have drawn much attention from the academia and many works have been done in the past years. A study of micro evolution on OSNs (Leskovec et al. 2008) captured the best fits of population growth in four different OSNs and showed that the growth tendency varies with time. Torkjazi et al. (2009) and Rejaie et al. (2010) observed S-shaped population growth: it experiences a slow growth in the beginning, following a period of exponential growth and finally a significant and sudden slow down in the growth of the population. However, most of these studies fail to provide a theoretical model to describe the population growth in OSNs. Besides, existing works study population growth only in the temporal dimension, and they lacks concern of the dynamics in the geographic scale.

In this paper, we investigate the population growth of OSNs from spatiotemporal scale-space. Our investigation is based on three data sets randomly sampled from the popular OSN website including Renren (2013), Twitter (2013) and Gowalla (2011), from which we explore their population distributions over various geographic locations and time-varying properties on population growth. We find out that in the spatial scale, the population size follows a power-law distribution over geographic locations. In the temporal scale, the population growth in the largest populated location is revealed to fit a power function increasing with time. The number of populated locations also increases as a power function as time. Based on these observations, we propose an aggregated population growth model by integrating the population growth over geographic locations and time. Theoretical analysis is presented to derive this model and comprehensive experiments are conducted to verify its effectiveness. It is shown that the proposed model fits well for population growth in large scale rapid growing OSNs such as Facebook (2013) and Sina Weibo (2013). As an application, we further utilize the model to predict population growth in three data sets, which illustrates that our model can achieve a prediction accuracy between 86.14% and 99.89%.

There are several applications of our work. It has significant meanings for Internet Service Providers (ISPs) to understand the population growth of OSN users, which will further reveal the user interaction patterns and network traffic patterns. It is also benign for the OSN web sites to deploy servers and cast advertisements on the base of population growth model. The third-party service providers can analyze the service market by the model and further optimize their resource deployment and investment.

Data

To conduct our analysis, we collect data from three online social network sites: *Renren*, a social-based application service, *Twitter*, a social-based media service and *Gowalla*, a location-based online social service. Renren, established in December 2005 and now with 160 million users, is a Chinese online social network which organizes users into membership-based networks representing schools, companies and geographic locations. It allows users to post short messages known as status, blogs and pictures. It also allows people to share contents such as videos, articles and pictures. Twitter, with over 500 million users, launched in July 2006, is known as its microblogging services by which users can write any topic within the 140 characters limit. Such kind of short message is known as tweet. A follower can follow any other users and receive any kind of tweets from his/her followings. Varied from above-mentioned two online social networks, Gowalla is a location-oriented online social service. People are allowed to check-in their visiting places via mobile devices. It is launched in 2007 and closed in 2012 with approximately 600,000 users.

We collect the Renren and Twitter data sets by crawling from their sites. We start our crawling with randomly selected users from the largest weakly connected component (WCC). Following friends' links in the forward direction in a breadth first search (BFS) fashion, we collect a sample of each social network. To eliminate the degree bias caused by BFS, we launch the BFS-bias correction procedure described in (Kurant et al. 2011). Furthermore, according to the estimation method of the size of social networks by Katzir et al. (2011), we believe the quality and quantity of our data sets are enough to reveal the laws of population growth in OSNs.

In order to capture the growth of population in different geographic locations, we need to know the account creation time and geographic information of each user. We trace user account creation time in Twitter from user profile. However, we cannot explicitly retrieve account creation time from user profile in Renren. To estimate the account creation time precisely, we use the time of a user's first activity such as updating status, posting a blog or interacting with friends as the time point when the account creates. Meanwhile, we seek users' geographic locations from user profiles and choose users with valid geographic information to compose our data sets.

The Gowalla data set, obtained from public source (Cho et al. 2011), contains more than 100,000 users, as well as their social relations and check-in histories. We find the user registration time by their first check-ins. To reveal users' geographic information, similar as (Cho et al. 2011), we infer a user's location by compartmentalizing the globe into 25 by 25 km cells and defining the location as the cell with the most check-ins.

The statistics of three data sets are shown in Table 1. Renren data set (Renren) contains around 1 million users and covers $10,039$ locations. It records user growth over the period of January 2006 to December 2010 (60 months). Twitter data set (Twitter) consists of more than 250 thousand users covering $8,929$ locations. We collect user population growth over the period of August 2006 and October 2010 (51 months). Gowalla data set (Gowalla) has around 100 thousand nodes, with $5,088$ populated locations. It contains the population growth ranging from February 2009 to October 2010 (21 months).

Table 1 Statistics of data sets

	Renren	Twitter	Gowalla
Nodes	997,849	257,929	107,089
Locations	10,039	8,929	5,088
Period	01/06 – 12/10	08/06 – 10/10	02/09 – 10/10

The table presents the detail information of Renren, Twitter and Gowalla data sets. It contains the number of nodes, the number of locations and the time period of the data set.

Methods

We present the methods for modeling of population growth in online social networks in this section.

Basic approach

To study the population growth in OSNs, we first illustrate the basic approach of modeling the population growth in spatiotemporal dimension.

The population in OSNs grows over locations and time. In spatial aspect, people from different geographic locations may register as users in an OSN, thus people from more and more locations join in the network. The OSN expanding from locations to locations, leads to the growth of population spatially. At the same time, the population in each geographic location grows in temporal scale. People in a geographic location may be attracted to join in the network from time to time and thus the location will have more and more people. Therefore, combining spatial and temporal effects, we model the population growth as the accumulation of populations in different geographic locations, while population in each location changes as a function of time. We describe the population growth in spatial and temporal dimensions as follows.

As the first step, we consider the population growth model in the spatial dimension. The aggregated population GP is a sum of populations in all populated locations on a certain time point. It can be formulated as an accumulation function:

$$GP = \sum_{x=1}^{M} S_x, \tag{1}$$

where S_x denotes the population size in location x. M is the total number of populated locations. To calculate this formulation, one needs to know the population size of every location, which is not scalable in large online social networks. We take one step further. Instead of numerating all population sizes, we use the proportion of the number of locations with population size s over the total number of populated locations, known as population distribution, which is denoted as P_s, the total number of populated locations L and the largest population size N to construct the formulation of aggregated population:

$$GP = \sum_{s=1}^{N} P_s * L * s, \tag{2}$$

This formulation describes the aggregated population on a certain time point in spatial aspect.

Now we consider the population growth process with temporal factor. The aggregated population growth is a dynamic process, with population in each location growing as a function of t. Therefore, we consider N, L and P_s are time-dependent functions, where $N = n(t)$, $L = l(t)$ and $P_s = P(s, t)$. The discrete accumulation of Eq. 2 can be approximated as integral as following:

$$GP(t) = \int_{1}^{n(t)} (P(s, t) * l(t) * s)ds. \tag{3}$$

The formulation reflects not only the spatial characteristics that it is aggregation of populations in different geographic locations, but also temporal factor that the population growth is a dynamic process as a function of time.

So far, we propose a population growth model in spatiotemporal perspective. To specify this model, we need to study three time-dependent functions: the dynamics of population distribution $P(s, t)$, the growth function of populated locations $l(t)$ and the growth function of the largest population $n(t)$ in the following subsections.

Population distribution: $P(s, t)$

The population distribution reveals the proportions of different population sizes in an OSN. We investigate the population distribution by drawing the log-log plots of probability density function (PDF) of populations in different geographic locations of three data sets as shown in Figure 1 with respect to different time points. The observation to the figures shows that the population distributions are close to each other in different time periods. In particular, the population distributions of Renren are close to each other from 10*th* month to 50*th* month, as shown in Figure 1a. The similar phenomenon can also be observed from Twitter from 8*th* month to 48*th* month in Figure 1b and Gowalla from 5*th* month to 20*th* month in Figure 1c. It allows us to use one distribution curve approximately fit all population distributions in one data set, which means the population distributions of various periods are roughly can be fitted with one identical distribution function. Therefore, $P(s, t)$ is a time-independent function. Then we find that the population distribution of each data set is seemingly a straight line in the log-log scale, which indicates that the distribution can be fitted with a power-law function. To confirm this observation, we conduct the hypothesis testing as described in (Clauset et al. 2009), which uses a *goodness-of-fit test* to determine plausibility of the power-law fit. It generates a *p*-value to quantify the plausibility. If *p*-value is close to 1, the fit is considered as plausible to the empirical data. Otherwise, it is considered as implausible. The results of the hypothesis tests show that $p = 1.0$ for Renren, $p = 0.90$ for Twitter and $p = 0.95$ for Gowalla. All three values are close to 1. Therefore, we say that the power-law function is plausible to fit three data sets.

Alternatively, we conduct alternative hypothesis testing regarding the population distribution by the *likelihood ratio test* (Clauset et al., 2009), which suggests that the distribution is a power-law if the likelihood ratio between the alternative and power-law distribution is positive. We calculated the likelihood ratio of exponential distribution compared with power-law distribution, which is 2.23, and the likelihood ratio of log-normal distribution compared with power-law distribution, which is 0.12. The results suggest that power-law is the best distribution to represent population distribution.

We fit each distribution in figures with maximal likelihood estimation (MLE) (Newman 2005, Clauset et al. 2009). The fitting results are shown in dashed lines. It shows that Renren data set has a power-law exponent of 1.4, Twitter has a power-law exponent of 1.78 and Gowalla has a power-law exponent of 1.4.

We give the mathematical model of the population distribution in different time periods as:

$$P(s, t) = \varphi * s^{-\lambda}, \tag{4}$$

where φ is the scaling factor and λ is the power-law exponent. The equation reveals that the population distribution in different time periods is a power-law function, and it is independent from time.

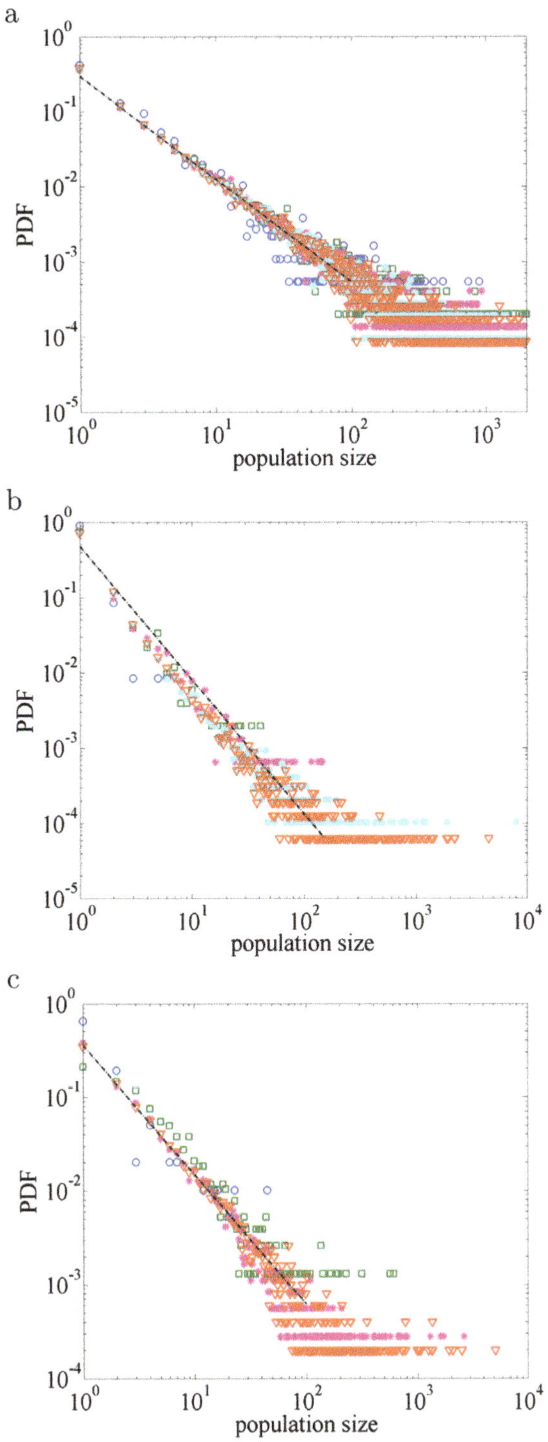

Figure 1 Population distribution of various periods in three data sets. (a) Renren, the distribution of population size in 10th month (circle), 20th month (square), 30th month (star), 40th month (solid circle) and 50th month (triangle). Five curves are close to each other and seemingly follow the same distribution. The dashed line is the power law distribution: $y \sim x^{-1.4}$. **(b)** Twitter, the distribution of population size in 8th month (circle), 18th month (square), 28th month (star), 38th month (solid circle) and 48th month (triangle). Five curves are close to each other and seemingly follow the same distribution. The dashed line is: $y \sim x^{-1.78}$. **(c)** Gowalla, the distribution of population size in 5th month (circle), 10th month (square), 15th month (star) and 20th month (triangle). Four curves are close to each other and seemingly follow the same distribution. The dashed line is: $y \sim x^{-1.4}$.

Populated locations: $l(t)$

To model the population growth in OSNs, one important aspect is to understand the growth of populated locations. In this subsection, we investigate the growth of populated locations.

The growth of populated locations is a function of time t and defined as $l(t)$. To formulate $l(t)$, we draw the number of populated locations of three data sets as a function of time in the log-log scale, as shown in Figure 2a, 2b and 2c. The numbers of populated locations are approximately straight lines. Again, we use MLE to fit them shown as dashed lines in figures. The fitting function can be formulated as:

$$l(t) = \eta * t^{\varepsilon}, \tag{5}$$

with scaling parameter η and the power exponent ε. The power exponent of Renren is 1.26, of Twitter is 1.96 and of Gowalla is 1.62.

In a summary, we find that the growth of the number of populated locations in OSNs is a power function of time.

The largest population: $n(t)$

As we model the population growth as an accumulation of populations in various locations, the largest population as the upper bound of the formulation also needs to be investigated.

To measure the largest population size, we select the location with the largest population sizes in $10th$ month from three data sets and study the growth of their populations as a function of time, respectively. We define the growth function of the largest population as $n(t)$. To obtain $n(t)$, we grab the largest population size from month to month in each data set and plot the growth of the largest population size as a function of time. The growth trends of them are shown in Figure 3a, 3b and 3c. Similar to the analysis of populated location growth, the growth of the largest population size can also be fitted using a power function as the following:

$$n(t) = a * t^{b} + c. \tag{6}$$

Besides the power component $a * t^{b}$, there is a constant number c added to the power function. We use this function to fit the largest population size in each data set shown as the solid lines in the figures. Specifically, the power parameter of Renren is 1.31, of Twitter is 2.97 and it is 1.61 for Gowalla.

Therefore, the growth function of the largest population size is a power function. The population growth of a location will affect the aggregated population growth. We will give the detailed model of the population growth in the following subsection.

The population growth model

Given the distribution of population, the growth function of populated locations and the largest population size, we insert the above expressions of $p(s,t)$, $l(t)$, and $n(t)$ as shown in Eq. 4, Eq. 5 and Eq. 6 into Eq. 3. Then we have

$$\begin{aligned} GP(t) &= \int_{1}^{a*t^{b}+c} ((\varphi * s^{-\lambda}) * (\eta * t^{\varepsilon}) * s) ds \\ &= \frac{\eta \varphi t^{\varepsilon}}{2-\lambda} [\, (a * t^{b} + c)^{2-\lambda} - 1\,]. \end{aligned} \tag{7}$$

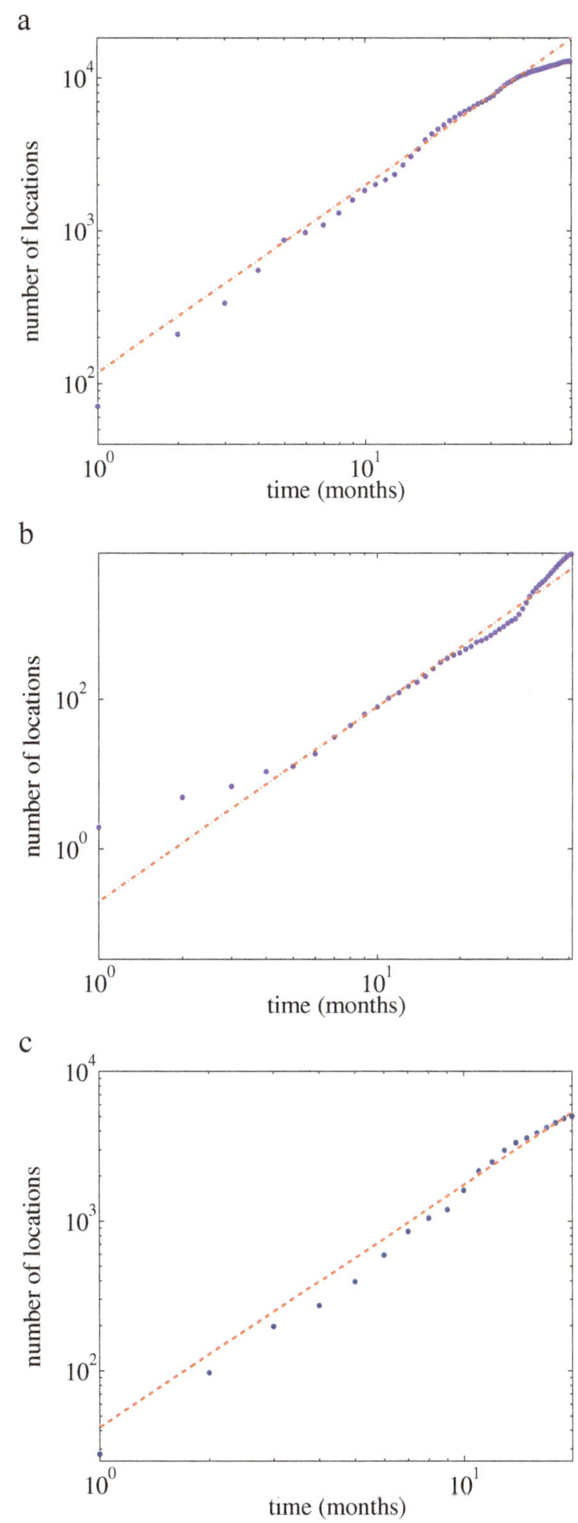

Figure 2 The growth of populated locations. (a) Renren, the growth of populated location as a function of time follows a power function $y \sim x^{1.26}$. **(b)** Twitter, the growth of populated location as a function of time follows a power function $y \sim x^{1.96}$. **(c)** Gowalla, the growth of populated location as a function of time follows a power function $y \sim x^{1.62}$.

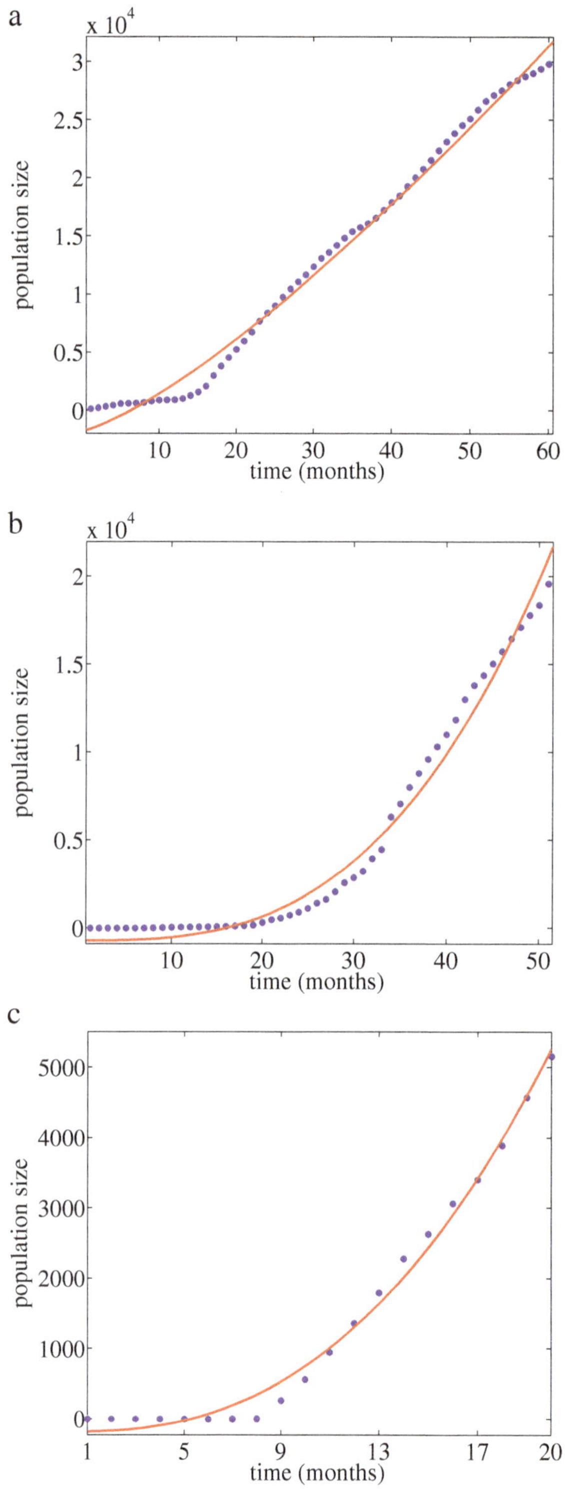

Figure 3 The population growth of the largest populated location. (a) Renren, the growth of population as a function of time in the largest populated location, follows a power function: $n(t) = 155.8t^{1.31} - 1706$. **(b)** Twitter, the growth of population as a function of time in the largest populated location, follows a power function: $n(t) = 0.19t^{2.97} - 678.8$. **(c)** Gowalla, the growth of population as a function of time in the largest populated location, follows a power function:$n(t) = 68.83t^{1.61} - 243.9$.

The above equation reveals that the population growth is a function of time, and it is similar to power function. The model describes the aggregated population growth of online social networks in both temporal dimension and spatial dimension.

Results

To present the effectiveness of our population growth model, we evaluate our proposed model from three aspects. First, we verify the model in the early stage population growth of three data sets. Then we evaluate the full population growth in Facebook (2013) and Sina Weibo (2013) by our model. Finally, as an application of our model, we use it to predict the population growth on the latter part populations of three data sets.

Model verification in three data sets

We verify our population growth model by estimating the early stages of population growth in three data sets (i.e. first 35 months' population growth of Renren, the first 40 months' population growth of Twitter and the first 14 months' population growth of Gowalla).

We use the population growth function as shown in Eq. 7 to fit the early stages of population growth in three data sets. We plot the values estimated by the model and the monthly ground-truth population values in three data sets as shown in Figure 4. The points in the figure denote the monthly ground-truth values of population size in each data set, and the dashed lines are the estimated value by the growth model.We compare the estimated value with the value from the data sets. Figure 4a shows the comparison of Renren. The minor difference indicates the model can represent population growth in Renren well. Although the modeled curve of Twitter has the difference with the data set value in the middle of the time period, as shown in Figure 4b, we observe that the model tends to fit the data set population very well at the end of the period. The model shows very good performance in Gowalla as shown in Figure 4c.

In a summary, the verification of the population growth model in three data sets validate the correctness of the model.

Model verification in full OSN populations

In this subsection, we verify our model in full population of Facebook (2013) and Sina Weibo (2013) to show its effectiveness in two OSNs with integral populations.

We conduct experiments on the full user populations in Facebook and Sina Weibo. Facebook was launched in 2004 and reached 800 million in September 2011. The population growth trend is shown in Figure 5a. Sina Weibo, a Chinese microblogging web site, was launched in 2009. Its population has reached 300 million in February 2012. The growth trend of the population is shown as Figure 5b. For each population growth, we use Eq. 7 to fit the population. We plot the value estimated by the model shown as dashed lines in Figure 5a and Figure 5b, respectively. By comparing the estimated values with the ground-truth population values, we find that the estimated values are close to the real values of populations, which suggests that our model is valid for population growth of Facebook and Sina Weibo.

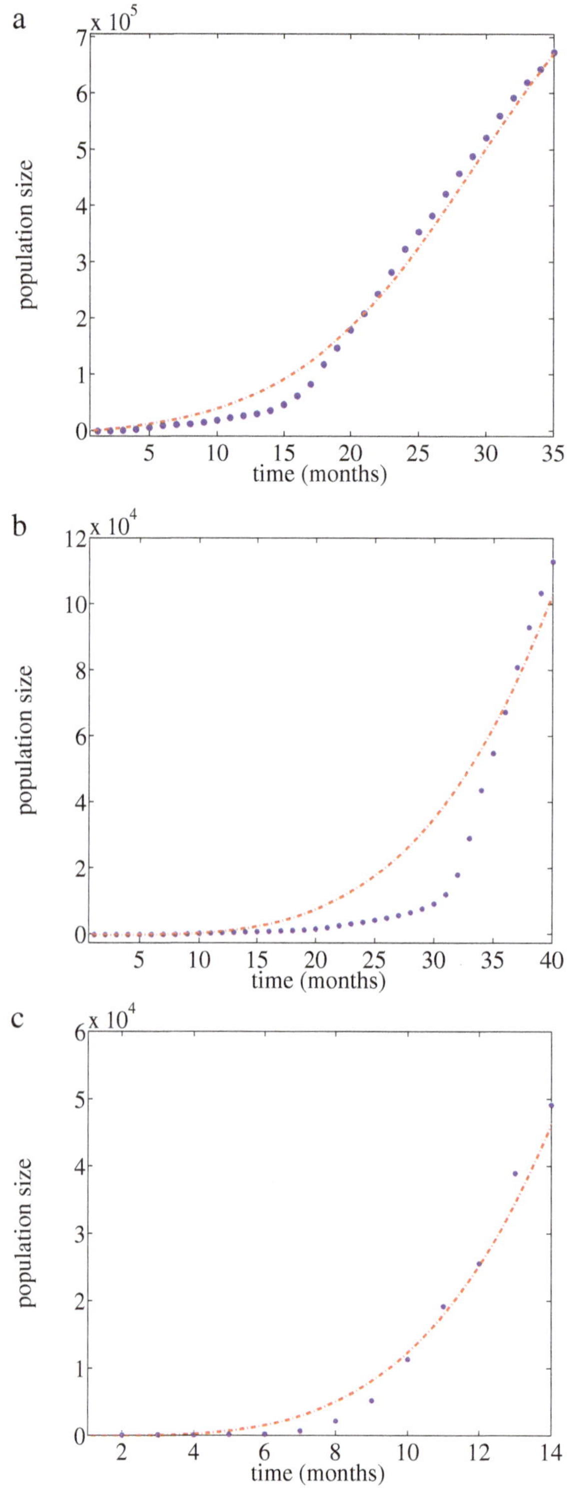

Figure 4 The verification of the model in three data sets. (a) Renren, the ground-truth value of the first 35 months' population size (as points) and the estimated population size by the model (as dashed lines). **(b)** Twitter, the ground-truth value of the first 40 months' population size (as points) and the estimated population size by the model (as dashed lines). **(c)** Gowalla, the ground-truth value of first 14 months' population size (as points) and the estimated population size by the model (as dashed lines).

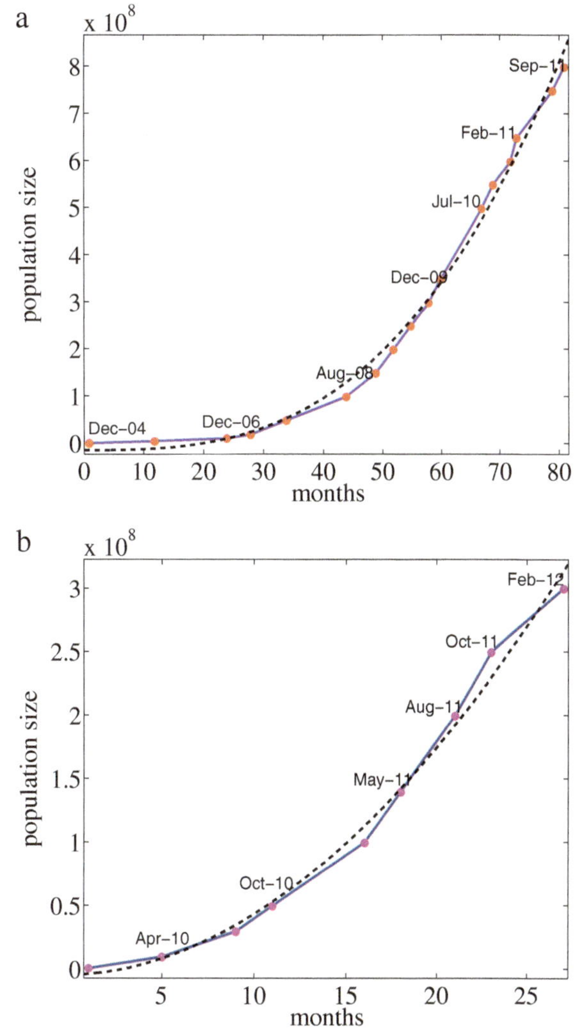

Figure 5 The verification of the model in full populations. (a) Facebook, the ground-truth value of population size (as pointed lines) and the estimated population size by the model (as dashed lines). **(b)** Sina Weibo, the ground-truth value of population size (as pointed lines) and the estimated population size by the model (as dashed lines).

Using the model for population prediction

As an application, we use our model to predict the latter part population growth of three OSN data sets.

Through estimating the early stage population size in three data sets, we have built the population growth model with parameters. Now, we use the established multi-step ahead model to predict the population growth from the 36*th* to 60*th* month in Renren, from the 41*st* to 51*st* month in Twitter, and from 14*th* to 20*th* month in Gowalla. We plot Figure 6 to show the ground-truth values of three data sets and their prediction results. The points are the population size of each data set at different times. The dashed lines are the predicted population size. Figure 6a presents the results of Renren. The predicted population fits the data set population well at the beginning. The data set population growth becomes slower than the predicted value as time changes. The prediction results of Twitter and Gowalla are also close to real populations as shown in Figure 6b and Figure 6c.

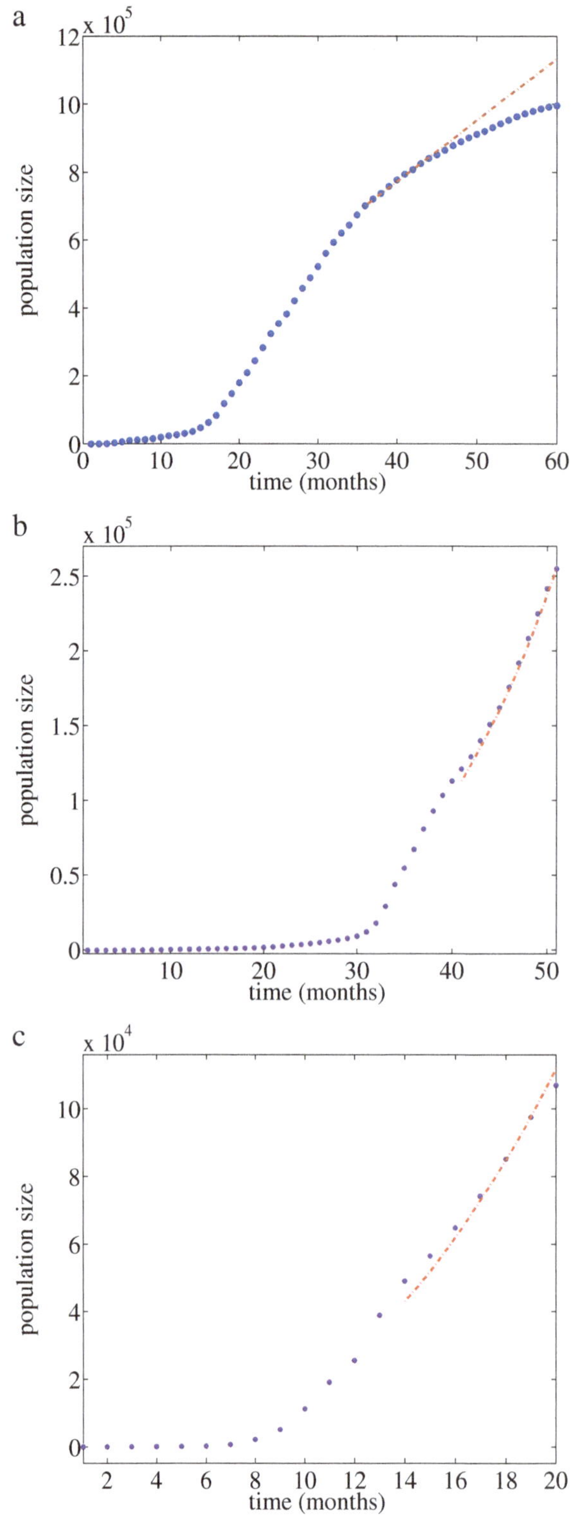

Figure 6 The predicted population in three data sets using population growth model. (a) Renren, the ground-truth value of the population size between 36th month and 60th month (as points) vs. the predicated value by the model (as dashed lines). **(b)** Twitter, the ground-truth value of the population size between 41th month and 51th month (as points) vs. the predicated value by the model (as dashed lines). **(c)** Gowalla, the ground-truth value of the population size between 15th month and 20th month (as points) vs. the predicated value by the model (as dashed lines).

To quantify the accuracy of the prediction, We use prediction accuracy (PA) and define it as follows:

$$PA = 1 - \left|\frac{predicted\ value - actual\ value}{actual\ value}\right| \tag{8}$$

According to Eq. 8, we present the prediction accuracy of our model in several months in Table 2. The results in the table show that our model achieves 94.30% average prediction accuracy for Renren, 98.94% average prediction accuracy for Twitter, and 97.73% average prediction accuracy for Gowalla. These results suggest that our model can perform very well in OSN population prediction.

Table 2 The prediction accuracy of the population growth model

Renren		Twitter		Gowalla	
40th	99.36%	42th	96.90%	16th	95.55%
45th	98.71%	44th	99.04%	17th	98.25%
50th	95.50%	46th	99.35%	18th	99.62%
55th	91.77%	48th	99.89%	19th	99.80%
60th	86.14%	50th	99.53%	20th	95.43%

The table is the prediction accuracy of the population growth model of selected months in three data sets.

Discussion

In this section, we discuss the impact of methods used for data collection and processing, the effects of populations acquirement in OSNs and the scope of our growth model.

Data collecting and processing

In this paper, we use BFS started from random selected nodes to collect the data samples from online social networks. The population growth model based on random sampling data may cause inaccurate population estimation and prediction. To avoid this issue, we conduct several actions to make the data sets fair enough. First of all, we launch a BFS-bias correlation procedure (Kurant et al. 2011) to eliminate the biases caused by random walking. Secondly, according to the estimation method of social network sizes by random sampling presented in (Katzir et al. 2011), we argue that the quality of data sets cannot be affected by random selected nodes. Finally, we use full population size in Facebook and Sina Weibo to validate the effectiveness of the proposed population growth model. All these efforts are made to let the data samples collected from online social networks be accurate enough for modeling population growth.

Population acquirement in OSNs

We use registered users as the population in OSNs. Thus, we count every registered user in the network as a member of the total population. It contains both active users and inactive users. In our model, we consider these inactive users as one part of the aggregated population for the following reasons: (1) Inactive users are also one part of the population. We cannot say a user who is not active in the network does not belong to this network. (2) Detecting active users is a complex process which cannot be done simply. For example, people may find the active users from the activity that the user conducts on the web site. However, many people may only browse the web site without any explicit activity. They perform inactive in interacting with others in OSNs, but they are also active users. By these two reasons, we consider registered users instead of active users as the population of an OSN.

Scope of the growth model

Our population growth model focuses on the growing stage of OSNs. We do not intend to track the life circle of an OSN. When an OSN's population stops growing, our model will not take effect on it. To specify the growing stage of an OSN, we define the population monthly growth rate as

$$r(t) = \frac{population(t) - population(t-1)}{population(t-1)},$$

if $r(t) > 0$, we consider the population is in growing stage between time $t-1$ and t. We say an OSN is in growing stage if its monthly growth rates are all greater than 0 in the observed time period. Actually, our crawled data sets and two full population OSNs (Facebook and Sina Weibo) are all in population growing stage, which adapt to our study. Besides, most popular OSNs (such as Facebook and Twitter) are currently still in the growing stage. Therefore, our population model focuses in the stage of population growth in OSNs.

Conclusions

In this paper, we propose a population growth model for online social networks. We investigate the population growth in spatiotemporal perspective. By studying the population growth over locations and time in three data sets of Renren, Twitter and Gowalla, we find out the population distribution is a power-law function over various locations. The growth of populated locations and the largest population are both power functions of time. By integrating the temporal and spatial characteristics of population growth, we conduct the general population growth model. Extensive experiments show that our model can fit the population growth in Facebook and Sina Weibo. As an application, we use the model for population predication in three data sets, and it can achieve a prediction accuracy between 86.14% and 99.89%.

Competing interests

The authors declare that they have no competing interests.

Authors' contributions

All authors contributed equally to the paper. All authors read and approved the final manuscript.

Acknowledgements

The authors acknowledge the funding from Alexander von Humboldt Foundation and DAAD Foundation. We would like to thank Mr. Cong Ding for the help to crawl Renren data sets. We also appreciate the comments from anonymous reviewers for improving the quality of the paper.

Author details

[1]Institute of Computer Science, University of Goettingen, Goettingen, Germany. [2]State Key Laboratory for Novel Software and Technology, Nanjing University, Nanjing, China.

References

Clauset, A, Shalizi CR, Newman MEJ: **Power-law distributions in empirical data.** *SIAM Rev* 2009, **51**(4):661—703.

Cho, E, Myers SA, Leskovec J: **Friendship and mobility: user movement in location-based social networks.** In *Proceedings of the 17th ACM SIGKDD International Conference on Knowledge Discovery and Data Mining (KDD '11).* New York: ACM; 2011:1082–1090.

Facebook; 2013. [http://www.facebook.com]

Katzir, L, Liberty E, Somekh O: **Estimating sizes of social networks via biased sampling.** In *Proceedings of the 20th International Conference on World Wide Web (WWW '11)*. New York: ACM; 2011:597—606.
Kurant, M, Markopoulou A, Thiran P: **Towards unbiased BFS sampling.** *IEEE J Selected Areas Commun* 2011, **29**(9):1799–1809.
Leskovec, J, Backstrom L, Kumar R, Tomkins A: **Microscopic evolution of social networks.** In *Proceedings of the 14th ACM SIGKDD International Conference on Knowledge Discovery and Data Mining (KDD '08)*. New York: ACM; 2008:462–470.
Newman, MEJ: **Power laws, Pareto distributions and Zipf's law.** *Contemp Phys* 2005, **46**(5):323–351.
Rejaie, R, Torkjazi M, Valafar M, Willinger W: **Sizing up online social networks.** *IEEE Netw: Mag Global Internetworking* 2010, **24**(5):32–37.
Renren net; 2013. [http://www.renren.com]
Sina weibo; 2013. [http://www.weibo.com]
Torkjazi M: **Hot today, gone tomorrow: On the migration of MySpace users.** In *Proceedings of the 2nd ACM Workshop on Online Social Networks (WOSN '09)*. New York: ACM; 2009.
Twitter; 2013. [http://www.twitter.com]

4

An agent-based framework for performance modeling of an optimistic parallel discrete event simulator

Aditya Kurve[1*], Khashayar Kotobi[1] and George Kesidis[2]

*Correspondence: ack205@psu.edu
[1] EE Department, The Pennsylvania State University, University Park, PA 16802, USA
Full list of author information is available at the end of the article

Abstract

Purpose: The performance of an optimistic parallel discrete event simulator (PDES) in terms of the total simulation execution time of an experiment depends on a large set of variables. Many of them have a complex and generally unknown relationship with the simulation execution time. In this paper, we describe an agent-based performance model of a PDES kernel that is typically used to simulate large-sized complex networks on multiple processors or machines. The agent-based paradigm greatly simplifies the modeling of system dynamics by representing a component logical process (LP) as an autonomous agent that interacts with other LPs through event queues and also interacts with its environment which comprises the processor it resides on.

Method: We model the agents representing the LPs using a "base" class of an LP agent that allows us to use a generic behavioral model of an agent that can be extended further to model more details of LP behavior. The base class focuses only on the details that most likely influence the overall simulation execution time of the experiment.

Results: We apply this framework to study a local incentive based partitioning algorithm where each LP makes an informed local decision about its assignment to a processor, resulting in a system akin to a self organizing network. The agent-based model allows us to study the overall effect of the local incentive-based cost function on the simulation execution time of the experiment which we consider to be the global performance metric.

Conclusion: This work demonstrates the utility of agent-based approach in modeling a PDES kernel in order to evaluate the effects of a large number of variable factors such as the LP graph properties, load balancing criteria and others on the total simulation execution time of an experiment.

Keywords: Agent-based modeling, Parallel simulation, Self organizing system, Game theory, Load balancing

Background

Parallel and distributed simulation techniques allow us to create and run large scale models that also allow fine grained modeling of the simulated entities. For example, the rapid growth of large-scale communication networks, particularly the Internet and the on-line social networks it supports, has brought an increased emphasis on the need to model and simulate them in order to understand their macroscopic behavior and to design and test defenses against network attacks e.g., denial-of-service and Border Gateway Routing Protocol (BGP) attacks (Sriram et al. 2006). It has been observed largely that in many cases the results of such experiments are affected significantly by the level of abstraction that is chosen. For example, (Chertov and Fahmy 2011) studied the need to model forwarding devices such as switches or routers and the sensitivity of the experimental results to their level of detail in the model. Parallel and distributed simulation distributes the simulation tasks across multiple machines and, in this way, exploits their combined resources. A challenge here is to maintain synchronization among collaborating machines in order to preserve the causality of processed events and at the same time attain a significant speed up.

Most parallel simulation mechanisms use the concept of a logical process (LP) which is a software component that runs in parallel with other LPs via processor time sharing or by running simultaneously on different processors in the case of multiprocessor systems. Each LP has a set of local variables that define its state. The state of the simulation at a given time is defined by the combined state of all LPs at that time. LPs communicate with each other using inter-process communication mechanisms such as shared memory or pipes. For the specific case of network simulation, a simulator kernel creates a distinct LP as a computational unit for each element of the simulated network model (Nicol and Fujimoto 1994). For example, all the properties and methods associated with a simulated router are executed and maintained locally in a distinct LP that represents it. This aids in the distribution of computation load across machines and also restricts the communication of an LP with only its "neighboring" set of LPs (*i.e.,* neighbours in the network topology under simulation). In this way, we can represent the LPs and their interdependencies with a graphical model.

Extensive studies have been conducted to optimize the performance of parallel simulations by focusing on different aspects such as methods of synchronization, effect of roll-backs, LP-to-machine assignment methods and so on. New methods that promise to speed up parallel simulations are evaluated directly on hardware platforms such as cluster computers. Although the performance of any proposed method should be benchmarked ultimately using a hardware platform, this work-flow also presents its own share of shortcomings. For instance, limitations posed due to available hardware and the simulated scenario in the experiment makes the method susceptible to fine-tuning to the hardware and the simulated experiment. An alternative is to create a computational or numerical model of the parallel simulator that factors in known dependencies (both deterministic and probabilistic) and then evaluate the proposed methods using this computational or numerical testbed. This allows us the additional luxury of testing the proposed methods under varying system environment variables since they are represented in software. The parameters that affect the total simulation time can be abstracted by choosing appropriate methods of representation. For instance, the particular scenario that is simulated can be represented as a graph of logical processes or LPs, and each LP is represented with its behavioral model of event generation and processing as shown in Figure (1). Note that other details can be ignored since they do not have consequence on the total simulation execution time of the experiment. The Discrete Event System Specification formalism (Zeigler et al. 2000) describes a hierarchical and modular method of system modeling where model of the specific scenario to be simulated is separated from the simulation engine using formal means of representing the simulated model as an

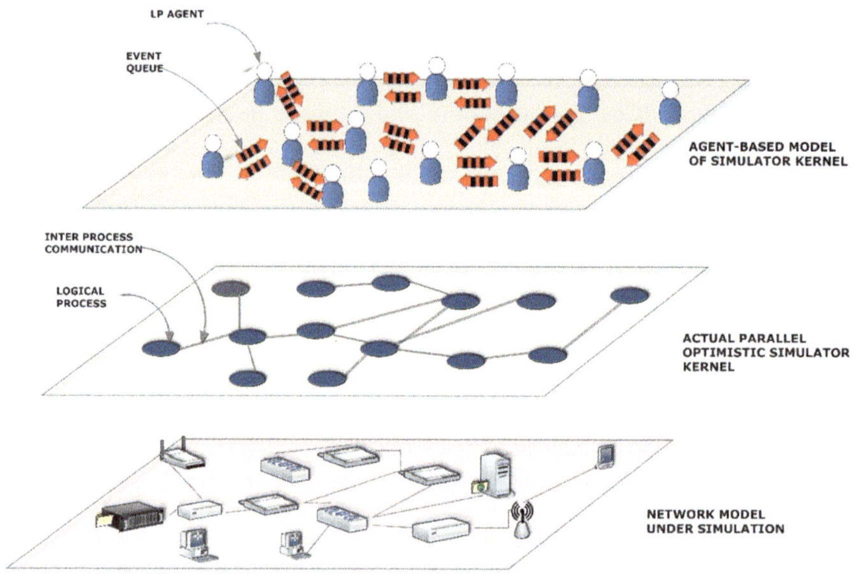

Figure 1 Network simulation illustration: Mapping to an agent-based model of PDES simulator.

input to the simulator. However since we are interested only in the simulation execution time of the experiment, we use a much simpler technique of representing the simulated scenario. One can extend the functionality of the LP agent class to include additional details.

In this work, we use an agent-based framework to simulate the behavior of a PDES (Parallel Discrete Event based Simulator) kernel wherein each LP is represented by an autonomous agent with its own set of distinct local variables. The agent-based modeling paradigm of LPs is novel and appropriate especially in the context of optimistic synchronization among LPs that requires each LP to maintain a local simulation state and use messages or events to communicate and synchronize with other LPs. Agent-based modeling allows us to study the resulting global performance metrics of the simulator kernel. The total time required for simulation is a non trivial and a complex (in many cases nondeterministic) function of various system parameters. The dynamics of an optimistic PDES kernel are easier to describe at the level of an LP in terms of the event generation and processing behavior. These local dynamics, their causal effects and the effect of state rollbacks have a direct consequence on the global performance metrics of the system, which in this case happens to be the total simulation wall-clock time of the experiment. We use this framework to evaluate an LP assignment scheme where each LP makes its own informed choice of a processor based on a novel incentive based cost criterion. The agent based modeling framework also makes it convenient to implement our assignment scheme since it is based on a local cost criterion at the level of LPs which are agents in our model. This is akin to a self organizing network of LP agents that form clusters across the machines.

Background

In the following subsections we detail some important background ideas and then develop our agent-based model of the simulator in the sections thereafter.

Need for parallel and distributed simulation

Networks in the form of social connections, information-carrying links, collaborative associations and reputation systems are ubiquitous. Large and complex networks result from the association of entities distinct in character, preferences, and inherent design. Network modeling is essential for understanding the behavior of a combined system of heterogeneous individual nodes that interact with each other. It is used, e.g, to test routing and forwarding protocols, defenses against network attacks and to study community structures in social networks. A typical communication network may contain tens of thousands of nodes with a large number of links. The nodes in reality can be whole routers, CPUs, or switches, individual router ports, firewalls, or end systems. (Sriram et al. 2006) studied the large scale impact of BGP peering session attacks that can cause cascading failures, permanent route oscillations or the gradually degrading behavior of the routers, using an Autonomous System (AS) level topology which was down-sampled from a typical AS-level topology consisting of 23,000 ASs and 96,000 BGP peering links (Zhang 2013). There were a fidelity/complexity tradeoff when simulating such a network. Some methods scaled down the network under study (Psounis et al. 2003) by omitting some intricate details and cleverly choosing the parameters that were expected to maximally affect the simulation results so as to minimize the loss of fidelity as much as possible. Recent papers such as (Gupta et al. 2008) introduced new methods to represent groups of nodes by a single node while (Carl and Kesidis 2008) studied path-preserving scaled-down topologies for large scale testing of routing protocols. Dimitropoulos et al. (2009) suggested use of annotations to a simple unweighted, undirected graph to represent the original network. However, the reliability of scaled-down methodologies depends largely on the assumption of low sensitivity of the outcome of the experiment to certain microscopic factors which may be at best crudely modeled by scale-down. An important macroscopic behavior of a network might be due (in an *a priori* unknown way) to a microscopic behavior ignored in the simulated model, thus resulting in inaccurate simulation results.

Parallel and distributed simulation: synchronization for causality

With the advancement of distributed processing systems, computing power was increased and one could thus feasibly represent networks using more refined models. Simultaneously, the theory of distributed simulation developed along with practical implementation of simulators such as PARSEC (Bagrodia et al. 1998). In the case of network simulation, one can represent each node in a network by a logical process (LP). An LP is an object that consists of a set of variables known as its "state" along with functions specific to the type of node being modeled. It normally has an event list containing time-stamped events to be executed. The local variables of an LP change when an event is processed. The value of these variables at any given time defines the state of the LP at that time. An LP maintains a local (simulation) time variable that contains the time stamp of the event currently being processed (if busy) or the most recent event processed (if idle). Events are stamped with their time of execution (in simulation time). Each LP maintains a list of events to be processed.

The methods to synchronize event execution between LPs running in parallel can be classified into two major types: conservative (Chandy KM, Misra 1981) and optimistic, e.g. Time Warp (Jefferson 1985). In the conservative methods, LPs strictly follow time-causality, *i.e.*, all events are processed in the strict order of their time stamps. Each LP tries to ensure, before processing an event A, that no other event, say B, with time stamp less than that of A, will arrive at the LP subsequent to the processing of event A. This time causality is followed by assuming that the graph topology of LPs is fixed and known beforehand. LPs communicate via messages and each message-carrying link ensures that they are sent in the order of their time stamps. Synchronizing null messages are exchanged between LPs to assure each other that no event with time stamp less than a specified time will be sent. Optimistic synchronization allows for non-causal event execution. An LP is allowed to process events "ahead of time" without any kind of synchronization assurance. Each LP maintains its own local time and processes the event with lowest time stamp in the event list. If an LP receives an event time-stamped which is less than its local time, it rolls back in time to the time stamp of the event. Rolling back means restoring a state prior to the time stamp of the event that triggered the rollback. In order for it to rollback to a prior state, an LP should archive the past history of its states and events. The combined system of LPs maintains a global variable called *global-time* which is equal to the minimum local time across all the LPs. Hence the global time is indicative of the overall progress of the simulation.

Using agent-based framework for performance modeling of a PDES kernel

The system modeled using an agent-based framework is characterized or described in terms of the actions, behavior, beliefs, etc. of an individual autonomous agent in a group of large number of such autonomous agents. To a large extent, the paradigm of agent-based modeling can be said to use the bottom-up approach to characterize, represent, predict or recreate a system or a phenomenon. However, agents have been used to drive large scale software systems (Jennings 2001) where the architecture is specified using the top-down approach. A common underlying theme motivating agent based modeling is that most complex phenomena observed in the physical world are the consequence of a much simpler set of rules that govern the dynamics of a large number of constituent entities that it is comprised of. For example, the Axelrod's model of social dissemination (Axelrod 1997) tries to explain the consensus towards cultures and the simultaneous existence of different cultures in the society using a simple set of rules of interaction between individuals or behavioral models that explain the flocking of birds (Reynolds 1987) or ant colony optimization (Dorigo et al. 2006). Agent-based modeling has received a large amount of interest in the last decade even from the non-computing research communities in areas such as social sciences and ecology (Niazi and Hussain 2011). It is a comparatively new method of modeling and a large number of existing models can be extended under this paradigm. Borschev and Filippov (2004) illustrates how the classic predator-prey model can be enriched by an agent-based model that makes more realistic assumptions without any significant addition to the model complexity. A large number of agent-based modeling toolkits such as NetLogo Tissue and Wilensky (2004) and Repast (North et al. 2007) have been designed in the last decade. A useful comparative evaluation of the existing toolkits has been provided in (2007). A high-level approach to performance prediction of simulators has been done previously using probabilistic models for event thread generation. These statistical methods use analytical means instead of simulations to predict the performance or speed up of synchronized iterative algorithms on multiprocessors. For instance, Agrawal and Chakradhar (1992) uses maximum order statistics on random variables that affect the speed-up on parallel machines. Xu and Chung (2004) proposes similar analytical methods for performance prediction of synchronous simulation. Agent-based modeling has always been explored as an alternative to equation-based modeling (Agent-based modeling vs. equation-based modeling). We argue that in the case of optimistic synchronization, analytical models are not feasible due to the complex and time-varying inter-dependencies of various parameters and hence we resort to agent-based approach. In the following sections we will demonstrate the application of agent-based modeling to performance modeling of an optimistic PDES kernel. The model is used to compute the simulation wall-clock time of the kernel for an experiment described by an event initialization list based on the dynamics and interactions at the level of LPs. The agent-based model can be used to evaluate and compare different cost criteria for LP assignment. In this work we focus on the additive cost framework described in (Kurve et al. 2011a) (See Appendix B) and the alternative cost framework (See Appendix C). This model can serve as a testbed to evaluate other cost criteria that can be more complex than the simple additive relationship in modeling effects of computation load imbalance and communication delays on the number of roll backs and consequently the simulation execution time.

LP assignment and simulation time

The total simulation time of the experiment is sensitive to the assignment of LPs to machines. In our model we focus on two aspects that are direct consequence of the LP assignment: the resulting load imbalance across the machines and the inter-processor communication[a]. We study the effect of optimization of these parameters on the total simulation time of the experiment. Our bi-criteria optimization algorithm assumes an additive relationship, *i.e.,* the total simulation time is represented mathematically as the addition of the load balancing cost and weighted inter-process communication cost. In the case of an optimistic PDES, both the parameters induce synchronization overheads in the form of rollback events that slows down the advance of the global simulation time. The graph partitioning problem is a well-known NP-complete problem that considers both these aspects. It is formally defined as follows.

Let $G = (V, E)$ be an undirected graph where V is the set of nodes and E is the set of edges. Suppose the nodes and the edges are weighted. Let w_i represent the weight of the ith node and let c_{ij} represent the weight of the undirected edge $\{i, j\}$. Then the K-way graph partitioning problem aims to find K subsets $V_1, V_2, .., V_K$ such that $V_i \cap V_j = \emptyset \; \forall \; i, j$ and $\bigcup_{i=1}^{K} V_i = V$, $\sum_{j \in V_i} w_j = \frac{\sum_k w_k}{K} \; \forall \; j$ and with the sum of the weights of edges whose incident vertices belong to different subsets minimized.

Heuristics to solve the graph partitioning problem primarily make use of spectral bisection methods (Pothen et al. 1990) or multilevel coarsening and refinement techniques (Karapis and Kumar 1996). Spectral bisection methods calculate the second smallest eigenvector, known as Fiedler vector, of the modified adjacency matrix of the graph. These methods by far give the best results. However, finding the Fiedler vector is computationally very expensive. For "geometric" graphs in which coordinates are associated with the vertices, geometric methods are available which are randomized and are quicker than spectral bisection methods. Multilevel partitioning algorithms are by far the most popular techniques. All of these discussed methods use centralized computational implementations involving access to global state information. A periodic refresh of partition is needed in most cases because of the highly time varying nature of the load generated across different LPs. The exchange of a chunk of nodes that are identified using a sparse cut metric was studied in Kurve et al. (2011b). In our work (Kurve et al. 2011a), we present an alternative to this, where decisions are made at the LP or node level. We model a game theoretic framework where the local cost for each LP incentivizes the LP to choose a processor so as to minimize the total simulation time of the experiment (represented by the "social welfare"). LPs can make distributed decisions eliminating the need for a separate software or hardware resource to refresh the partition across the machines. We are particularly interested in the class of games known as "potential games" (Monderer and Shapley 1996) which guarantee a descent in a global cost function (indicative of the system performance) with each decision made at the local (machine or node) level. The global state required to make an accurate choice of LP is independent of the total number of LPs and scales linearly with the number of neighbors of an LP. Please refer to Appendix A and B for more details on the local cost criteria. Note that the local incentive based criteria can be applied in other scenarios which involves the two competing incentives for load balancing and clustering. For instance, Kurve et al. (2013) has studied the application of the local incentive criteria in super-peer based peer-to-peer (P2P) networks.

Method

We designed the model using Netlogo (Tissue and Wilensky 2004). Our model is based on simulating the LPs as agents with individual characteristics and then observing the system level performance of the combined graph of nodes in terms of the simulation execution time. Some of the key features of the model are listed below:

1. We do not consider any specific simulation scenario and both the LP agent class and the class of events are generic to accommodate different particularities. The LPs can represent a router or a switch in the case of network simulation or logic gates in gate-level simulation of VLSI circuits. We describe the LP agent and event classes in detail in the following subsections.
2. One of the inputs to the model is a list of events to which event lists of the LPs is initialized before the simulation begins.
3. We abstract the simulation model across different scenarios using a graph of LPs and an event generation model defined by the initial event list and the cause-effect relationships between different events resulting in the event execution thread across LPs during simulation.
4. The parallel processing hardware is modeled as simulated artifact in terms of the number of processors, relative speed of each processor and the mean intra and inter processor communication delays.
5. We focus our attention on the total simulation time and the synchronization overhead in terms of the number of rollbacks, which depends largely on the event generation and processing model rather than an actual specification of the simulation scenario.

To illustrate the abstraction mechanism, consider an oversimplified scenario of N peers sharing content using a peer-to-peer (P2P) system such as Gnutella. We would like to simulate the query generation and resolution mechanism in the system. Suppose the queries are resolved using random walk over the overlay network graph. Using our method of abstraction, we can represent this using a graph of LPs which is similar to the graph of the overlay network. Each peer is represented by a unique LP in the graph. The events for each LP correspond to queries generated by the LP itself and those forwarded by its neighbors. In this case the event list for an LP is initialized based on the query generation rate of each peer. The query forwarding mechanism is represented using limited hop event forwarding and time to process each event. Note that our abstraction is agnostic about the actual changes that happen to a local state of an LP (representing a peer) in the simulation, because this does not directly affect the total simulation time.

As shown in Figure 2, the inputs to the model are the graph of LPs, the initial event list for each LP and the LP assignment algorithm and the output is the simulation execution time.

The LP agent class

We now describe some important properties and methods that define the base class of an LP agent. The LP agent base class is based on optimistic synchronization and uses rollback to synchronize its event execution sequence. It supports two basic event types: regular and rollback. One can easily derive more specific classes of LP agents that support a richer class of events. An LP has a set of local variables which are specific to the scenario under simulation. We do not factor in these scenario-specific local variables in the total simulation time and hence in our base LP agent class, this is an empty set. It has an *event-list* that holds the list of pending events that it needs to process. Each event has an associated time stamp and the agent processes each event in the order of the time stamps starting with the least. The variable *local-time* is equal to the time stamp of event being processed currently if the agent is busy or the time stamp of last processed event if not busy. When *busy?* is true, the agent performs no operation and simply decrements the *busy-tick* timer until it becomes zero. It also contains a list of event history with the local state of the LP archived before processing each event. In the case of a rollback event or a non-causal event (event whose time stamp is less than the local simulation time), the state of the LP is restored back to the time stamp of the non-causal event using the archived states from the *event-list-history*. Tables 1 and 2 shows a list of some important properties and member functions of the LP agent base class.

The *process_causal_event()* member function in the base class simply loads the *busy-tick* counter and sets the *busy?* bit true and possibly generates events in the neighboring LPs when *busy-tick* becomes zero. This is because we are simply interested in the time

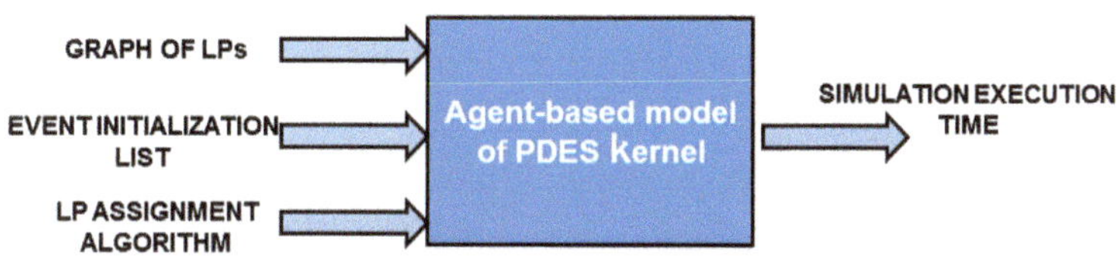

Figure 2 Software based model of a simulator.

Table 1 Properties of an LP agent base class

local-time	Contains time stamp of event being processed
state	Local variables of the LP
event-list	List containing pending events
event-list-history	List containing the archived local state of LP before each processed event in the past
busy-tick	The time remaining in wall-clock units to finish processing the current event
busy?	A boolean variable indicating if the LP is busy processing an event or idle.
my-machine	Current machine assigned to the LP
sim-time	Time in wall-clock ticks needed for processing the current event

of execution of the events. As shown in algorithms 1 and 2 the difference between the two member functions that process non-causal and rollback events is that in the case of former, the LP state restoration is followed by the execution of the current event. Whenever an LP rolls back in simulation time, it sends rollback events to all its neighboring LPs that it had sent events in the past until the time of rollback. If both the regular event and its corresponding rollback event are present in the *event-list* of an LP then they nullify each other and both are deleted. Algorithm 3 shows the pseudo code for the *simulate()* member function which is called for every tick of wall-clock time.

Event generation and processing model

The properties of the base event class are listed in Table 3.

In our model, each thread of events has a unique number denoted by *event-thread-number*. We define thread as a sequence of events where one event is generated by a previous event in the sequence. Each event in the event initialization list starts its own thread of events across LPs. This helps us track the spread of events in a limited-scope flooded packet-flows scenario. The time stamp of the events are stored in the *event-time* variable. The time stamp is the simulation time when the event is supposed to occur. The *event-type* variable tells the LP what its reaction to the event should be. Generally, there is a specified procedure call for every type of event. A rollback event is one of the types by default. In order to model delay in communication (in wall-clock time) between two LPs, we use the variable *event-tick*, which is programmed to a value by the LP that generates the event. At every tick of wall-clock time, *event-tick* is decremented by one unit until it reaches zero. An event can be processed by an LP only if its corresponding *event-tick* variable is zero. The variables *event-list-history* are the lists containing the information the

Table 2 Member functions of an LP agent base class

process_causal_event()	Member function to process causal event
process_noncausal_event()	Member function to process noncausal event
process_rollback_event()	Member function to process rollback event
generate_event(agent-list)	Generate new events for the thread of currently processing event in each of the agents in *agent-list*
create_neighbor_list ()	Returns a agent-list which is a subset of neighbors
simulate()	Member function called at every tick of wall-clock time
get_busy_time()	Returns the time needed to process the current event. Depends on the type of event, current processor load and processor speed
compute_node_weight()	Computes node weight
compute_edge_weight()	Computes the weight of edges with all its neighbors

Algorithm 1: *process_noncausal_event()*

```
for each event in event-list-history
 {
   if event-time > time of current-event then
     Restore event data event-list, event-time, event-type, event-tick, event-count.
     Delete event data from event history lists
     for each neighboring node
      {
        if current-event exists in event-list of neighbor
           Delete the event
        else if current-event exists in event-list-history of neighbor
           Send rollback event to the neighbor
        end if
      }
   end if
 }
Put all data of current-event in event history lists
Call process_causal_event() for the current-event
```

events already processed by the LP along with the contemporary state of the LP before the event was processed. In the case of a rollback, the LP restores its prior state from these variables. The LP regularly flushes out the data of events from the history with time stamps less than the global time. This is because the LP will never have to rollback to a state prior to the global time. Note that in such a model of the simulator, the functions that actually process events are generic. All that is needed is the time in wall-clock ticks stored in the variable *sim-time,* required for processing the event and the neighboring nodes where new events will be created as a result of processing this event. The *sim-time* is a function of the speed of the machine on which the LP resides given by *my-machine,* which in turn depends on the number of LPs that reside on that machine. We use a simple limited scope event forwarding using the generic framework of the simulator model. In this model, events are generated at random times by randomly chosen LPs and these events traverse the network for a limited number of *hops, i.e.,* each node that receives such a packet forwards it to a randomly chosen neighbor, provided the hop count given by *event-hop-count* is not zero. In our experiments, we randomly initialize the *event-list* of each LP that generate such a thread of events in such a way that the load generated

Table 3 Properties of an event base class

event-thread-number	Thread number of the event
event-time	Time stamp of the event
event-type	Type of event: regular or rollback
event-tick	Waiting time in wall-clock units the event is ready to be processed. This is used to model communication delays between LPs
event-hop-count	The maximum hop count of the event thread. Event threads survive for limited hops equal to this value

across the LPs becomes highly dynamic during the course of simulation. More specifically, we generate "hot spots" of traffic or a cluster of LPs that generate large amounts of traffic over a short period of (simulation) time. The locations of these hot spots change regularly.

LP assignment algorithm

Initial static LP-to-Machine assignment

We argue that due to initial uncertainty of computation and communication costs and their dynamic nature, an optimum partition is not possible *a priori*. As a result, we employ a simple initial partitioning method in which each machine chooses an initial "focal node" from among the nodes of the graph and then expands hop-by-hop to include neighboring nodes. The idea is to have connected sub-graphs within each partition so as to minimize inter-process communication. To avoid contention between partitions, we require that each machine wait a random amount of time after every hop and check for a semaphore before claiming ownership of new nodes. Unit edge and node weights are assumed during initial partitioning. The choice of the focal nodes is important to ensure a high probability of a good initial partition. As will be described later, the iterative partition refinement algorithm converges to a local minimum. Hence, a good initial partition might improve the chances of converging to the global minimum or a good local minimum. Refer to Appendix A for additional details.

Iterative partition refinement

The partition refinement algorithm utilizes the cost framework developed in the game theoretic study. During the course of the simulation, each LP can decide to switch processors either synchronously or asynchronously. In synchronous transfers which we refer to as a refinement step, LPs take turns to evaluate their costs and the prospective new machine based on the cost framework. In asynchronous transfers, LPs decide to switch without a global synchronization scheme. However, asynchronous transfers might lead to

Algorithm 2: *process_rollback_event()*

```
for each event in event-list-history
{
  if event-time > time of current-event then
    Restore event data event-list, event-time, event-type, event-tick, event-count.
    Delete event data from event history lists
    for each neighboring node
    {
      if current-event exists in event-list of neighbor
        Delete the event
      else if current-event exists in event-list-history of neighbor
        Send rollback event to the neighbor
      end if
    }
  end if
}
```

Algorithm 3: *Simulate()*

> Remove events from *event-list-history* with *event-time* < *global-time*
> Remove rollback events from *event-list-history* if *event-time* > *local-time*
> **if** *busy?* = idle **then**
>   *current-event* = event in *event-list* with lowest time stamp and *event-tick* = 0
>   **if** *current-event* ≠ rollback and time stamp of *current-event* ≥ *local-time* **then**
>   *busy?* = true
>   *busy-time* = *get_busy_time()*
>   Put all data of *current-event* in event history lists
>   **else if** *current-event* ≠ rollback and time stamp of *current-event* < *local-time*
>   *process_noncausal_event()*
>   **else if** *current-event* = rollback **then**
>   *process_rollback_event()*
>   **end if**
> **else**
>   *busy-time* = *busy-time* - 1
>   **if** *busy-time* = 0
>   *busy?* = false
>   **if** *event-count* of *current-event* ≠ 0 **then**
>   *generate_event*(*create_neighbor_list ()*)
>   **end if**
>   **end if**
> **end if**
> **for** each event in *event-list*
> {
>   **if** *event-tick* ≠ 0 **then**
>   *event-tick* = *event-tick* - 1
>   **end if**
> }

inconsistent decisions due to the likelihood of simultaneous transfers of more than one LPs. Note that the inconsistencies can be eliminated if the transfers occur between two different pairs of machines. We thus assume the use of a processor *mutex* in the case of asynchronous transfers that allows only single transfer for every processor. The transfer rates in our model are controlled by the global parameter *partition-refine-freq* which determines the frequency when the the node weights corresponding to the computation cost of each LP and the edge weights corresponding to the communication cost between LPs are recomputed and the LPs are triggered to evaluate processor transfers.

In the case of synchronous transfers, LPs are triggered to compute their "destination processor" based on the current state of the system. Then the "most dissatisfied" LP in the system is allowed to make transfer to its destination machine. The "most dissatisfied" node is the one which would benefit the most in terms of its cost by changing its machine. The cost function used here can be either of the two defined in (3) or (6). In our experiments we make comparative evaluation between the two cost frameworks. Hence $C_i(k)$

Algorithm 4: Simulator engine pseudocode for limited scope flooded packet flow

Initialize global variables.
ask nodes
{
  Initialize local variables
  Initialize the *event-list*
  initial-partition()
}
wall-clock-time = 0
while *event-list* of all nodes not empty **do**
wall-clock-time = *wall-clock-time* + 1
ask nodes
{
simulate()
}
global-time = minimum *local-time* and time stamps of events in *event-list* across all nodes
if remainder *sim-time*/*partition-refine-freq*= 0
ask nodes
{
  compute_node_weight()
  compute_edge_weight()
}
refine_partition
end if
end while

represents the cost of the ith node if it were to be assigned to the kth machine and $C_i(r_i)$ represents the current cost of the ith node. We define the dissatisfaction of the ith node as

$$\Im(i) = C_i(r_i) - \min_k C_i(k) \tag{1}$$

Hence the most dissatisfied node is the one with maximum value of $\Im$. The most dissatisfied nodes in the system having $\Im = 0$ indicates that the algorithm has converged to a local minimum. We can prove that this algorithm converges to a local optimum of the social welfare function (Kurve et al. 2011a).

Simulation engine

Finally, the pseudocode for the entire simulation engine of the agent-based model is shown in algorithm 4. It begins with the initialization of the global and local variables including the event-list of all the LP agents. At each tick of *wall-clock-time, simulate()* member function for all the LP-agents is invoked exactly once. The simulation is continued and the *sim-time* is incremented until all the LP agents have an empty *event-list. partition-refine-freq* defines how frequently the LP assignment algorithm is invoked. Particularly, it is the interval between two invocations of the LP assignment algorithm. Before every invocation of the LP assignment algorithm, the node and edge weights are recomputed. *compute_node_weight()* and *compute_edge_weight()* are used to estimate the computation cost of an LP and the communication cost over the edge with its neighboring LPs. There can be many different ways of defining these functions. In our experiments, we use the length of the *event-list* as being indicative of the computation cost that the LP will generate in the near future. Hence, node weight is calculated as the length of a weighted *event-list* where each event in *event-list* is weighed by *event-weight*, *i.e.*, with a weight which is the number of machine cycles needed to process the event. Communication cost should quantify the amount of "event traffic" between two LPs. We estimate edge weight from the history of event traffic across the edge. Specifically, we calculate the number of events generated across the edge during a fixed interval of time given by *estimation-window*. We then multiply it by 100 to make its magnitude comparable to the computation cost and then assign it to the weight of the link.

Results and discussion

We evaluated the iterative LP assignment algorithms on the agent based model described above to compare the speed up in the simulation of an experiment. We also observed the performance under different settings of system such as synchronous and asynchronous LP transfers and for two different models of random graph generation. As described previously, one of the inputs to the model was a graph of LPs. The first model of random graphs that we used to generate an instance of the LP graph was the preferential-attachment model which is scale-free and, according to some studies, can be used to model the Internet topology at the level of Autonomous System (AS) (Bu and Towsley 2002). The second model incorporates geometric information along with the degree of a node, *i.e.*, each node has associated coordinates in two dimensions: while forming links, each node randomly chooses another from among a set of 15 closest nodes (in terms of distance with respect to the coordinate axes). Our NetLogo based model of LP graph consisted of 300 LP nodes generated from one of the models above each of which was an instance of the LP agent class described in Section 3.1. The event initialization list was created keeping in mind the highly dynamic nature of the load generated during a network simulation, *i.e.*, we created two genres of event threads: one that was uniformly distributed across all nodes with a relatively large value of *event-hop-count* and second one that had shorter hop count and was localized to a closely connected set of nodes. The former created a wider scope of event interdependencies between nodes that were far apart in geodesic distance while the later generated hot spots of traffic that were localized in space and time.

get_busy_time() as listed in Table 2 gave the total time in wall-clock units needed to process the *current-event*. We used the following relationship to compute the *busy-time*.

$$busy{-}time = \frac{weight \text{ of } current - event \times Number\ of\ LPs\ sharing\ same\ processor \times 100}{processor - speed}$$

Note that the *processor-speed* is the normalized speed of the processor that the LP resides on. We set the inter processor communication delay as 10000 wall-clock ticks uniformly across all the links between processors, *i.e.*, any event that is created by an LP in another LP that resides in a different processor other than its own processor would experience a delay of 10000 wall-clock ticks. We set the value of *estimation-window* described in the previous section to 10000 clock ticks.

In our first set of experiments, we tuned the value of μ (relative weight given to inter-machine potential rollback-delay cost), *i.e.*, we tried to optimize the total simulation execution time of the experiment over different values of μ. The search space for μ is unbounded and hence we exponentially increased μ by powers of 10. For each setting, we kept all other parameters of the model constant, *i.e.*, the LP graph, initial partition before the simulation and sources of randomness such as event weights. We repeated this for ten different realizations of the LP graph and event thread initialization, and computed the average simulation time. As seen in Figure 3, the optimum values of μ for the two cost frameworks are different: about 100 for the first method and about 1000 for the second method. We can also observe that increasing the value of μ beyond a certain limit does not change the total simulation execution time.

The above experiment was performed for the geometric model of LP graph. In our next set of experiments, we further fine tune the value of μ to optimize the simulation execution time. In this case we varied μ from 0.1 to 1000 with an exponent of 2 and observed the optimal value of μ for the first method to be between 6.4 to 25.6, whereas for the second method it was between 400 and 800. However, as seen in Figure 4, the curves are not monotonous suggesting the presence of a large number of locally optimum points.

We also plotted similar curves using preferential attachment model to realize the random LP graph. Considering Figure 5, we observe quite a few differences from the graph in Figure 3. Most notable among them is the fact that the optimum values of μ for the two methods are different. This suggests that the additive rule that we used in both the cost frameworks, and the technique to estimate the node and edge weights, renders the efficacy of the partitioning algorithm sensitive to the properties of the LP graph. This is an important observation in the context of assigning the value of μ. Secondly we observe that the simulation execution-time rapidly increases as we increase μ beyond its optimum value unlike in the case of geometric model where it tends to stagnate after some value of μ. Similar to Figure 4, we plotted curves for the preferential attachment model

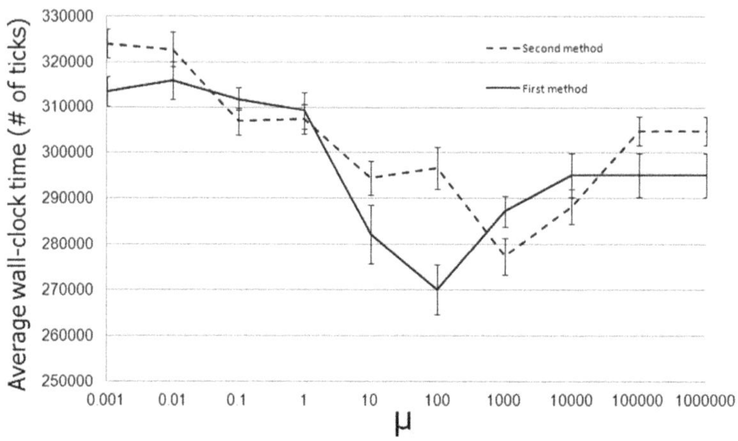

Figure 3 Finding optimum balance between computation and communication costs: For geometric model of random graph.

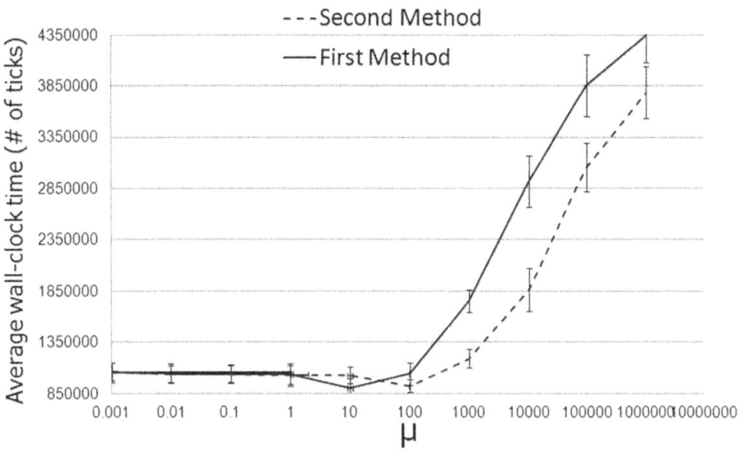

Figure 4 Resolving μ further: For geometric model of random graph.

in Figure 6. We can see that at this scale of resolution of μ, it is difficult to find the exact value of optimal μ due to the presence of large number of local minimum points in the curve. In all these cases it was observed that the first method generally performed better than the second.

In Figures 7 and 8, we plot the effect of the partition refinement interval on the simulation execution time. We observed that there is no significant gain in reducing the refinement interval below 5000 wall-clock ticks for both the LP graph generation models. However, this value depends on several different factors such as the dynamics associated with the events generated and the processing time of an event by the LP. These graphs further reiterate the need for partition refreshing in the case of a dynamic network simulation and that a static partitioning scheme is highly sub-optimal in this case.

From Figures 9 and 10, we can observe the effect of partition refinement interval on the distribution of rollback counts. We plotted the number of rollback events across five processors with the progress of simulation execution. We can observe that partition

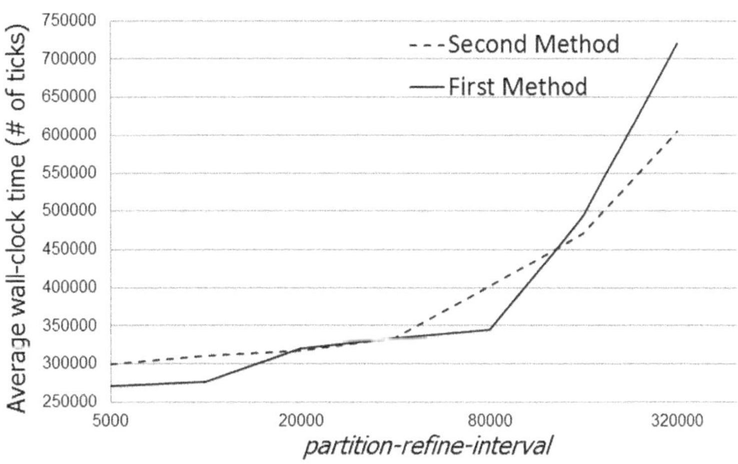

Figure 5 Finding optimum balance between computation and communication costs: For preferential attachment model of random graph.

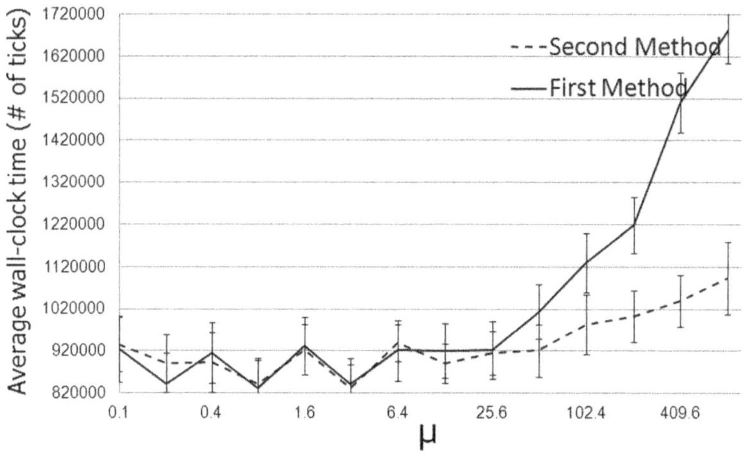

Figure 6 Resolving μ further: For preferential attachment model of random graph.

refinement reduces the variance of the number of rollback events across the processors as seen from Figure 10 when the interval is 20000 clock ticks as compared to Figure 9 when the interval is 100000 clock ticks.

In order to study the scalability of our results to variations in the LP graph size, we also plotted curves similar to Figure 5 in Figures 11 and 12 wherein the number of LP nodes in the graph were 100 and 500, respectively . We observed similar trends as in Figure 5 in both cases.

Conclusion

We presented a method of agent-0based modeling of an optimistic parallel discrete-event simulator. The agent-based modeling paradigm is novel to the study of a PDES performance and greatly expedites the evaluation of performance optimization techniques with different system settings. The model was used to evaluate and compare two LP-to-Machine assignment algorithms. In this process, we got several insights into the

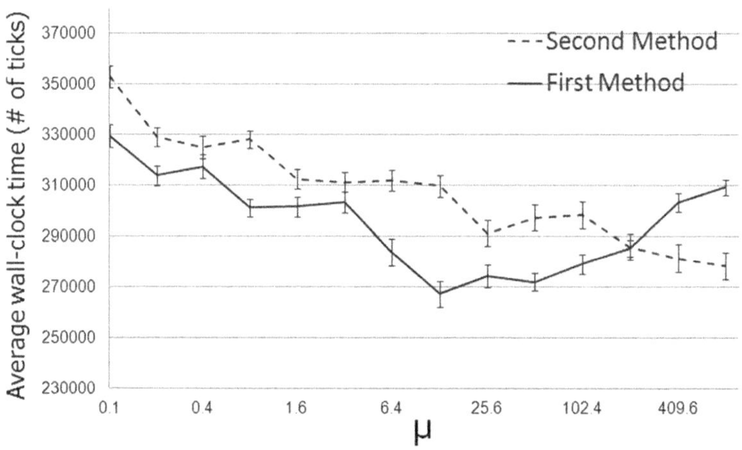

Figure 7 Effect of partition refinement frequency on simulation time: For geometric model of random graph.

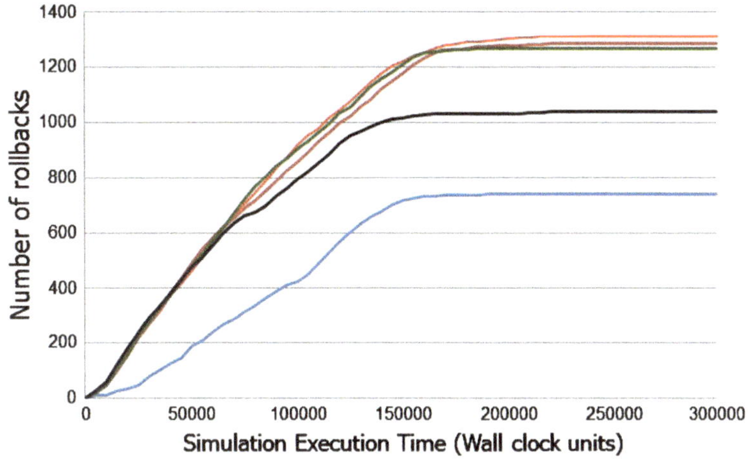

Figure 8 Effect of partition refinement frequency on simulation time: For preferential attachment model of random graph.

complex relationship between the LP agents and the simulation time. We observed that the comparative weight given to computation and communication delay cost in the LP assignment scheme depends on the properties of the LP graph. We studied the sensitivity of the simulator performance to other parameters such as partition refinement frequency, and the graphical properties of the network under simulation. The need for dynamic load balancing was further emphasized from the results of our experiments by studying the distribution of the rollback count curves with the advance of simulation execution time. One can use the proposed agent based modeling framework to study different LP to machine assignment techniques that characterize system performance using different objective functions. This framework can also be extended to study different rollback strategies and methods to estimate LP load and communication patterns based on the knowledge of current and past events.

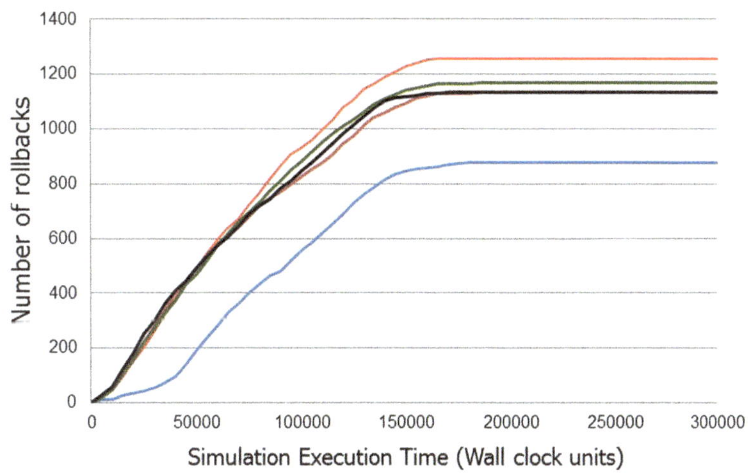

Figure 9 Rollback event distribution across processors with refresh interval of 100000.

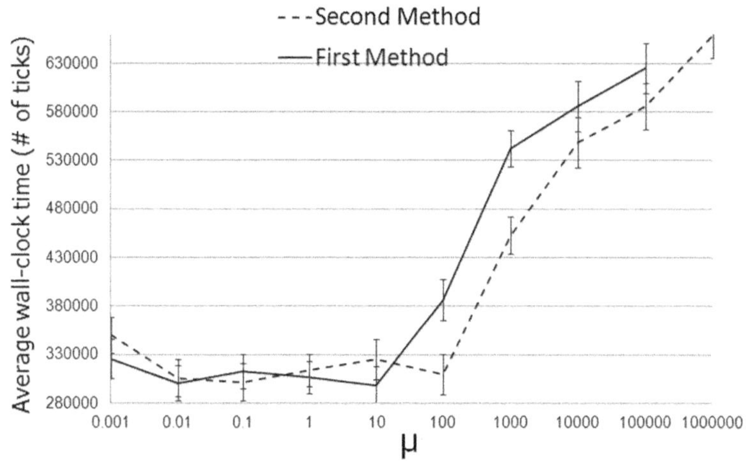

Figure 10 Rollback event distribution across processors with refresh interval of 20000.

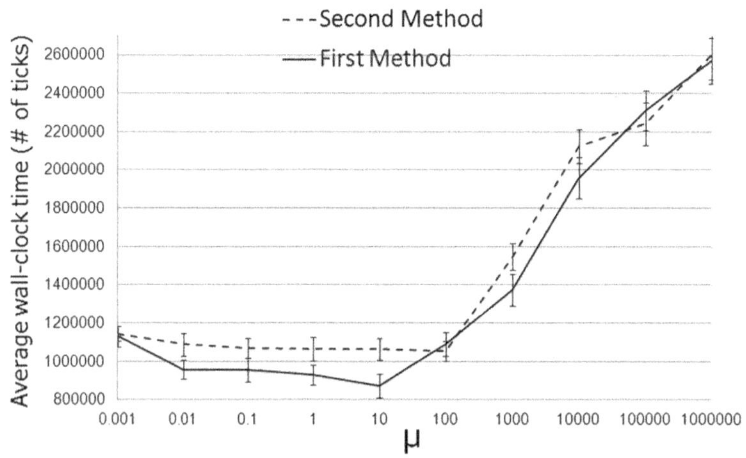

Figure 11 Finding optimum balance between computation and communication costs: For preferential attachment model of random graph of 100 nodes.

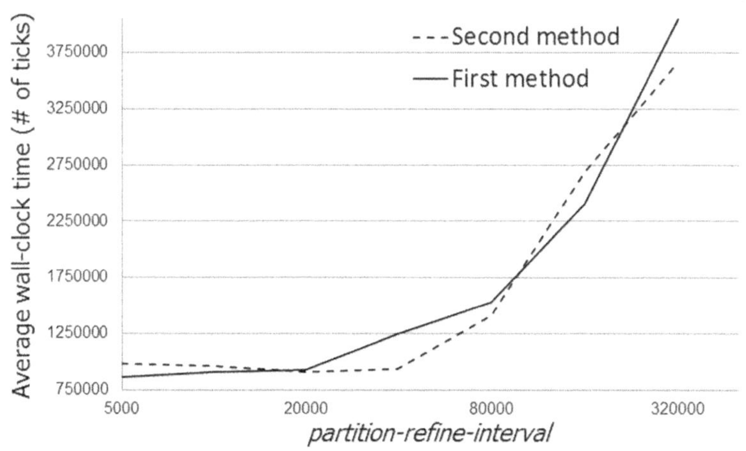

Figure 12 Finding optimum balance between computation and communication costs: For preferential attachment model of random graph of 500 nodes.

Appendix

Appendix A: initial partitioning algorithm

In this appendix, we describe the initial partitioning algorithm. As mentioned previously, the choice of focal nodes is important to get a good initial partition. The focal nodes are chosen so that they are at maximum geodesic distance from each other. Ideally we should find focal nodes such that each one of them is at least $2N_{\frac{|V|}{K}}$ geodesic distance away from the others, where $N_{\frac{|V|}{K}}$ is the mean number of hops that cover $\frac{|V|}{K}$ nodes. In this case, we should know the properties of the underlying graph for calculating $N_{\frac{|V|}{K}}$. We find this value for the example of an Erdos-Renyi random graph model.

Theorem 1. *For an Erdos-Renyi random graph $G = (V, E)$ with p as the probability of existence of an edge between any two nodes, if N_k and N_{k-1} are the expected number of nodes collected into a cluster at the kth hop and $(k-1)^{\text{th}}$ hop respectively, then the expected number of nodes in the cluster by the $(k+1)^{\text{th}}$ hop is:*

$$N_{k+1} = N_k + (|V| - N_k)(1 - (1-p)^{N_k - N_{k-1}}) fork > 0,$$
$$N_{k+1} = 1 fork = 0.$$

Proof. Suppose after the kth hop, the graph is divided into two sets of nodes: the set A of nodes obtained by the kth hop and set A' of nodes not yet obtained. Let $|A| = N_k$ and $|A'| = |V| - N_k$. Denote the set of newly acquired nodes in the kth hop by B, where $|B| = N_k - N_{k-1}$.

For any node $a \in A'$:

P(a is not connected to any node in B)= $(1-p)^{N_k - N_{k-1}}$ and
P(a is connected to at least one node in B) = $1 - (1-p)^{N_k - N_{k-1}}$.

The number of nodes acquired during the $(k+1)^{\text{th}}$ hop is binomial, with $(|V| - N_k)$ trials and the probability of success being $1 - (1-p)^{N_k - N_{k-1}}$. Hence the expected number of nodes acquired during $(k+1)^{\text{th}}$ hop is $(|V| - N_k)(1 - (1-p)^{N_k - N_{k-1}})$ □

More practically, we would like to have the focal nodes as far as possible (in geodesic distance) from each other. This will ensure that the partitions formed after hop-by-hop expansion are somewhat equal in number of nodes per partition. Suppose $F = \{f_1, f_2, ..., f_K\}$ is the set of focal nodes. Hence we would like to have

$$F = \underset{H \subseteq V \text{ s.t. } |H|=K}{\arg\max} \ \min_{h,l \in H: l \neq h} d_G(h, l), \tag{2}$$

where d_G is the geodesic distance between the two nodes. We can attempt to find such nodes using heuristics. For example, start by assigning an arbitrary set of distinct nodes to each machine. In round-robin fashion, each machine takes a turn at finding a node from the set of nodes that are neighbors of its current focal node that increases the minimum of its geodesic distance with focal nodes of other machines. This becomes the new focal node for that machine. This process is iterated until there is no further improvement possible. We iterate this process over multiple initializations of the focal node set and the best set of focal nodes is identified. In the next phase, starting at the focal nodes, the partitions try to collect-in nodes in their neighborhood, thus expanding their clusters. We can use random waiting time between two successive hops and semaphores to deal with contention issues, *i.e.*, when two or more machines try to claim ownership of the same node.

Appendix B: a local cost function

We will now describe our iterative local incentive for the partitioning algorithm (Kurve et al. 2011a) in detail. Consider an undirected weighted, graph $G = (V, E)$ representing the network model to be simulated. As described previously, the graph can be interpreted as a model of a network of LPs as vertices or nodes, with weights assigned to the nodes and edges of G. Thus, we want to first estimate the computational load generated by each node and the amount of communication between the logical processes associated with each node to assign node weights and edge weights, respectively, to the graph. Second, we wish to find a distributed technique to equitably load-balance among the machines while also taking into consideration the amount of inter-machine communication, the latter reflecting the risk of rollback. We will address the first problem in our following sections and focus on the second problem now.

- Let K be the number of machines ($K < |V|$). The graph G is partitioned among K machines or less, since in some cases where the cost of inter-processor communication is high, partitioning the workload among less than K machines might be optimal, *i.e.*, some machines may be not be assigned any LPs.
- Let b_i represent the computational load of ith LP.
- Let c_{ij} denote the cost of communicating over the edge $\{i, j\}$, representing the average amount of traffic between node i^{th} and j^{th} LP .
- Let $r_i \in \{1, 2, ...K\}$ be the partition chosen by the ith LP.
- Let the normalized capacity or speed of the kth machine be:

$$w_k = \frac{s_k}{\sum_{j=1}^{K} s_j},$$

where s_j is the speed of the jth machine.

Then according to (Kurve et al. 2011a) the cost function at the ith LP is given by.

$$C_i(r_i, \mathbf{r}_{-i}) = \frac{b_i}{w_{r_i}} \sum_{\substack{j:r_j=r_i \\ j\neq i}} b_j + \frac{\mu}{2} \sum_{j:r_j\neq r_i} c_{ij}, \tag{3}$$

where $\mathbf{r}_{-i}$ denotes the vector of assignment of all the nodes in $\mathbf{r}$ except that of the ith node and μ denotes the relative weight given to the inter-machine potential rollback-delay cost. This can vary across simulation platforms. For example, if the participating machines are remotely connected, then this cost will be higher than if locally connected. In the function above, the computational cost of a node intuitively depends on two factors: the existing load on the machine assigned to the ith LP, *i.e.*,

$$\sum_{j:r_j=r_i, j\neq i} b_j,$$

and the computational load that the ith LP will bring to the machine, *i.e.*, b_i. If the computational load generated by the node is zero ($b_i = 0$), then the computational part of the cost should be zero, thus motivate multiplication by b_i. This cost incentivizes the node to choose a machine that has relatively less existing load, thus encouraging load balancing at the system level. For example, if $\mu = 0$ then a node i which is currently assigned to machine r_i would choose to relocate to machine r_i^* only if

$$\frac{1}{w_{r_i}} \sum_{j:r_j=r_i, j\neq i} b_j > \frac{1}{w_{r_i^*}} \sum_{j:r_j=r_i^*} b_j. \tag{4}$$

In this way the load balancing mechanism is implicitly manifested in the local cost function. The second term in the sum represents the weight of edges that connect the ith node with nodes in other partitions. This term incentivizes the node to choose a partition with which it is well connected.

In (Kurve et al. 2011a), we also showed that this cost function allows for the existence of a Nash equilibrium which means that there is a fixed point in the strategy space of all the LPs. A strategy profile $\mathbf{r} = (r_1^*, r_2^*, ..., r_N^*)$ is a Nash equilibrium if and only if

$$C_i(r_i^*, \mathbf{r}_{-i}^*) \leq C_i(r_i, \mathbf{r}_{-i}^*) \quad \forall r_i \in \{1, 2, ..., K\} \; \forall i \in V. \tag{5}$$

Thus, at Nash equilibrium every node, say node i, will not be able to improve its cost by unilaterally changing its current processor r_i^*, *i.e.*, provided that the decisions of all other nodes are given by the assignment vector $\mathbf{r}_{-i}^*$.

Appendix C: an alternate cost framework

Let us consider an alternate local cost function for the graph partitioning problem.

$$\tilde{C}_i(r_i, \mathbf{r}_{-i}) = \frac{b_i^2}{w_{r_i}^2} + \frac{2b_i}{w_{r_i}^2} \sum_{\substack{j:r_j=r_i \\ j\neq i}} b_j - \frac{2b_i}{w_{r_i}} \sum_j b_j + \frac{\mu}{2} \sum_{j:r_j\neq r_i} c_{ij}. \tag{6}$$

Note that this cost function also has the property of machine-to-machine level overhead similar to the cost function described previously.

Next, we state a centralized graph partitioning problem that reasonably models the total simulation execution time by minimizing the load variance across machines as well as the potential inter-machine rollback-delay cost. Let $\underline{\mathbf{X}}$ be a $K \times |V|$ assignment matrix such that $x_{ki} = 1$ if node i belongs to machine k; otherwise $x_{ki} = 0$. We require $\sum_k x_{ki} = 1$ $\forall i = 1, 2, ..., |V|$, *i.e.*,

$$x_{ki} = 1 \quad \text{when } r_i = k. \tag{7}$$

Let w_k be the normalized capacity of the kth machine so that $\sum_k w_k = 1$. Then, the centralized LP assignment problem is:

$$\min_{\underline{\mathbf{X}}} \; \tilde{C}_0(\underline{\mathbf{X}}) = \sum_{k=0}^{K} \left(\frac{\sum_{j\in V} x_{kj} b_j}{w_k} - \sum_j b_j \right)^2 + \frac{\mu}{2} \sum_{i,j} c_{ij} x_{ki} (1 - x_{kj}) \tag{8}$$

$$\text{subject to} \sum_k x_{kj} = 1 \; \forall \, j \text{ and } x_{kj} \in \{0, 1\} \; \forall \, k, j.$$

The above standard formulation of the graph partitioning problem (Van Den Bout and Thomas Miller 1990) is a quadratic integer programming problem the convexity of which depends on the network graph. In most cases it will not be convex. We can think of decomposing this problem into a set of K subproblems each of which is solved by a single partition. However, with the constraints $\sum_k x_{kj} = 1$, $\forall \, j \in V$, such a decomposition is difficult to realize. So, instead we study the effect of sequential node-by-node transfer on (8). We can prove that for the local node cost function (6), Nash equilibria exist at the local optima (minima) of the centralized cost function (8). Thus, we can define a new cost function for each node as given by (6). Note that at each of the locally optimal points of (8), none of the nodes can improve their costs (6) given that the node assignments of all other nodes remain constant. Hence, the assignment vectors at these points are the Nash equilibria for this game. And since the node decisions always perform descent on (8), there is convergence guaranteed to one of the equilibrium points.

Endnote

[a]High volume inter-LP communication is not just overhead or delays, but also indicates the threat of rollback.

Competing interests

The authors would like to declare that there are no competing interests.

Authors' contributions

The authors of this paper contributed to this work in the following way: AK and GK were involved in the conceptualization of the agent based paradigm for PDES. AK and KK contributed to the coding and software implementation of the model in NetLogo. KK contributed to the method of modeling rollbacks and designing experiments. All authors read and approved the final manuscript.

Acknowledgements

The local incentive cost-based LP assignment method was presented at the International Workshop on Modeling Analysis and Control of Complex Network (CNET) 2011 at San Francisco. This work is supported in part by the NSF under grant number 0916179.

Author details

[1]EE Department, The Pennsylvania State University, University Park, PA 16802, USA. [2]CSE and EE Department, The Pennsylvania State University, University Park, PA 16802, USA.

References

Agent-based modeling vs. equation-based modeling: A case study and users guide. In *Proceedings of the, First International Workshop on Multi-Agent Systems and Agent-Based Simulation*: Springer-Verlag; 1998:10–25.

Agrawal, VD, Chakradhar ST: **Performance analysis of synchronized iterative algorithms on multiprocessor systems.** *IEEE Trans, Parallel Distributed Syst* 1992, **3**(6):739–746.

Axelrod, R: **The dissemination of culture a model with local convergence and global polarization.** *J Confl Resolution* 1997, **41**(2):203–226.

Bagrodia, R, Meyer R, Takai M, Chen Y, Zeng X, Martin J, Song HY: **Parsec: a parallel simulation environment for complex systems.** *Computer* 1998, **31**(10):77–85.

Borshchev, A, Filippov A: **From system dynamics and discrete event to practical agent based modeling: reasons, techniques, tools.** In *Proceedings of the 22nd International Conference of the System Dynamics Society*; 2004. number 22.

Bu, T, Towsley D: **On distinguishing between internet power law topology generators.** In *Proceedings of the 21st, Annual Joint Conference of the IEEE Computer and Communications Societies IN-FOCOM*: volume 2. IEEE; 2002:638–647.

Carl, G, Kesidis G: **Large-scale testing of the Internet's Border Gateway Protocol (BGP) via topological scale-down.** *ACM Trans Model, Comput Simul (TOMACS)* 2008, **18**(3):1–30.

Chandy, KM, Misra J: **Asynchronous distributed simulation via a sequence of parallel computations.** *Commun ACM* 1981, **24**(4):198–206.

Chertov, R, Fahmy S: **Forwarding devices: From measurements to simulations.** *ACM Trans Model and, Comput Simul (TOMACS)* 2011, **21**(2):12.

Dimitropoulos, X, Krioukov D, Vahdat A, Riley G: **Graph annotations in modeling complex network topologies.** *ACM Trans Model, Comput Simul (TOMACS)* 2009, **19**(4):17.

Dorigo, M, Birattari M, Stutzle T: **Ant colony optimization.** *Comput Intell, Mag, IEEE* 2006, **1**(4):28–39.

Gupta, D, Vishwanath KV, Vahdat A: **Diecast: testing distributed systems with an accurate scale model.** In *Proc. 5th USENIX Symposium on Networked Systems Design and Implementation*: USENIX Association; 2008:407–422.

Jefferson, DR: **Virtual time.** *ACM Trans Program, Languages Syst (TOPLAS)* 1985, **7**(3):404–425.

Jennings, NR, Commun ACM: **An agent-based approach for building complex software systems.** 2001, **44**(4):35–41.

Karypis, G, Kumar V: **Parallel multilevel k-way partitioning scheme for irregular graphs.** In *Proc. 1996 ACM/IEEE Conference on Supercomputing*: IEEE; 1996.

Kurve, A, Griffin C, Kesidis G: **A graph partitioning game for distributed simulation of networks.** In *Proceedings of the 2011 International Workshop on, Modeling, Analysis, and Control of Complex Networks*: ITCP; 2011a:9–16.

Kurve, A, Griffin, C, Kesidis, G: **Iterative partitioning scheme for distributed simulation of dynamic networks.** In *Proc. 16th IEEE International Workshop on Computer Aided Modeling and Design of Communication Links and Networks (CAMAD)*: IEEE; 2011b:92–96.

Kurve, A, Griffin C, Miller DJ, Kesidis G: **Optimizing Cluster Formation in Super-Peer Networks via Local Incentive Design.**: Springer; 2013:Accepted.

Monderer, D, Shapley LS: **Potential games.** *Games Econ, Behav* 1996, **14:**124–143.

Niazi, M, Hussain A: **Agent-based computing from multi-agent systems to agent-based models a visual survey.** *Scientometrics* 2011, **89**(2):479–499.

Nicol, D, Fujimoto R: **Parallel simulation today.** *Ann Oper, Res* 1994, **53**(1):249–285.

Nikolai, C, Madey G: **Anatomy of a toolkit: A comprehensive compensium of various agent-based modeling.** In *Proceedings of the Agent 2007 Conference on Complex Interaction and Social Emergence*; 2007:87–92.

North, M, Howe T, Collier N, Vos J: **A declarative model assembly infrastructure for verification and validation.** In *Advancing Social, Simulation: The First World Congress*: Springer; 2007:129–140.

Pothen, A, Simon HD, Liou KPetal: **Partitioning sparse matrices with eigenvectors of graphs.** *SIAM J Matrix Anal Appl* 1990, **11**(3):430–452.

Psounis, K, Pan R, Prabhakar B, Wischik D: **The scaling hypothesis: Simplifying the prediction of network performance using scaled-down simulations.** *ACM SIGCOMM Comput Commun, Rev* 2003, **33**(1):35–40.

Reynolds, R: **Flocks, herds and schools: a distributed behavioral model.** In *ACM SIGGRAPH Comput Graph*: volume 21. ACM; 1987:25–34.

Sriram, K, Montgomery D, Borchert O, Kim O, Kuhn DR: **Study of BGP peering session attacks and their impacts on routing performance.** *IEEE J Selected Areas, Commun* 2006, **24**(10):1901–1915.

Tisue, S, Wilensky U: **Netlogo: A simple environment for modeling complexity.** In *Proc. International Conference on Complex Systems*: Citeseer; 2004:16–21.

Van Den Bout, DE, Thomas Miller III TK: **Graph partitioning using annealed neural networks.** *IEEE Trans Neural Netw* 1990, **1**(2):192–203.

Xu, J, Chung MJ: **Predicting the performance of synchronous discrete event simulation.** *IEEE Trans Parallel, Distributed Syst* 2004, **15**(12):1130–1137.

Zeigler, B, Praehofer H, Kim TG: **Theory of modeling and simulation: integrating discrete event and continuous complex dynamic systems.** *Academic Pr* 2000.

Zhang, L: **Internet topology data.** 2013. [http://irl.cs.ucla.edu/topology]

5

A theoretical formalism for analyzing agent-based models

Michael J North[1,2]

Correspondence: north@anl.gov
[1]Decision and Information Sciences Division, Argonne National Laboratory, 9700 South Cass Avenue, Lemont, IL 60439, USA
[2]University of Chicago, 5801 South Ellis Avenue, Chicago, IL 60637, USA

Abstract

Purpose: Following Holland, complex adaptive systems (CASs) are collections of interacting, autonomous, learning decision makers embedded in an interactive environment. Modeling CASs is challenging for a variety of reasons including the presence of heterogeneity, spatial relationships, nonlinearity, and, of course, adaptation. The challenges of modeling CASs can largely be overcome by using the individual-level focus of agent-based modeling. Agent-based modeling has been used successfully to model CASs in many disciplines. Many of these models were implemented using agent-based modeling software such as Swarm, Repast 3, Repast Simphony, Repast for High-Performance Computing, MASON, NetLogo, or StarLogo. All of these options use modular imperative architectures with factored agents, spaces, a scheduler, logs, and an interface. Many custom agent-based models also use this kind of architecture. This paper's contribution is to introduce and apply a theoretical formalism for analyzing modular imperative agent-based models of CASs. This paper includes an analysis of three example models to show how the formalism is useful for predicting the execution time and space requirements for representations of common CASs.

Method: The paper details the formalism and then uses it to prove several new findings about modular imperative agent-based models.

Results: It is proven that the asymptotic time and space performance of modular imperative agent-based modeling studies is computationally optimal for a common class of problems. Here 'optimal' means that no other technique can solve the same problem computationally using less asymptotic time or space. Modular imperative agent-based models are shown to be universal models, subject to the correctness of the Church-Turing thesis. Several other results are also proven about the time and space performance of modular imperative agent-based models. The formalism is then used to predict the performance of three models and the results are found to compare closely to the measured performance.

Conclusions: This paper's contribution is to introduce, analyze, and apply a theoretical formalism for proving findings about agent-based models with modular agent scheduler architectures. Given that this kind of modeling is both computationally optimal and a natural structural match for many modeling problems, it follows that it is the best modeling method for such problems.

Keywords: Agent-based modeling; Computational complexity theory; Random access stored-program machine

Background

Introduction

Complex adaptive systems (CASs) are collections of interacting, autonomous, learning decision makers embedded in an interactive environment (Holland 1992; Holland 1999; Holland 2006). CASs are common in both nature and society. Examples include ecosystems composed of organisms, industrial supply chains consisting of companies, social networks of formed by people, and even the human body's multitudinous cells. The contributing decision makers in each of these CAS, and all others, have properties and behaviors. The decision makers interact with and influence each other; learn from their experiences; and change their behaviors so they are better suited to their environment.

Modeling CASs is challenging for several reasons. The details matter, so averages often fail to properly represent these systems. Heterogeneity is the norm, so decision makers must be modeled individually. Adaptation is expected, so models must be highly dynamic. Space is central, so models must be intensively spatial. Scale matters, so models must often include large numbers of decision makers. Nonlinearity is routine, so models must track specific decision maker attributes against relevant thresholds. As an example, consider human tissue growth at the cellular level. In this situation, averages over the number of cells and extracellular materials do not take into account the unique, heterogeneous situation of each cell. Cells adapt to their circumstances and substantially change their behavior in nonlinear ways. The relative locations of cells and extracellular materials are critical for predicting long-term outcomes. Furthermore, scale matters since large numbers of cells are needed to form tissues. Consider further an example of tissue engineering where human tissues such as bone are grown in controlled environments. Now, there are nonlinear and even adaptive environmental inputs. If vascularization (e.g., blood vessel formation) is one of the goals then appropriate growth factors may be periodically added to the growth media. If growth optimization is the goal then additional factors and nutrients may also be included.

The challenges of modeling CAS such as those for tissue growth and engineering can largely be overcome by using the individual-level focus of agent-based modeling (ABM) (Bonabeau 2002; Macal and North 2010; North and Macal 2007). By modeling CAS decision makers as individual agents, the full effects of the diversity that exists among the decision makers with respect to their attributes and behaviors can be observed as they give rise to the dynamic behavior of the system as a whole. Like CAS decision makers, ABM agents have properties and behaviors, interact with and influence each other, learn from their experiences, and adapt their behaviors so they are better suited to their environment. Agent-based modeling has been used successfully to model CAS in many disciplines, including archaeology, biology, ecology, supply chains, consumer market analysis, military planning, and economics (North and Macal 2007; North and Macal 2009). Many of these models were implemented using agent-based modeling software such as Swarm (Accessed 2014), Repast 3 (Accessed 2014), Repast Simphony (Accessed 2014), Repast for High-Performance Computing (Accessed 2014), MASON (Accessed 2014), NetLogo (Wilensky 2014), and StarLogo (Accessed 2014). All of these options use modular imperative agent-based modeling architectures. Many custom agent-based models also use this kind of architecture. Three examples are reviewed later in the section on example models. These examples to show how the formalism introduced in this paper is useful for predicting the execution time and space requirements for representations of common CASs.

There are many possible theoretical foundations for computational agent-based modeling. The value of each candidate depends on the questions to be answered by the theoretical approach. For example, the theoretical foundation in North (2012) focuses on how information is actually processed in agent-based models. Other candidates are reviewed in the section on related work. This paper extends the implementation-oriented theoretical foundation in North (2012) and applies it to study a variety of new questions.

This paper's contribution is to introduce and apply a theoretical formalism for analyzing agent-based models. This paper details the formalism and then uses it to prove several new findings about agent-based models. It is proven that the asymptotic time and space performance of agent-based modeling studies is computationally optimal for common classes of problems. Here 'optimal' means that no other technique can solve the same problem computationally using less asymptotic time or space. Asymptotic approaches are likely to become increasing attractive as the sizes of agent-based models grow (Parker and Epstein 2011; Murphy 2011; Collier and North 2012). Agent-based models are also shown to be universal models, subject to the correctness of the Church-Turing thesis (Kleene 1943). Here 'universal' means that any computational model can be expressed as an agent-based model. Several other results about agent-based model time and space performance are also proven. The formalism is then used to predict the performance of three agent-based models and the results are found to compare closely to the measured performance.

The formalism contributes several things to the modeling of CAS and is therefore needed by CAS researchers from a range of disciplines. First, the formalism provides fundamental results about the time and space performance of one of the most common ways to investigate CASs, namely modular imperative agent-based models. Second, the formalism provides a way to predict the runtime and memory required to execute modular imperative agent-based models of CASs. This will become increasingly important as the number of agents commonly represented in models rises. Third, the formalism provides a straightforward way for CAS model developers to compare the time and space performance of modular imperative CAS agent-based models to that of traditional approaches such as optimization. Fourth, the formalism anchors modular imperative agent-based models, and their associated CASs, within the algorithmic hierarchy of traditional computer science complexity theory.

Related work

There is a range of published work that uses formal systems to prove selected properties of agent-based models or related kinds of models. This work includes the use of equation-based modeling, game theory, discrete systems, classifier agents, and an earlier random access stored-program (RASP) machine-based formalism. Each of these candidates is reviewed in the next sections.

Equation-based modeling

Parunak et al. (1998) compare agent-based modeling and equation-based modeling. They conclude in part that agent-based modeling is best for spatial problems and individual choices, while equation-based modeling is best for geographically concentrated problems driven by well-defined mathematical rules. Specific equation-based formalisms for analyzing simulations include partial recursive functions, difference equations, and dynamical systems.

Partial recursive functions

Epstein (2007) invalidates the commonly cited dichotomy between computational agent-based models and equation-based models by proving, albeit informally, that all computational agent-based models can be expressed using partial recursive functions. Epstein begins his proof by observing that because any standard computer program is effectively Turing complete, then programs for computational agent-based models must be so as well. Epstein then notes that there is an equivalent partial recursive function for every Turing machine. Therefore, there are sets of partial recursive functions for every computational agent-based model. These partial recursive functions are equations that fully specify any given computational agent-based model. Epstein also uses the partial recursive functions approach to argue informally that computational agent-based models, particularly those in the social sciences, are deductive in nature.

Considering the partial recursive functions approach further, Epstein (2007) discusses how unnatural or "unrecognizable" this expression of an agent-based model is likely to be for most agent-based modelers and how far this formalism is from real implementations. This suggests that partial recursive functions are not presently a suitable implementation-oriented theoretical foundation for agent-based modeling.

Difference equations

Leombruni and Richiardi (2005) introduce an equation-based formalism that can describe both agent-based and analytical models. Their formalism is intended to bridge the conceptual gap between equation-based economics and agent-based modeling.

Leombruni and Richiardi (2005) define agent attributes using vectors of state variables and then employ time-indexed difference equations to specify how each agent's state changes from one time step to the next. Naturally, analytical models that can be reduced to difference equations can also be represented with their formalism. Once they define their formalism, they then consider the question of how to estimate the relevant parameters in the difference equations.

Leombruni and Richiardi (2005)'s formalism has the advantage of including both agent-based and analytical models. Unfortunately, difference equations can be an extremely awkward way to specify the behavior of agents with complicated algorithmic behavior. Furthermore, they do not use their formalism to prove properties of either kind of model or show how this could be done.

Algebraic dynamical systems

Hinkelmann et al. (2011) extend Grimm et al.'s (2006) Overview, Design Concepts, and Details (ODD) protocol by adding an algebraic field to describe agent behavior. First we review ODD then consider Hinkelmann et al.'s (2011) extension of it.

ODD describes models using a three-part approach involving an overview, concepts, and details. The model overview includes a statement of the model's intent, a description of the main variables, and a discussion of the agent activities. The design concepts include a discussion of the foundations of the model. The details include the initial setup configuration, input value definitions, and descriptions of any embedded models.

Hinkelmann et al.'s (2011) ODD extension maps each agent's state into a set of state variables. A set of time-indexed polynomial discrete dynamical equations that update the values of each agent's state variables are then defined in such a way as to allow the state variables to form an algebraic field. They then use the rich body of existing algebraic techniques to prove system-level properties of two simple demonstration models.

Hinkelmann et al.'s (2011) approach has the advantage of offering a straightforward way to prove properties of agent-based models while leveraging a large body of existing research. Unfortunately, expressing complicated agent behaviors using time-indexed polynomial discrete dynamical equations can become prohibitively complex. Also, Hinkelmann et al.'s (2011) note that the algebraic techniques they seek to leverage sometimes have serious weakness of their own, such as an inability to scale to realistic model sizes.

Reaction–diffusion dynamical systems

Dynamical systems are sets of equations that specify a path through a space over time (Smale 1967). Selected agent-based models can be expressed as dynamical systems models or can be approximated by them. For example, Dorofeenko and Shorish (2002) used this approach to study a variation of the prisoner's dilemma game. It is interesting to note that even though the prisoner's dilemma is a classic game theoretic model, Dorofeenko and Shorish did not choose analytical or algorithmic game theory as the primary method for their study.

Dorofeenko and Shorish (2002) map Epstein's (1998) demographic prisoner's dilemma game to a reaction–diffusion dynamical system metamodel. Epstein's original game used a 30×30 square lattice populated by reproducing agents with fixed prisoner's dilemma strategies. Dorofeenko and Shorish investigated a one-dimensional version of the game and proved that certain conditions produce regions of long-term cooperation. Their findings are interesting, but are limited to the specific model that they are investigating. In particular, they do not provide a method for generalizing their findings or applying dynamical system analysis to a wider range of agent-based models. This absence illustrates the general conclusion that dynamical systems have limited potential as a theoretical foundation for agent-based modeling, as there is no known general mapping between computational agent-based models and dynamical systems models.

Game theory

Game theory is a mathematical approach for studying multiparty decision problems in interactive environments (Myerson 1991). Some of the findings that are the most relevant for this paper are that many games have at least one stable outcome (i.e., Nash equilibrium) and that many games are nondeterministic polynomial time (NP)-complete (Nisan et al. 2007). Nisan et al. (2007) show that games with exactly one Nash equilibrium are in a lower complexity class than are NP-complete problems, whereas games with two or more Nash equilibriums are NP-complete.

Game theory can be used to prove properties about agent-based models to the extent that such models can be expressed as games. There are several approaches to solving game theoretic formulations including analytic game theory and algorithmic game theory. Each will be considered separately.

Analytic game theory (Nisan et al. 2007) applies mathematical proof techniques to deduce facts about game theoretic formulations. The previously mentioned proof that certain games are NP-complete is an example. Unfortunately, only a few kinds of games are amenable to analytical proof techniques (Nisan et al. 2007). Therefore, most practical problems do not have a corresponding analytically solvable game. This constraint prevents analytic game theory from being a useful general theoretical foundation for computational agent-based modeling.

Algorithmic game theory (Nisan et al. 2007) uses computational tools to solve game theoretic problems. Its computational nature allows it to address a much wider range of cases than does analytic game theory. However, this added flexibility often comes at the expense of theoretical generality.

There are several major differentiators between algorithmic game theory and agent-based modeling. Algorithmic game theory tends to focus on specific types of outcomes, such as stable equilibrium or various kinds of oscillatory dynamics (Nisan et al. 2007). Agent-based modeling tends to address much broader outcomes (North and Macal 2007). Algorithmic game theory uses extended game formulations to specify models, whereas agent-based modeling typically uses more scalable software engineering design tools. Unlike algorithmic game theory formulations, agent-based modeling software is often designed to represent detailed geographies, rich interaction topologies, nuanced agent learning, and complex agent behaviors. Therefore, algorithmic game theory is not an appropriate theoretical foundation for agent-based modeling because of the limited kinds of outcomes that are represented, the constraints on scaling, and the substantial practical differences in how models are specified.

Discrete systems

Quite a few different approaches to representing agent-based models using discrete systems have been published. These approaches include cellular automata and the Discrete Event System Specification (DEVS). These approaches will be considered in the following sections.

Cellular automata

Cellular automata (Christos and Lafortune 2008) are grids of cells, each of which has a state transition diagram. Uniform cellular automata have a single state transition diagram for every cell. Non-uniform cellular automata can have state transition diagrams that vary by cell. Brown et al. (2005) discuss cellular automata in relation to spatial agent-based modeling. They observe that cellular automata tend to yield Eulerian (i.e., field-oriented) models versus Langrangian (i.e., individual-oriented) representations. Here we conclude that this limits cellular automata as a foundation for proving properties about agent-based models, since many are domains are naturally Langrangian.

DEVS

DEVS is one of the best developed theoretical formalisms for analyzing complicated simulations. DEVS describes simulations as hierarchical Moore machines with timing information added to the state transitions (Zeigler et al. 2000). Classic DEVS is the core abstraction. It is a structure that consists of the following:

- A set of external events that can be input to the DEVS,
- A sequence set of states,
- A set of internal results that can be output from the DEVS,
- An internal transition function that specifies how timeouts change states,
- An external transition function that specifies how inputs change states,

- A time advance function that specifies how long before each state times out, and
- An output function that specifies the results for each state.

Coupled DEVS extends classic DEVS by allowing hierarchical nesting of the Moore machines. An example of a hierarchical DEVS implementation is described by Bolduc and Vangheluwe (2002). Extensions that add other functionality to classical DEVS and restrictions that limit the functionality have also been studied.

DEVS provides a theoretical foundation for analyzing some critical computational properties of agent-based models such as decidability. If the Church-Turing thesis is correct, then any computing system as complex as a Moore machine can be used to calculate any function. Therefore, in principle, DEVS can represent any computable model. Nonetheless, this finding by itself does not mean that DEVS is an ideal theoretical foundation for agent-based modeling because the same conclusion can be reached about any Turing-complete system.

Unfortunately for agent-based modeling, DEVS imposes an awkward boundary between state and behavior that makes it difficult to represent dynamically generated agent behaviors. The DEVS boundary also does not correspond to common agent-based model architectures. In addition to dynamically generated behaviors, agent-based models often have state structures that are a dynamic function of the events in the models. Dynamic Structuring DEVS (Shang and Wainer 2006) attempts to address these weaknesses, but it also adds new constraints that similarly limit its utility.

In addition, DEVS lacks important constraints commonly found in agent-based models. Agent-based models have extremely wide-ranging properties, yet exhibit characteristic forms of state storage and agent behavior. These conventions place specific theoretical limitations on the computational properties of agent-based models. DEVS does not include these conventions. An appropriate theoretical foundation should represent these conventions and allow analysts to reason about them rigorously. These agent-based modeling conventions are detailed in later sections.

Classifier agents

Holland (2006) discusses the modeling of complex adaptive systems using a variety of techniques including agent-based modeling. He briefly reviews the capabilities of control theory, economic modeling, biological cell modeling, and game theory. Holland concludes that although the formalisms he has reviewed have unique strengths, they all face the problem of relying on differential calculus in general and on partial differential equations in particular. He then argues that the usefulness of partial differential equations is undermined by the inherent nonlinearity of the domains being modeled. For Holland, the answer is to develop exploratory computer-based models that have agent rules defined using individual classifier systems. Holland's approach does not include enough explicit simulation engine details to derive time or space performance bounds. Furthermore, his approach is a special case of the formalism presented in this paper because it can be implemented using this formalism. Unlike this paper, Holland's paper does not include time or space bounds derived using his recommended approach.

Formal methods

Niazi and Hussain (2010) use the Z specification language to represent an agent-based model of wireless sensor networks. Z is an International Standards Organization standard that specifies computations using set theory and predicate calculus operating on typed expressions (Spivey 1992). Z and its derivatives offer correctness-proving tools and model-based checking capabilities. Z specifications are often detailed enough to be translated into executable software.

Niazi and Hussain (2010) show how to use Z to specify a wireless sensor network model by defining an interlocking set of Z schemas. They report that the resulting specification is useful for demonstrating or proving the correctness of the wireless sensor network model relative to well-defined requirements. Also, they show that the Z specification provides a way to document detailed expectations for model behavior. This matches well with two of Z's purposes, namely system documentation; and system verification and validation. However, the Z language does not specifically address model execution time or space performance analysis. For example, Niazi and Hussain (2010) do not use their Z specification to predict the execution time or space performance of their wireless sensor network model.

Earlier RASP formalisms

North (2012) introduces a RASP-based formalism for the analysis of agent-based model performance and uses this formalism to prove that the total asymptotic time and space performance of a specific Mars Rover model is computationally optimal. The current paper substantially extends North (2012) in several ways. First, the formalism presented in this paper is both different and more general than that of North (2012). The formalism in this paper includes a discrete event scheduling option, a wider range of interaction spaces, a discussion of user interfaces, analysis of parallel execution, and provisions for modeling studies. Second, broadly applicable computational optimality conditions are proven in this paper. Third, several other results about agent-based model time and space performance are also proven. Fourth, this paper provides performance analysis examples for three real models and compares these results to the theoretical predictions.

Methods

Common features of agent-based modeling libraries

An implementation-oriented theoretical foundation for computational agent-based modeling should take into account the conventions commonly seen in applied modeling. This is not to say that current agent-based modeling efforts represent the expected final form of computational agent-based modeling research. Agent-based modeling is younger than techniques such as game theory. It is anticipated that agent-based modeling will evolve rapidly in the future. As this growth proceeds, it is hoped that agent-based modeling theory and practice will advance together. This paper is intended to contribute to this evolution.

Seven free and open-source agent-based modeling libraries and environments with published descriptions will be used to illustrate the common structural conventions of agent-based models. The examples are Swarm, Repast 3, Repast Simphony, Repast for

High-Performance Computing (HPC), MASON, NetLogo, and StarLogo. Each of these options has a modular imperative agent-based modeling architecture.

Swarm is an agent-based modeling library with both Objective-C and Java bindings (Minar et al. 1996). Repast 3 (North et al. 2006) is a family of three agent-based modeling libraries, two in Java and one in C#, each with architectures inspired by Swarm. Repast Simphony (North et al. 2013) is a Java-based library and environment that generalizes Repast 3. Repast HPC (Collier and North 2012) is a C++ Message Passing Interface agent-based modeling library designed to run Repast Simphony's core feature set efficiently on supercomputers.

MASON is an agent-based modeling library that seeks to provide a focused set of core services that can be easily used with other libraries (Luke et al. 2005). Its structure is related to that of Swarm.

NetLogo (Wilensky 1999) is an educational agent-based simulation environment that uses a modified version of the Logo programming language (Harvey 1997). Repast Simphony and Repast HPC include a ReLogo module and environment inspired by Logo in general and NetLogo in particular. StarLogo (Resnick 1994) is another Logo environment related to NetLogo.

When considering the commonalities between these libraries and environment, a short list of features emerges. These features include agents, a scheduler, interaction spaces, a random number source, logging, and a user interface. These features are discussed in detail in the sections that follow. It should be noted that these features are significantly different than those suggested by the previously discussed DEVS formalism.

Agents

Agents are the decision makers in all of the selected libraries and environments. Methods for representing agents vary. For example, the simulation systems discussed above are all object-oriented except for NetLogo and StarLogo. At a minimum, agents have some type of a unique identifier (e.g., index code or memory reference), behaviors that can be activated, and attributes that can be modified. Every agent can have unique behaviors and attributes, although agents are often grouped into types, classes, or breeds. Some libraries or environments offer additional features, such as dynamically modifiable behaviors.

Initialization of agents is completed using a bootstrap module that is called before the model begins executing. Some libraries or environments also offer persistent storage using text files, extensible markup language (XML) files, binary files, databases, or clouds.

Scheduler

The scheduler is responsible for representing the flow of time in a simulation. Scheduling can be time stepped or can use discrete events.

Time-stepped agents activate once for each scheduler time increment. The time counter normally increases in fixed increments of one unit for each step. Schedulers often implement time-step scheduling by repeatedly scanning through a list of agents. This list may be constant or may be randomized over time to allow the agent activation order to vary.

With discrete event scheduling, each agent action occurs at a unique time. Time advances from event to event. Some systems allow multiple events at a given time.

These pseudo-parallel systems usually execute simultaneous events at a given time in a randomized sequential order.

Schedulers often implement discrete event scheduling by maintaining a list of pending, time-stamped agent behaviors sorted by time of activation. The next behavior is activated whenever a running behavior is completed. If any ties occur, they are often broken with random draws as previously discussed.

Interaction spaces

Several different kinds of agent interaction spaces are usually available. All of the previously listed libraries and environments offer several basic interaction spaces, except for StarLogo, which does not have networks:

- A finite, one- (or more) dimensional integer lattice with one computable real value for each lattice point provides a simple way to represent scalar fields. The number of dimensions for each scalar field is fixed when the grid is created.
- A finite, one- (or more) dimensional integer lattice with multiple occupancy and computable real number coordinates for agents provides a simple representation of physical space. The computable real numbered coordinates allow each agent to take on any location within the range supported by the grid. The integer lattice allows agents to find other agents who have similar coordinates (i.e., those agents with the same location when their coordinates are truncated to an integer). As with scalar fields, the number of dimensions for each grid is fixed when the grid is created. This entity will be called a "grid."
- A basic network stored as a weighted graph between agents provides a way to represent connections between agents. Both directed and undirected graphs are supported.

Each model can have zero or more of each kind of interaction space. Generally, the number of interaction spaces is a small constant relative to the number of agents. By default, every agent in each model is present somewhere in each of the model's interaction spaces.

Random number streams

Random number streams are used to generate reproducible sequences of pseudorandom numbers. Several probability distributions are usually provided, including uniform random sources.

Logging

Logging provides a mechanism for recording values within an executing model for later analysis. Data collection is usually performed at regular intervals during model execution both for the overall model and for sets of agents. Data collection is usually performed automatically based on user input. User-programmed logging will be used in this paper.

User interface

Agent-based models usually run in one of two modes. The first is an interactive mode that presents the user with a graphical interface. The second is a 'headless' batch mode that reads input files, the runs model, and then stores the results into output files. Databases, data warehouses, or clouds may also be used in place of files.

Agent-based models are often run in interactive mode during model development and scenario creation. Graphical user interfaces (GUIs) for agent-based models are usually composed of a main window containing a nested set of subwindows. The main window is called the 'frame'. Sometimes a frame is not used and the subwindows appear independently. In either case, the subwindows contain various simulation controls, one or more views of the spaces, and one or more displays of the properties of selected individual agents. The agent property subwindows are called 'probes'. The properties often dynamically update as the underlying model executes. An overview schematic is shown in Figure 1.

Displaying a GUI often consumes a large amount of computing time and space. Most of these resources are not needed during batch runs. Therefore, batch mode is commonly used to generate large numbers of runs once the model and scenarios are finalized. Batch mode is also typically used for individual model runs with large numbers of agents. Batch mode will be the main focus of this paper because asymptotic performance analysis of agent-based models implies large numbers of agents.

Abstract machine for modular imperative agent-based modeling

This section provides an abstract machine specification for the modular imperative agent-based modeling architecture identified in the previous section. In this paper, we use the RASP architecture for our abstract machine. In keeping with the implementation-oriented theme discussed earlier, the RASP architecture was selected because it has a closer

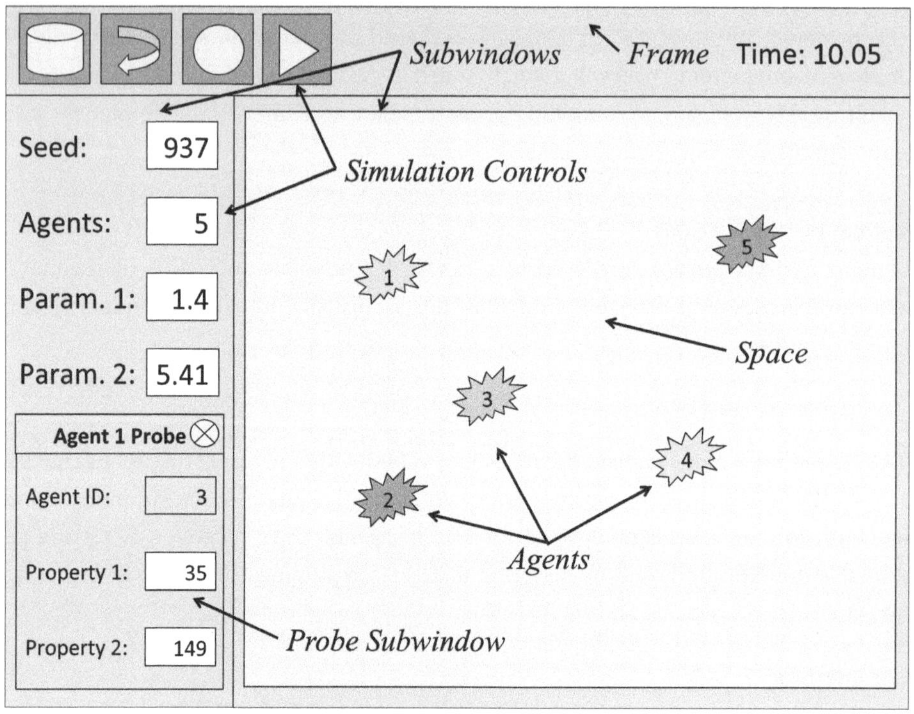

Figure 1 Schematic diagram of an agent-based model GUI.

correspondence to real computer systems than do other typical options, such as Turing machines or the λ-calculus.

A RASP is an enhanced random access machine (RAM) (Cook and Reckhow 1973). Several variations exist (Elgot and Robinson 1964; Cook and Reckhow 1973). This paper uses the formulation by Cook and Reckhow (1973).

A RAM consists of an infinite set of memory registers that are used by a finite program. To better match the real implementations in this paper, a large but finite number of registers will be used instead of an infinite number. Each register can hold an integer. Register i is denoted X_i for a nonnegative integer.

Memory will be managed using a doubly linked free list with a fixed block size combined with a free boundary reference. The free boundary reference points to the first free register beyond which all registers are unallocated. To avoid external memory fragmentation, the block size will be set to be the maximum of any the dynamically allocatable uses of memory. Memory allocation will use the first block in the free list, if one is available. If the list is empty, then a block will be created by using and then incrementing the free boundary reference. Memory deallocation will insert the freed memory at the head of the list. In this case, the free boundary reference will remain the same.

By default, a RASP only processes integers. Finite computable real numbers will also be included in our modified machine. The operations for computable real numbers will work in the same fundamental way as the corresponding integer operations defined later in this section.

Cook and Reckhow (1973) scale the sizes of their registers independently to match the values held in storage. Modern computers rarely allow such variation. Therefore, we will set the register size for each RASP run to be a RASP-wide constant value, w. Naturally, w defines the maximum value of any integer or computable real number that can be directly processed in the given program run. w will vary between runs to reflect the widths of the numbers used in the problems being solved. The size of w is discussed further, once agents are introduced.

A RAM has access to a sequential input tape and a sequential output tape. The RASP enhancements add an accumulator (AC) that contains an integer; an instruction counter (IC) that contains the address of the next instruction; the ability to run programs by sequentially reading, decoding, and executing the values in the registers pointed to by IC; and an ALGOL-like programming language that can be translated into instructions for the underlying RAM. The RASP instructions are shown in Table 1.

For large w and constants c_k, all of the operations shown in Table 1 take $O(c_k w) = O(w)$ time. In particular, processing integers and computable real numbers requires $O(w)$ time and space. Operations that allocate new registers take $O(0)$ time and $O(w)$ space.

Cook and Reckhow (1973) show that a RASP has higher computing power than a standard multi-tape Turing machine—but only by the square root of the time. The RASP architecture was chosen for this paper because it is amenable to complexity analysis while still retaining an architecture closely resembling commonly used computers.

In the next sections, the abstract machine's program and memory layout is presented. The outline intentionally uses standard data structures and algorithms to insure that the expected code is straightforward.

Table 1 RASP instructions adapted from Cook and Reckhow's (1973) Table 2

Instruction	Interpretation	Code	Time	New space
`LOD c`	Load constant **c** into **AC**	1	2w	0
	Increment **IC**			
`ADD X_i`	Add the value in register X_j to **AC**	2	4w	0
	Increment **IC**			
`SUB X_i`	Subtract the value in register X_j from **AC**	3	4w	0
	Increment **IC**			
`STO X_i`	Store the value in **AC** to register X_j	4	3w	w, if X_j was not allocated else 0
	Increment **IC**			
`BPA c`	If **AC** > 0, then set **IC** to **c**	5	3w	0
	If **AC** ≤ 0, increment **IC**			
`RD X_i`	Load the next input tape value into X_j	6	3w	w, if X_j was not
	Increment **IC**			allocated else 0
`PRI X_i`	Output the value in X_j to the output tape	7	3w	0
	Increment **IC**			
`HLT`	Halt execution	-∞ o 0 and 8 to ∞-	2w	0

Agents

Agents will be represented using a set of stored programs and a set of registers for data values, as shown in Table 2. h is the number of different types of agents in a given model. n_i is the maximum number of active agents for agent type i in a given model run. The n_i values count agents that execute at least one behavior, have attributes that are accessed at least once for each agent, or are included in at least one calculation. Completely inactive agents are not included in the n_i counts. We then have the total number of agents, N, as the sum of n_i from 1 to h. We set $w \geq \lceil \log_2(N) \rceil$ to insure that agent counts can be properly calculated in all cases. To minimize time and space requirements, we then have $O(w) = \max(O(\lceil \log_2(N) \rceil), Z)$, for Z the widest numeric value to be directly processed by a given model.

Agent definitions are read from input files. The input files have either individual agent descriptions or distributions. If the agents are stored individually, then there will be at least one input per agent. If distributions are used, then the needed agents are created using random draws from the given distributions.

Table 2 Agent storage layout

Register	Contents
`I`	The first of `j` registers needed to store the agent attributes.
`i + j - 1`	The last of `j` registers needed to store the agent attributes.
`i + j`	The head of a linked list used to store dynamically generated agent behaviors, if any.
`i + j + 1`	The first of **k** registers needed to store the heads of doubly linked lists used to store network links for network `1`.
`i + j + k`	The last of **k** registers needed to store the heads of doubly linked lists used to store network links for network **k**.
`i + j + k + 1`	The address of the next agent in the agent list.
`i + j + k + 2`	The address of the previous agent in the agent list.

Agents will be stored using the layout shown in Table 2. The default identifier for each agent is its lowest memory address.

Agents can have both statically defined and dynamically generated behaviors. Statically defined behaviors are included in the code for a given RASP instance. Dynamically generated behaviors are stored in an agent-specific linked list as shown in Table 2.

A doubly linked list of agents is maintained by the system. Inserting or removing agents updates this list. This list is used for activities discussed later such as time scheduling.

Basic operations to be provided by the agent-based modeling library are as follows:

- Creating an agent requires initializing the agent register storage, as shown in Table 2. This operation has time $O(r_1\lceil \log_2(N)\rceil)$ complexity and $O(r_1\lceil \log_2(N)\rceil)$ space complexity for r_1, the number of registers used by the agent. r_1 is a constant relative to N.
- Disposing of an agent requires adding the agent's space back to the free list, which takes $O(\lceil \log_2(N)\rceil)$ time.
- Changing the value of an agent attribute for a given agent j uses O(0) space and $O(r_2\lceil \log_2(N)\rceil)$ time for r_2, the required number of registers. r_2 is again a constant relative to N.
- Modifying an agent's behavior involves writing a new routine for the agent. This action has $O(r_3\lceil \log_2(N)\rceil)$ time complexity and $O(r_3\lceil \log_2(N)\rceil)$ space complexity for r_3, the required number of write operations. Statically compiled agent behaviors are stored with the main model program. Dynamically generated agent behaviors are added to an agent-specific linked list, as noted in Table 2. r_3 is yet again a constant relative to N.
- Executing an agent behavior takes time and space as determined by the contents of the corresponding routine.

The agents are accompanied by information on their surrounding environment. This information includes the statically compiled agent behaviors and other pieces of overall control information stored with the main model program. This information is in addition to a description of the interaction spaces that are present, which are discussed in a later section. The space needed to store information on the surrounding environment is always small and constant compared to that needed to store information about agents for any substantial number of agents. Therefore, environments use $O(A\lceil \log_2(N)\rceil)$ space and take $O(B\lceil \log_2(N)\rceil)$ access time for constants A and B. The need for a large amount of information on the surrounding environment implies that the agent-based model has been designed incorrectly and that agents have been blended into the environment.

Scheduler

Both time-stepped and discrete event scheduling are available. Each model will use one of these options exclusively.

The time-stepped scheduler uses the agent list and an integer time counter. This scheduler simply picks elements sequentially from the agent list and executes the next behavior. Every agent runs for each time step. Each time-stepped agent has a default 'step' routine that is called for each time step. A given agent's step routine can be empty if no behavior is needed for a given time step. This default is taken to be the first element in the linked list of dynamically generated agent behaviors, if any. If an agent's dynamic procedures list is empty, then a call is made to the first routine in the list of statically compiled functions for the corresponding agent type. Whichever way the step routine is found, it can call other routines as needed.

The discrete scheduler uses a time counter and an event time slot list stored as a B-tree sorted by time stamp. Each event time slot list entry includes a time value and the head pointer of a singly linked list of events. Each event, in turn, contains a time stamp, the address of the routine to be invoked, a single register value that can contain an optional parameter, and a link to the next list element. Decoding the parameter is solely the responsibility of the routine to be invoked. Inserting a new event time slot into the event time slot list, in the worst case, requires $O(\lceil \log_2(N) \rceil \log v)$ time for a list with v unique time slots. Finding an event time in the worst case takes $O(\lceil \log_2(N) \rceil \log v)$. Deleting an event when it is completed in the worst case takes $O(\lceil \log_2(N) \rceil \log v)$. The B-tree will use space $O(v\lceil \log_2(N) \rceil)$. In the worst case inserting, reading, or deleting an event once the corresponding event time has been found or inserted takes $O(\lceil \log_2(N) \rceil)$ time.

A model is said to have asymptotically linearithmic[a] scheduling in time or space when the time or space performance is $O(N\lceil \log_2(N) \rceil)$. Time step scheduling is always asymptotically linearithmic in both time and space. Discrete scheduling is asymptotically linearithmic in time and space when the number of pending unique discrete event time slots, referred to as v previously, is constant with respect to N.

Regardless of the kind of scheduling that is employed, the number of time steps or unique discrete event times is usually independent of N. In other words, models are rarely run for more time steps or unique discrete event times because they have more agents. We thus assume that the length (i.e., time steps or slots) of each run is a constant relative to N, and therefore run length does not affect asymptotic time or space performance. We will briefly return to this assumption in a later section.

Interaction spaces

Three kinds of interaction spaces are supported. Each model can have any number of these spaces, although in practice the number of different interaction spaces is usually quite small.

Scalar fields will be stored as multidimensional arrays indexed in row major order. Each d-dimensional scalar field with maximum dimension range g will use space $O(g^d\lceil \log_2(N) \rceil)$. Each grid point holds a computable real number or an integer. An individual value can be read or written in $O(\lceil \log_2(N) \rceil)$ time.

Grids will also be stored as multidimensional arrays indexed in row major order. Each array element will store the head pointers of doubly linked lists of agent records and a counter of the length of each list. There is one head pointer and counter pair for each agent type. Each agent record holds the reference of the associated agent and the agent's floating point coordinates. Agents will also keep a reference to their own records. The doubly linked list can be of any length between zero and N. It is up to each modeler to process these lists in an appropriately efficient way. Each d-dimensional grid with width g will use space $O(g^d\lceil \log_2(N) \rceil)$. An individual value can be read or written in $O(\lceil \log_2(N) \rceil)$ time.

Networks will be stored using a doubly linked list of edges for each agent. The head references for each agent are shown in Table 2. Each linked list element contains the identifier of the linked agent, a computable real number weight, a reference to the next element, if any, and a reference to the previous element, if any. The use of weights by agent behaviors is optional. Undirected networks contain linked lists with paired references (i.e., if agent A's list includes agent B, then B's list will also include A). Directed networks allow lists with asymmetric references. Each network will use $O(bN\lceil \log_2(N)\rceil)$ space for b the maximum number of links per agent. An agent can insert, read, or delete an entry in their list in $O(b\lceil \log_2(N)\rceil)$ time.

Random number stream

A pseudorandom number generation algorithm with $O(\lceil \log_2(N)\rceil)$ asymptotic time and space performance will be used to produce a uniformly distributed random number stream. The Mersenne Twister (Matsumoto and Nishimura 1998) is an example, at least until the number of random draws approaches the algorithm's period of 2^{19937}-1. Non-uniform distributions can be provided by applying appropriate constant time functions to the core uniform random number stream. Also, the random seed used for stochastic execution is an input parameter.

Logging

The log stores values to the output tape using the PRI instruction. The log will be unstructured. This will allow users to determine the format of the logs individually for each model. The PRI instruction uses $O(\lceil \log_2(N)\rceil)$ time and O(0) space for each value to be logged.

User interfaces

The abstract machine will not normally use a GUI. As previously discussed, the machine will typically execute in a headless batch mode, since we are analyzing asymptotic (i.e., large-scale) performance. Nonetheless, the performance impact of a GUI will be briefly analyzed.

The abstract machine model can accommodate a GUI by including an optional memory region on startup that represents the graphics buffer. The machine will then include routines to draw agents into the memory locations representing the graphics buffer.

Drawing an agent-based model GUI usually involves rendering the frame and subwindows shown in Figure 1. The complexity of the overall frame and simulation control subwindows are independent of the number of agents. Thus, these components run in constant time and use constant space relative to the number of agents.

Users normally invoke probes manually rather than programmatically. As such, there is a fixed upper bound on the number of agents that can realistically be shown with probes in a GUI. Therefore, probes also run in constant time and use constant space relative to N.

Drawing the interaction spaces usually involves rendering many, if not all, of the agents individually. Generally, it will take $O(\lceil \log_2(N)\rceil)$ time to draw each view of the agents. Optimizations that reduce the number of agents considered for drawing may improve the time complexity. For example, simply not showing interaction spaces reduces the complexity of this task to O(0).

Drawing can be performed on an individual basis with intermediate results accumulating in the graphics buffer (i.e., graphics memory). Graphics buffers have a fixed size that is related to the screen resolution and is independent of the number of agents. Thus, drawing takes constant storage space.

The remainder of this paper will focus on batch mode model runs. The analysis that follows can be augmented using the results in this section, as needed.

Parallel execution

There are multiple ways to use parallel computing to improve simulation performance Perumulla (2006), Parker and Epstein (2011) and Collier and North (2012) present examples. Two major paths are parallel parameter sweeps and distributed execution of individual simulations. Obviously, these approaches can be combined. Both paths will be considered in the next sections.

Parallel parameter sweeps

Simultaneously executing each run of a parameter sweep is sometimes called an "embarrassingly parallel" approach because of the relative simplicity of its implementation compared to more sophisticated alternatives (Foster 1995). The speedup factor is just the number of simultaneous runs discounted for any overhead required for job control. The overhead for this type of implementation is generally quite low, as implied by the "embarrassingly parallel" moniker. The major limiting factor is the number of processors, p, available for executing model runs. For even the largest computers, p is constant relative to N. Therefore such runs do not affect the asymptotic time requirements. Similarly, the memory requirements for a parallel parameter sweep will be the size of a model run times the number of simultaneous runs, again with additions for job control requirements and with discounts for possible shared memory. These factors, combined with the constant number of processing nodes, imply that parallel parameter sweeps do not affect the asymptotic space requirements.

Distributed model runs

Executing a single model using many processing nodes offers the potential to speed up the delivery of individual results. Here we will assume an ideal implementation. In this case, the asymptotic space requirements remain unchanged, given that the same number of agents needs to be tracked as with a sequential model. The best speedup for asymptotic time is 1/p for p processors because no computing work can be ignored. Unfortunately, as previously discussed, p is a constant relative to the number of agents, so there is no asymptotic time speedup.

Modeling studies

Agent-based modeling studies often require multiple model runs to cover different scenarios and to account for stochastic variation. The time and space complexity results for individual runs can be used to estimate the requirements for each needed type of run and then multiplied by the corresponding number of runs and summed, as appropriate.

In many situations, even though the number of model runs needed for a given study is a factor of many input variables, it is independent of the number of agents. In these cases, the asymptotic time and space requirements, of course, are the same as for an individual model run.

The abstract machine reads the first input tape value to determine the number of input sets that are present. If more that one input set is present then they are simply concatenated in the desired order of execution.

The abstract machine has constant time routines to reset itself and start the new model runs, as needed. The output from each run is simply appended to the output tape in order of execution. It is the responsibility of modelers to store appropriate delimiters or use other encodings to differentiate output run results.

The abstract machine has an optional provision for executing a routine to aggregate a set of runs after they have completed. The performance of this routine is included in the overall performance of the data aggregation system. In general, if the number of model runs is constant with respect to N and the model-level data aggregation is linearithmic or better in asymptotic time and space performance then the parameter sweep-level data aggregation routine will also be linearithmic or better in performance.

Results and discussion

Best cases for any model

This section presents results that apply to all models. Considering memory usage, there must be an assignment operation for every register that is used beyond the size of the basic machine and model program, both of which are constant with respect to N. Therefore, for maximum model memory space usage S_{max}, we have:

(1) For any model, $O(T_{total}) \geq O(S_{max})$.

Considering output logs, there must be a PRI instruction for each output value. For total time T_{total} and total output log space usage S_{log} we therefore have:

(2) For any model, $O(T_{total}) \geq O(S_{log})$.

Best cases for asymptotically thorough and incompressible models

Two important definitions are presented in this section. These definitions are then used to analyze the best cases for time and space.

Asymptotically thorough models

A model with N agents is said to be *asymptotically thorough* when individual values from O(N) agents are needed from at least two different representative time periods to correctly calculate the logged results. Representative means that the span between logged time periods includes typical agent behaviors. Logging a value from each agent for two or more normal time steps is generally sufficient to meet this standard.

Asymptotically incompressible models

A model with N agents is said to be *asymptotically incompressible* when a minimum of O(N) space is needed to store the agents and recalculating an attribute for an agent takes at least as much space as storing it between uses. In other words, the best compression ratio achievable for agent storage is constant relative to N for the general case for a given model. Agent attributes that can take on a wide range of values are generally sufficient to meet this standard.

Analysis

We will now consider the best cases for time and space performance for asymptotically incompressible and asymptotically thorough models.

Time

For N decision makers, any asymptotically thorough model must access the attributes for at least O(N) decision makers for a minimum of two different time periods. Thus, the output aggregation must use at least O(N) operations, each of which takes $O(\lceil \log_2(N) \rceil)$ time. Then we have the following:

(3) The minimum asymptotic time needed for asymptotically thorough models is in Ω $O(N\lceil \log_2(N) \rceil)$ for variable word size computers.

Space

Asymptotically thorough models must log individual values from O(N) agents from at least two different representative time periods to correctly calculate the logged results. Any asymptotically incompressible model must either store the relevant decision maker attributes between the first and second time periods or use at least as much space recreating them for the second logging period. Either way, this requires $O(N\lceil \log_2(N) \rceil)$ space. Therefore, we have the following:

(4) The minimum asymptotic space needed for asymptotically thorough and incompressible model is in Ω $(N\lceil \log_2(N) \rceil)$ for variable word size computers.

Time-stepped, boundedly rational modular imperative agent-based model worst cases

In this section, we will consider the worst cases for time and space for asymptotically thorough and incompressible batch-mode modular imperative agent-based models with large numbers of individual, boundedly rational, decision makers; input parameter ranges and counts that do not increase with the number of decision makers; asymptotically linearithmic scheduling; and output aggregation that is linearithmic or faster in time and space. Obviously, it is assumed that the agent-based model in question is suitable for the problem, correctly written, and efficiently implemented relative to the constraints and requirements of the question to be answered.

Time

The N agents in a model with, at most, s unique event times (i.e., times steps or unique discrete event times) and d operations in the agent routine with the greatest time complexity requires, at most, $O(Nsd\lceil \log_2(N) \rceil)$ time. s does not increase with N. Thus, s is a constant with respect to N. Bounded rationality means in part that the amount of computation that can be completed by any one agent in a single time step is strictly limited. This limitation, in turn, means that d must be a constant relative to N because N is unbounded. All of the agent's operations are either programmed using the instruction set given in Table 1 or are interaction space access calls. Thus, the agents themselves require, at most, $O(N\lceil \log_2(N)\rceil)$ time plus access time for any interaction spaces that might be used.

As discussed above, environments use $O(\lceil \log_2(N)\rceil)$ time. There is only one environment. Therefore, the total time N agents spend accessing information in the environment is, at most, $O(N\lceil \log_2(N)\rceil)$.

Scalar fields can be accessed in $O(\lceil \log_2(N)\rceil)$ time. The number of scalar field accesses in each agent behavior must be bounded due to bounded rationality. Therefore, the total time N agents spend accessing scalar fields is, at most, $O(C_1N\lceil \log_2(N)\rceil) = O(N\lceil \log_2(N)\rceil)$, where C_1 is the maximum number of accesses in an agent behavior.

Grids are stored as simple multidimensional arrays. The access time is just the time needed for array indexing, namely, $O(\lceil \log_2(N)\rceil)$ per access. The number of linked list checks and overall grid accesses in each agent behavior must be bounded due to bounded rationality. Therefore, the total time N agents spend accessing grids is, at most, $O(C_2N\lceil \log_2(N)\rceil) = O(N\lceil \log_2(N)\rceil)$, where C_2 is the maximum number of grid accesses in an agent behavior.

Each network has $O(b\lceil \log_2(N)\rceil)$ access time for b, the maximum number of links per agent. Bounded rationality implies that the number of network links that can be remembered by any one agent is strictly limited. This constraint, in turn, means that b must be a constant relative to N, given that N is unbounded. The number of accesses must also be bounded due to bounded rationality. Therefore, the total time N agents spend accessing networks is, at most, $O(C_3bN\lceil \log_2(N)\rceil) = O(N\lceil \log_2(N)\rceil)$, where C_3 is the maximum number of network accesses in an agent behavior.

Output aggregation uses linearithmic or less time. The outputs, therefore, require, at most, $O(N\lceil \log_2(N)\rceil)$ time in addition to that used by the rest of the model.

We have the following for the total execution time, T_{total}:

$$T_{total} \le T_{agents} + T_{environment} + T_{fields} + T_{grids} + T_{networks} + T_{outputs}$$

The count of each type of interaction space will be a constant relative to N as previously discussed. Substituting in the findings thus far produces the following:

$$T_{total} \le O(N\log_2(N)) + O(N\log_2(N)) + O(N\log_2(N)) + O(N\log_2(N)) + O(N\log_2(N)) = O(N\log_2(N))$$

Therefore, $T_{total} \in O(N\log_2(N))$.

The input parameter ranges and counts do not increase with the number of decision makers. This constraint means that the number of model runs, M, needed to solve a particular problem instance is, at worst, constant relative to N. Thus, if one run requires, at most, $O(N\lceil \log_2(N)\rceil)$ time, then a proper design of experiments for a study to answer a specific question requires $O(MN\lceil \log_2(N)\rceil) = O(N\lceil \log_2(N)\rceil)$ time. Thus:

(5) The maximum asymptotic time for studies done with asymptotically thorough and incompressible batch mode modular imperative agent-based models with large numbers of individual, boundedly rational, decision makers; input parameter ranges and counts that do not increase with the number of decision makers; asymptotically

linearithmic scheduling; and output aggregation that is linearithmic or better in time and space is Ω $(N\lceil \log_2(N)\rceil)$ for variable word size computers.

Space

Modular imperative agent-based models read all of their agent inputs, create the corresponding agents, and then store them. The combined requirements of asymptotic thoroughness and incompressibility prevent agent data from being recreated for each time step to save space. Thus, $O(N\lceil \log_2(N)\rceil)$ space is required.

Environments use $O(C_4\lceil \log_2(N)\rceil)$ space for constant C_4. There is one environment. Therefore, environments use, at most, $O(\lceil \log_2(N)\rceil)$ space.

Each scalar field requires $O(g^d\lceil \log_2(N)\rceil)$ space. d will be a constant relative to N, since the number of space dimensions normally does not depend on the number of agents. The growth rate for the maximum dimension range, g, is tied to the agent space density, y, given by $y = N/g^d$, as covered by these two cases:

1. g grows asymptotically faster than $N^{1/d}$. In this case, the agent density will drop and eventually either the agents will cluster or most agents will be isolated from one another. In either case, the model's use of the space becomes trivial. The space can then be replaced with either a set of smaller spaces when most of the agents cluster or with a network when most of the agents are isolated from one another. Thus, this case is not allowed.
2. g grows at the same rate or asymptotically slower than $N^{1/d}$. The agent density is constant or rising so the agent interactions can continue to be mediated via the field. The overall space usage for the grid is then $O(g^d\lceil \log_2(N)\rceil) \leq O(((N/y)^{1/d})^d \lceil \log_2(N)\rceil) = O(((N/y)\lceil \log_2(N)\rceil) \leq O(N\lceil \log_2(N)\rceil)$

The number of scalar fields must be bounded due to bounded rationality. Therefore, scalar fields use $O(C_5N\lceil \log_2(N)\rceil) = O(N\lceil \log_2(N)\rceil)$ or less space, for C_5 scalar fields.

Each grid requires $O(g^d\lceil \log_2(N)\rceil)$ space. d will be a constant relative to N. The doubly linked grid point membership lists can individually vary in size, but in total they take up $O(N\lceil \log_2(N)\rceil)$ space for N agents. The growth rate for g is tied to the agent space density, y, given by $y = N/g^d$, as covered by these two cases:

1. g grows asymptotically faster than $N^{1/d}$. In this case, the agent density will drop and eventually either the agents cluster or most agents will be isolated from one another. In either case, the model's use of the space becomes trivial. The space can then be replaced with either a set of smaller spaces when most of the agents cluster or with a network when most of the agents are isolated from one another. Thus, this case is not allowed.
2. g grows at the same rate or asymptotically slower than $N^{1/d}$. The agent density is constant or rising so the agent interactions can continue to be mediated via the field. The overall space usage for the grid is then $O(g^d\lceil \log_2(N)\rceil) \leq O(((N/y)^{1/d})^d \lceil \log_2(N)\rceil) = O(((N/y)\lceil \log_2(N)\rceil) \leq O(N\lceil \log_2(N)\rceil)$

The number of grids must be bounded due to bounded rationality. Therefore, grids use, at most, $O(C_6N\lceil \log_2(N)\rceil) = O(N\lceil \log_2(N)\rceil)$ or less space, for C_6 grids. Of course, agents must be selective in how they access the grid point membership lists to maintain bounded rationality.

Each network requires $O(bN\lceil \log_2(N)\rceil)$ space. b has been previously shown to be a constant relative to N. Thus, each network requires $O(N\lceil \log_2(N)\rceil)$ space. As with the other kinds of spaces, the number of networks must be bounded due to bounded rationality. Therefore, networks use, at most, $O(C_7N\lceil \log_2(N)\rceil) = O(N\lceil \log_2(N)\rceil)$ space, for C_7 networks.

Output aggregation uses linearithmic or less space. The outputs, therefore, require at most $O(N\lceil \log_2(N)\rceil)$ space in addition to that used by the rest of the model.

The input parameter ranges and counts do not increase with the number of decision makers. This means that the number of model runs, M, needed to solve a particular problem instance is at worst constant relative to N. Thus, if one run requires at most $O(N\lceil \log_2(N)\rceil)$ space, then a proper design of experiments for a study to answer a specific question requires $O(MN\lceil \log_2(N)\rceil) = O(N\lceil \log_2(N)\rceil)$ space. Thus:

(6) The maximum asymptotic space for studies done with asymptotically thorough and incompressible batch mode modular imperative agent-based models with large numbers of individual, boundedly rational decision makers; input parameter ranges and counts that do not increase with the number of decision makers; asymptotically linearithmic scheduling; and output aggregation that is linearithmic or faster in time and space is Ω $O(N\lceil \log_2(N)\rceil)$ for variable word size computers.

Computational optimality

Next, we consider the conditions under which modular imperative agent-based modeling studies may be considered to have optimal asymptotic time or space performance. As previously stated, (optimal) means that no other modeling technique can solve the same problem as an agent-based model can using fewer resources — in this case, time and space. The strategy is straightforward because both the lower bounds on all modeling techniques and the upper bounds on a common class of agent-based models have been proven. Using this background, we will now prove the following:

(7) Studies with modular imperative agent-based models are computationally optimal in asymptotic time and space performance if the models are asymptotically incompressible; asymptotically thorough; have large numbers of individual, boundedly rational, decision makers; have input parameter ranges and counts that do not increase with the number of decision makers; use batch mode operation; have asymptotically linearithmic scheduling; and have output aggregation that is linearithmic or faster in time and space for variable word size computers.

(3) and (4) show that the best case time and space bounds for any asymptotically incompressible and thorough model are both in Ω $(N\lceil \log_2(N)\rceil)$. We also showed that individual, boundedly rational decision makers in an asymptotically incompressible and thorough agent-based model with individual inputs requires $O(N\lceil \log_2(N)\rceil)$ time and space, plus the time and space needed for accessing the interaction spaces. We additionally showed that the agent-based modeling interaction spaces require, at most,

$O(N\lceil \log_2(N)\rceil)$ time and space for each space for N boundedly rational agents. This formulation yields $O(N\lceil \log_2(N)\rceil) + O(C_8N\lceil \log_2(N)\rceil) = O(N\lceil \log_2(N)\rceil)$ total asymptotic time and space performance for C_8 spaces. Finally, we showed that modeling studies and parallel runs do not change the asymptotic time and space complexity of agent-based models. Therefore, (7) is proven.

Given (2) we also have:

(8) Modular imperative agent-based models are computationally optimal in asymptotic time performance if $O(T_{total}) = O(S_{log})$.

Computational power of modular imperative agent-based models

In this section, we use the formalism to investigate other properties of agent-based models. Here we focus on characterizing the computational generality of modular imperative agent-based modeling.

Modular imperative agent-based models are computationally complete

Any RASP algorithm can be stored as the sole behavior for a single agent. Executing the behavior can be the sole scheduled event for the model. The inputs for the algorithm can be stored as the input data for the solitary agent in the model. All of these mappings take constant time and space relative to the problem size. Memory allocation time and space are at worst a constant multiple of register assignment costs. The algorithm can thus use a standard RASP approach and produce results and performance metrics asymptotically identical to a default RASP. RASPs have been proven to be a general model of computation, subject to the correctness of the Church-Turing thesis. The RASP formulation of agent-based models thus has the same computing power and performance as a general RASP. Therefore:

(9) Modular imperative agent-based models are computationally complete, subject to the correctness of the Church-Turing thesis.

Hartmanis (1970) and Luginbuhl (1970) discuss the computational complexity of RASPs in detail. Naturally, agent-based models have the same computational complexity as general RASPs.

Modular imperative agent-based models are universal models

Result (9) says in part that if the Church-Turing thesis is correct, then any computable model output can be computed with modular imperative agent-based modeling. Therefore:

(10) Any computable model can be expressed as a modular imperative agent-based model.

For example, Macal (2010) presents an algorithm to convert any system dynamics model to a modular imperative agent-based model. Of course, result (10) does not mean that modular imperative agent-based modeling should be automatically used to solve all modeling problems. The practical circumstances surrounding a given modeling problem need to be taken into account as well.

Modular imperative agent-based model complexity classes: ABM, ABMSPACE, and ABMTIME

Let ABMTIME be the set of all modular imperative agent-based models that have optimal time performance. Let ABMSPACE be the set of all modular imperative agent-based models that have optimal space usage. Then let ABM be the set of all modular imperative agent-based models that have optimal time and space performance. Naturally:

(11) ABM = ABMTIME ∩ ABMSPACE

ABMTIME

What can we say about ABMTIME? We can construct a set of models. Let g(N) be a valid complexity class that is linearithmic or larger. Then let $g'(N) = (N^{-1})g(N)$. We have $g'(N) \geq 1$ since g(N) is linearithmic or larger. Define a modular imperative agent-based model with N agents and no interaction spaces. Assign each agent a random number 'tag' so that the set of agents is asymptotically incompressible. Program each agent to have a constant time interaction with $\lceil g'(N) \rceil$ other randomly selected agents for each of a fixed number of time steps. The output is a log entry of each interaction so the model is asymptotically thorough. This model is optimal according to result (7). It has $O(g'(N)) = O(N(N^{-1})g(N)) = O(g(N))$ asymptotic time performance. Therefore:

(12) ABMTIME includes modular imperative agent-based models of all complexity classes linearithmic or larger.

ABMSPACE

Similarly, what can we say about ABMSPACE? We can again construct a set of models. Begin with the model used for the proof of (12). Change the agents so they store the tag for each interaction from the previous time. Also, have the agents compute a new tag as the sum of the tags from the last interaction. This is $O(g'(N)) = O(N(N^{-1})g(N)) = O(g(N))$ asymptotic incompressible space usage As before, this model is optimal according to result (7). Thus:

(13) ABMSPACE includes modular imperative agent-based models of all complexity classes linearithmic or larger.

Toward real computers

In the proceeding sections, we have assumed an abstract model of computation that scales with the size of N. In this section, we will briefly relax this assumption to better approximate real computers. Even in this section, the relaxation will only be partial, since we want the formalism to accommodate the full range of typical computers. The relaxation here will occur along two axes, namely word size and run length.

Fixed word size

Real computers generally have fixed word sizes. In this case, the $\lceil \log_2(N) \rceil$ term we previously used to account for increasing word size for increasing N becomes 1. The results in the proceeding sections can then be simplified by replacing the $\lceil \log_2(N) \rceil$

term with 1 and replacing the linearithmic requirements with linear requirements. In particular, result (7) becomes the following:

(14) Studies with modular imperative agent-based models are computationally optimal in asymptotic time and space performance if the models are asymptotically incompressible; asymptotically thorough; have large numbers of individual, boundedly rational, decision makers; have input parameter ranges and counts that do not increase with the number of decision makers; use batch mode operation; have asymptotically linear scheduling; and have output aggregation that is linear or faster in time and space on fixed word size computers.

Substantial run length

The performance bounds in the proceeding sections included the observation that the number of time steps or unique discrete event times is generally independent of N. Therefore, these factors do not affect asymptotic time or space performance. For real computers, a substantial constant may still affect performance. In this case, the time performance results should be multiplied by s, the number of time steps or slots. The space performance results remain the same for boundedly rational agents with linear output aggregation on fixed word size computers. Here bounded rationality limits the agent's ability to accumulate memory, and linear output aggregation storage constrains output aggregation space.

Examples

Returning to the cellular-level human tissue growth example discussed in the introduction, if N is the number of cells, each cell's behavior is boundedly rational, cell density is finite, and the cells exist in three dimensional grid then the model will be O(N) on fixed word size computers.

In the following sections we analyze the asymptotic time and space performance of three example modular imperative agent-based models. The examples were chosen from the set of published agent-based models, so as to cover a range of different designs and a spectrum of varying uses. These examples show how to apply both the formalism and the optimality conditions to several different kinds of real agent-based models. These examples also demonstrate how the formalism is useful for predicting the execution time and space requirements for representations of common CASs.

A simple synthetic model

The Lotka-Volterra (Lotka 1925; Volterra 1926) predator–prey model is a classic differential equations model that is commonly implemented using system dynamics. Here we will consider a three species 'predator–prey-plant' variant. In this model, there are predators that eat prey and prey that eat plants. The animals (i.e., the predators and prey) have an energy level and die if their energy goes to zero. Plants regrow at a specified rate. N is the sum of the three species' populations. By convention, the output is a population count time series for each tracked species or compartment. The input parameter ranges and counts do not increase with N.

Theoretical analysis

The modular imperative agent-based predator–prey-plant implementation has one agent type for each species and is time-stepped. Since it is time-stepped, it has asymptotically linear scheduling. A selected Repast Simphony implementation is used for reference.

For the Repast Simphony implementation, animals reproduce probabilistically using random draws checked against reproduction rate thresholds. The animal agents move on a two dimensional toroidal grid, with plant agents at each point. During each time step, the animals move to a random point near their current location and then eat if the appropriate food is available. This is a simple boundedly rational $O(a\lceil \log_2(a)\rceil)$ time and space behavior for a animals on variable word size computers.

Plants can be living or dead. Live plants grow at a constant rate. Areas with dead plants lie fallow for a period of time and then life returns. Plants die due to excessive consumption by prey. This is a simple boundedly rational $O(b\lceil \log_2(b)\rceil)$ time and space behavior for b plants.

a and b can both be large. With $N = a + b$, we have N large, as well.

The three species each maintain individual energy levels that can take on any double precision value. As such, they are asymptotically incompressible.

Combining results (5) and (6) with the minimum requirements that were just discussed implies that this model is $O(N\lceil \log_2(N)\rceil)$ in time and space. This version of the agent-based model does not qualify for time or space optimality under result (14) because the output requires only compressible information from each agent (i.e., the population count) rather than specific attributes. In other words, the model's output aggregation is not asymptotically thorough. Is there a more efficient implementation?

The system dynamics predator–prey-plant implementation has one stock for each species and a set of flows between stocks. On variable word size computers, this model has $O(C_9\lceil \log_2(N)\rceil) = O(\lceil \log_2(N)\rceil)$ asymptotic time and space performance for constant C_9 measuring the time and space usage for the fixed stocks and flows. Obviously, this model has better time and space performance than the previous agent-based version.

Consider a variation of the model used to study predator and prey migration patterns. Now, the model must output the grid locations of each animal for every time step. This agent-based model is asymptotically thorough. It remains $O(N\lceil \log_2(N)\rceil)$ in time and space but it is now optimal under result (14).

What about the $O(\lceil \log_2(N)\rceil)$ system dynamics implementation? It cannot generate the needed outputs. A scalar field approach is also insufficient, since this only shows net flows, not long distance migrations. Modifications that allow the system dynamics model to track individual animal coordinates and energy levels results in a model with $O(N^2\lceil \log_2(N)\rceil)$ terms in each of the animal coordinate and energy level equations. The result is a model that requires $O(N^2\lceil \log_2(N)\rceil)$ space and $O(N^2\lceil \log_2(N)\rceil)$ evaluation time on variable word size computers. Furthermore, the model has awkward constraints on the range of model execution, since the formulation limits the maximum number of animals allowed in a given run. Wilson (1998) further explores the relationship between agent-based modeling and system dynamics for the two species variant of this model.

Experimental results

Results for 100 runs of the Repast Simphony predator–prey-plant model for varying grid sizes and 500 time steps are shown in Figures 2 and 3. The number of time steps is held constant with respect to N. The runs in Figures 2 and 3 were completed on a 1.8 GHz Intel Core i7 MacBook Air with 4 GB of memory.

For fixed word-size computers, we have $O(N\lceil \log_2(N)\rceil) = O(N)$, which is a linear relationship. As predicted, the test measurements of the predator–prey-plant model execution time are collectively linear versus N with an adjusted R^2 of 0.7794 and a p-value less than 2.200×10^{-16}. The space measurements are also collectively linear versus N with an adjusted R^2 of 0.5687 and a p-value less than 2.200×10^{-16}.

A research model

Strains of the bacterium community-associated methicillin-resistant *Staphylococcus aureus* (CA-MRSA), are responsible for potentially life-threatening infections of the skin, soft tissue, blood, bone, and other human tissues. CA-MRSA strains are resistant to standard antibiotics related to penicillins and have a high prevalence in the general community, as well as in healthcare facilities. Macal et al. (2012) have developed two fine-grained, modular imperative agent-based models of CA-MRSA for the Chicago metropolitan area. These hourly time-stepped models feature people as agents. Both models share a common design. We will consider the large-scale Repast HPC version here due to its ability to scale across a wide range of agent population sizes. We will use this ability to compare the theoretical predictions of our time and space performance model to measured results over several orders of magnitude for N. We will also compare this simulation to system dynamics epidemic simulations.

Theoretical analysis

The Repast HPC CA-MRSA modular imperative model has N person agents. The number of agents is constant throughout each simulation run. The number of agents is generally large because the model is used to study major parts or all of the Chicago area. The agents have activity schedules and a list of locations at which to act. The activities are coded by hour, location, and activity type. The agent activities can be interrupted when problems such as hospitalization or incarceration occur.

The model uses time step scheduling with a step size of one hour. Since it is time stepped, it has asymptotically linearithmic scheduling.

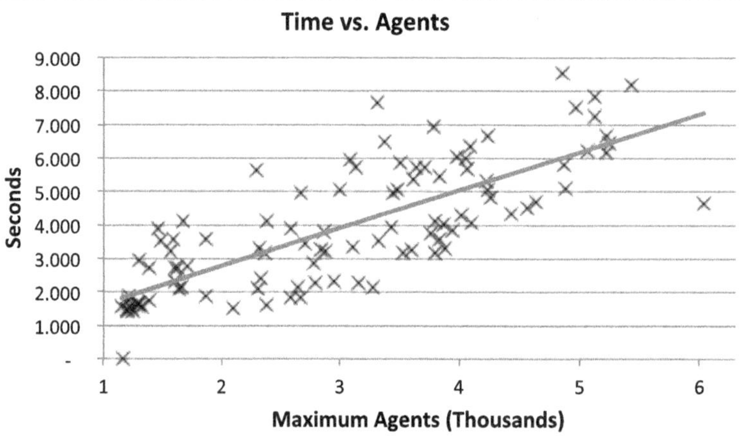

Figure 2 Predator–prey-plant model time versus N.

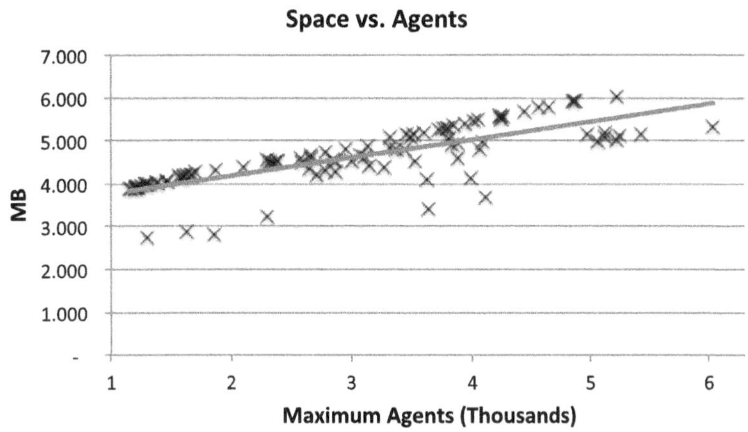

Figure 3 Predator–prey-plant model space versus N.

The location and activity type for each agent for each simulated hour contribute to the determination of the risk of disease transmission and the likelihood of disease progression. The disease states of the people in a location also contribute to the risks and likelihoods. The locations include residences such as homes or group quarters, workplaces, schools, gyms, hospitals, and jails.

The model tracks three disease states for each agent, namely uncolonized, colonized, and infected. Uncolonized people do not have the CA-MRSA. Colonized people have the disease but do not show symptoms. Infected people have symptoms and are believed to be more likely to spread the disease than are colonized people. Infected people may either treat themselves or seek medical care.

Individual behaviors involve selecting new activities from a personal list that is short and constant compared to N, moving to the selected activity location, and then interacting with the other agents at that place. The number of agents at each place is always small with respect to N. The individual, boundedly rational agent behaviors are thus $O(\lceil \log_2(N) \rceil)$ for each agent.

The people in the model have a disease state that can take on any one of three values for each individual. As such, they are asymptotically incompressible.

The model produces two kinds of outputs. The first kind includes an activity entry for each agent for each time step. This is asymptotically thorough. The second kind consists of a summary line for each time step. There is a fixed number of each kind of output. Both types of outputs use straightforward data collection that is linearithmic or less time and space.

The input parameter ranges and counts do not increase with the number of people in the simulation, since the parameters provide constants for the previously described bounded rational behavior. Therefore, modeling studies do not increase the asymptotic time or space complexity. This model implementation is thus $O(N\lceil \log_2(N) \rceil)$ in time and space. Its optimality depends on which outputs are needed to answer the modeling question at issue. What about alternative techniques?

Epidemics have long been studied using system dynamics modeling (Kermack and McKendrick 1927). Compartments are generally used to track the populations of interest. System dynamics epidemic models usually have a fixed number of stocks, often one for each compartment, and a fixed set of flows between stocks. These models have

$O(C_{10}\lceil \log_2(N)\rceil) = O(\lceil \log_2(N)\rceil)$ asymptotic time and space performance for constant C_{10} measuring the time and space usage for the fixed stocks and flows. Obviously, these models have better time and space performance than the previously discussed agent-based version as long as the number of compartments is constant relative to N and limited compartments are sufficient to answer the question at hand.

Consider the use of the model to study disease contagion networks. The needed model output will be the data needed to create a disease spread dendrogram showing who contracted the tracked disease from whom. Now, the model must output the identities of the interacting parties (e.g., agents) every time the disease spreads from person to person. The first model output previously mentioned, namely hourly activity information, meets this requirement. This agent-based model is asymptotically thorough. It remains $O(N\lceil \log_2(N)\rceil)$ in time and space, but it is now optimal under result (14).

What about the $O(\lceil \log_2(N)\rceil)$ system dynamics implementation? As with the previous predator–prey-plant model, it cannot generate the needed outputs. Modifications that allow the system dynamics model to track individual disease states and contacts result in a model with $O(N^2\lceil \log_2(N)\rceil)$ terms in each person's state equation. The result is a model that requires $O(N^2\lceil \log_2(N)\rceil)$ space and $O(N^2\lceil \log_2(N)\rceil)$ evaluation time. Furthermore, the formulation fixes the maximum number of people allowed in a given run. Rahmandad and Sterman (2008) further explore the relationship between agent-based and system dynamics disease dispersion models.

Experimental results

Results for runs of the CA-MRSA model for 10 years and five geographic regions are shown in Figures 4 and 5. The regions in order of increasing population are a single ZIP code, three ZIP codes, the South Side of Chicago, and Chicago itself. Figures 4 and 5 show 32 runs for each region. The number of time steps is held constant with respect to N.

The runs in Figures 4 and 5 were completed on the 320-node Argonne National Laboratory Fusion Linux cluster. Each Fusion computing node has two Nehalem 2.6 GHz Pentium Xeon processors with 36 GB of RAM.

For fixed word-size computers, we have $O(N\lceil \log_2(N)\rceil) = O(N)$, which is a linear relationship. As predicted, the test measurements of the CA-MRSA model execution

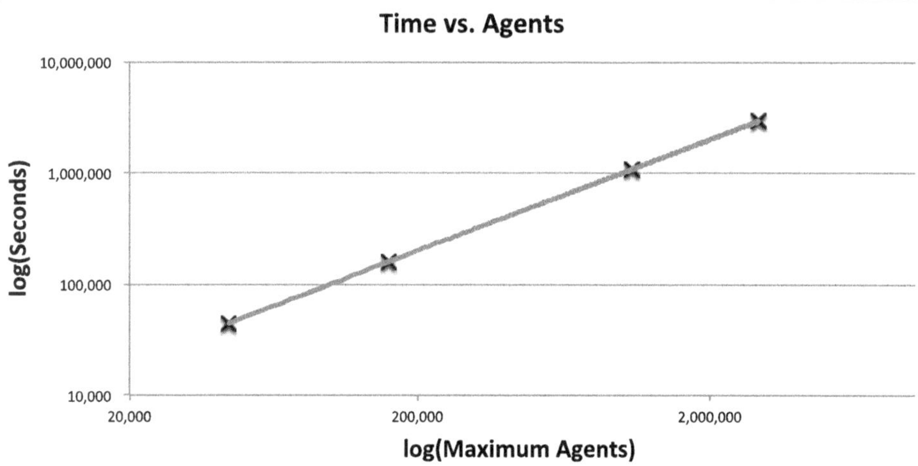

Figure 4 Measured MRSA model time versus N.

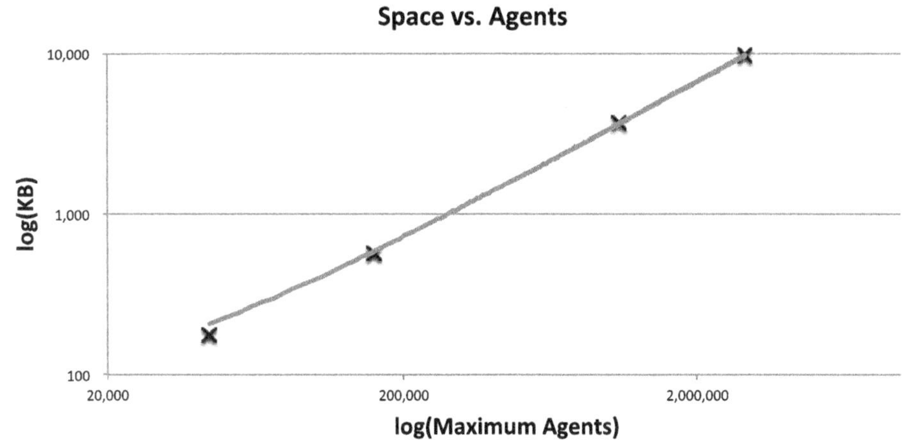

Figure 5 Measured MRSA model space versus N.

time are collectively linear versus N with an adjusted R^2 of 1.000 and a p-value less than 2.200×10^{-16}. The space measurements are also collectively linear versus N with an adjusted R^2 of 0.9999 and a p-value less than 2.200×10^{-16}.

An industrial model

The Virtual Market Learning Lab (North et al. 2010) is a large-scale modular imperative agent-based model of consumer markets co-developed by Argonne National Laboratory and Procter & Gamble (P&G). It represents the shopping behavior of consumer households and the business behavior of retailers and manufacturers in a simulated national consumer market. Argonne and P&G successfully calibrated, verified, and validated the resulting agent-based model using several independent, real-world data sets for multiple consumer product categories with more than 60 comparison tests per data set. First, Repast and then later Repast Simphony were used to implement the model. P&G has successfully applied the model to several challenging business problems, where it has directly influenced managerial decision-making and has produced substantial cost savings. The version of the model analyzed in this paper does not use social networking.

Theoretical analysis

The model uses time step scheduling with a step size of one day. Since it is time stepped, it has asymptotically linearithmic scheduling.

The agents in this model are consumer households (n_0), retail stores (n_1), retail regions (n_2), retailers (n_3), and manufacturers (n_4) (North et al. 2010). Here, N is the sum of n_i for i = 1 to 5. In principle, N scales with the number of any of the agent types. In practice, the number of consumers, n_0, is always orders of magnitude greater than the counts of the other agents. Thus, we have $N = n_0$. n_0 is generally large, so N is as well.

Each consumer household exists in a local neighborhood, buys products in local retail stores, and then uses those products. Buying products means selecting a store and then selecting a product from that store. Using products means selecting an item from inventory, reducing the stored amount of that item, and then updating the agent's memory about the product usage experience. These processes are $O((\lceil \log_2(N) \rceil)$ in time and space for each agent.

Each consumer household remembers varying attributes of both stores and products. The Miller (1956) range of seven plus or minus two bounds the sizes of the consumer attribute lists. The store and product selection processes depend on these attribute lists. These boundedly rational processes are $O(\lceil \log_2(N) \rceil)$ in time and space for each agent.

Households are randomly created based on draws from an input set of demographically indexed distributions. This process is $O(\lceil \log_2(N) \rceil)$ in time and space for each agent.

Retail stores stock their shelves; adjust prices; compare their prices to other stores in their neighborhood; and offer promotions including flyers, sales, volume discounts, and in-store displays. Retail stores are initialized using an input list of properties. All of these boundedly rational processes are $O(\lceil \log_2(N) \rceil)$ in time and space for each agent.

Retail regions track sales volumes and profits at their stores and provide feedback to their stores on sales goals. Retail regions are initialized using an input list of properties. All of these boundedly rational processes are $O(\lceil \log_2(N) \rceil)$ in time and space for each agent.

Retailers track sales volumes and profits in their regions; provide feedback to their regions on sales goals; advertise; determine stocking options for regions and stores; and receive, use, and distribute trade support from manufacturers. Retailers are initialized using an input list of properties. All of these boundedly rational processes are $O(\lceil \log_2(N) \rceil)$ in time and space for each agent.

Manufacturers maintain and update brand and product lists, advertise, run promotions, adjust suggested retail prices, and offer trade support incentives to retailers. Manufacturers are initialized using an input list of properties. All of these boundedly rational processes are $O(\lceil \log_2(N) \rceil)$ in time and space for each agent.

The agents in the model have a variety of attributes that can take on any double precision value as well as individualized lists, Boolean values, etc. As such, they are asymptotically incompressible.

The model produces two kinds of outputs. The first kind includes a result for each agent for each time step. The second kind consists of a summary line for each time step. There is a fixed number of each kind of output. Both types of outputs use straightforward data collection that is linearithmic or faster in time and space.

The input parameter ranges and counts do not increase with the number of consumers in the simulation, since the parameters provide constants for the previously described boundedly rational behavior. Therefore, modeling studies do not increase the asymptotic time or space complexity. This model implementation is thus $O(N\lceil \log_2(N) \rceil)$ in time and space.

This model fits into the requirements for result (14). Therefore, the Virtual Market Learning Lab is computationally optimal in asymptotic time and space performance.

Experimental results

For fixed word-size computers, we have $O(N\lceil \log_2(N) \rceil) = O(N)$, which is a linear relationship. As predicted, seven test measurements of the model execution time for a fixed number of time steps relative to N and a range of consumer population sizes were found to be collectively linear with an adjusted R^2 of 0.9963 and a p-value of 3.221×10^{-6}.

Conclusions and future work

Following Holland (1992, 1999, 2006), complex adaptive systems (CASs) are collections of interacting, autonomous, learning decision makers embedded in an interactive environment. CASs are common in both nature and society. Modeling CASs is challenging for a variety of reasons. The challenges of modeling CASs can largely be overcome by using the individual-level focus of agent-based modeling. This paper's contribution is to introduce, analyze, and apply a theoretical formalism for proving findings about modular imperative agent-based models. The use of the formalism is demonstrated with three example models. These examples also show how the formalism is useful for predicting the execution time and space requirements for representations of common CASs. We have proven results (1) to (14), including that modular imperative agent-based modeling studies are asymptotically optimal in computational time and space for a common class of modeling problems. Given that modular imperative agent-based modeling is both computationally optimal and a natural structural match for many modeling problems, it follows that modular imperative agent-based modeling is the best modeling method for these situations.

There are many next steps for future work. First, the theoretical formalism can be used to analyze additional applied models for computational complexity and optimality. Second, the range of modeling problems covered by the computational optimality results in this paper might be expanded. Third, the performance results for modular imperative agent-based models violating the computational optimality criteria might be compared to that of other modeling techniques by directly using the formalism. Fourth, the formalism may be used to prove other properties of modular imperative agent-based models. Fifth, the formalism might be extended as modular imperative agent-based modeling practice evolves.

Endnote

[a]The term "linearithmic" refers to relationships of the form ($x \log_2 x$).

Abbreviations

ABM: Agent-based model; AC: Accumulator; CA-MRSA: Community-associated methicillin-resistant *Staphylococcus aureus*; CAS: Complex adaptive system; DEVS: Discrete Event System Specification; GB: Gigabyte; GHz: Gigahertz; GUI: Graphical user interface; HPC: High-performance computing; IC: Instruction counter; NP: Nondeterministic polynomial time; ODD: Overview, Design Concepts, and Details; P&G: Procter & Gamble; RAM: Random access machine; RASP: Random access stored-program; XML: Extensible markup language.

Competing interests

The author declares that he has no competing interests.

Acknowledgments

The submitted manuscript has been created by UChicago Argonne, LLC, Operator of Argonne National Laboratory ("Argonne"). Argonne, a U.S. Department of Energy Office of Science Laboratory, is operated under contract number DE-AC02-06CH11357. Part of this work was supported by National Science Foundation Division of Information and Intelligent Systems (IIS) award number 1125412. Part of this work was supported by National Institutes of Health National Institute of General Medical Sciences Models of Infectious Disease Agent Study (MIDAS) grant number U01GM087729. The U.S. Government retains for itself, and others acting on its behalf, a paid-up nonexclusive, irrevocable worldwide license in said article to reproduce, prepare derivative works, distribute copies to the public, and perform publicly and display publicly, by or on behalf of the government.

The author wishes to thank Charles Macal, David Sallach, and William Rand for providing insightful reviews of this paper and perceptive suggestions for improvement. The author also wishes to thank the Repast team, the CA-MRSA modeling team, and the Virtual Market Learning Lab team for their contributions.

Any opinions, findings, conclusions, or recommendations are solely those of the author.

References

Bolduc JS, Vangheluwe H: *A Modeling and Simulation Package for Classic Hierarchical DEVS*. Montreal, Quebec, Canada: Modelling, Simulation and Design Lab, School of Computer Science, McGill University, Technical Report; 2002.

Bonabeau E: **Agent-based modeling: Methods and techniques for simulating human systems.** In *Proceedings of the National Academy of Sciences*, 3, Volume 99. Washington, D.C. USA: National Academy of Sciences Press; 2002:7280–7287.

Brown DG, Riolo R, Robinson DT, North MJ, Rand W: **Spatial process and data models: toward integration of agent-based models and GIS.** *J Geogr Syst* 2005, **7**(1):25–47.

Christos CG, Lafortune S: *Introduction to Discrete Event Systems.* 2nd edition. New York, NY USA: Springer; 2008.

Collier NT, North MJ: **Parallel agent-based programming with Repast for High Performance Computing.** *Simulation* 2012:1–21.

Cook SA, Reckhow RA: **Time-bounded random access machines.** *J Comput Syst Sci* 1973, **7**(4):354–375.

Dorofeenko V, Shorish J: *Dynamical Modeling of the Demographic Prisoner's Dilemma.* Vienna, Austria: Institute for Advanced Studies Economics Series; 2002.

Elgot CC, Robinson A: **Random-access stored-program machines, an approach to programming languages.** *J ACM* 1964, **11**(4):365–399.

Epstein JM: **Zones of cooperation in demographic prisoner's dilemma.** *Complexity* 1998, **4**(2):36–48.

Epstein J: *Generative Social Science: Studies in Agent-Based Computational Modeling.* Princeton, NJ USA: Princeton University Press; 2007.

Foster I: *Designing and Building Parallel Programs.* Reading, MA USA: Addison-Wesley; 1995.

Grimm V, Berger U, Bastiansen F, Eliassen S, Ginot V, Giske J, Goss-Custard J, Grand T, Heinz SK, Huse G, Huth A, Jepsen JU, Jørgensen C, Mooij WM, Müller B, Pe'er G, Piou C, Railsback SF, Robbins AM, Robbins MM, Rossmanith E, Rüger N, Strand E, Souissi S, Stillman RA, Vabø R, Visser U, DeAngelis DL: **A standard protocol for describing individual-based and agent-based models.** *Ecol Model* 2006, **198**(1–2):115–126.

Hartmanis J: **Computational complexity of random access stored program machines.** *Math Syst Theory* 1970, **5**(3):232–245.

Harvey B: *Computer Science Logo Style.* Boston, MA USA: MIT Press; 1997.

Hinkelmann F, Murrugarra D, Jarrah AS, Laubenbacher RC: **A mathematical framework for agent based models of complex biological networks.** *Bull Math Biol* 2011, **73**(7):1583–1602.

Holland JH: *Adaptation in Natural and Artificial Systems: An Introductory Analysis with Applications to Biology, Control, and Artificial Intelligence.* Cambridge, Mass: MIT Press; 1992.

Holland JH: *Emergence: From Chaos to Order.* Reading, Mass: Perseus Books; 1999.

Holland JH: **Studying complex adaptive systems.** *J Syst Sci Complex* 2006, **19:**1–8.

Kermack WO, McKendrick AG: **A contribution to the mathematical theory of epidemics.** *Proc Royal Soc A: Math, Phys Eng Sci* 1927, **15**(772):700.

Kleene SC: **Recursive predicates and quantifiers.** *Trans Am Math Soc* 1943, **53**(1):41–73.

Leombruni R, Richiardi M: **Why are economists skeptical about agent-based simulations?** *Phys A: Stat Mech Appl* 2005, **355:**103–109.

Lotka AJ: *Elements of Physical Biology.* Baltimore, Maryland, USA: Williams and Wilkins; 1925.

Luginbuhl DR: *Computational complexity of Random-Access Models. PhD thesis.* University of Illinois at Urbana-Champaign, Computer Science Department; 1970.

Luke S, Cioffi-Revilla C, Panait L, Sullivan K, Balan G: **MASON: A multiagent simulation environment.** *SIMULATION* 2005, **81:**517–527.

Macal CM: **To agent-based simulation from system dynamics.** In *Proceedings of the 2010 Winter Simulation Conference.* Edited by Johansson B, Jain S, Montoya-Torres J, Hugan J, Yücesan E. Baltimore, MD: IEEE/ACM; 2010.

Macal CM, North MJ: **Tutorial on agent-based modeling and simulation.** *J Simul* 2010, **4:**151–162.

Macal CM, North MJ, Collier N, Lauderdale DS, David MZ, Shumm P, Daum RS, Evans JA, Wilder JR, Dukic VM, Wegner DT: **Modeling the spread of community-associated MRSA.** In *Proceedings of the 2012 Winter Simulation Conference.* Edited by Laroque C, Himmelspach J, Pasupathy R, Rose O, Uhrmacher AM. Berlin: IEEE/ACM; 2012.

MASON: Accessed 2014 [http://cs.gmu.edu/~eclab/projects/mason]

Matsumoto M, Nishimura T: **Mersenne twister: A 623-dimensionally equidistributed uniform pseudo-random number generator.** In *ACM Transactions on Modeling and Computer Simulation*, Volume 8. 1st edition. New York: ACM; 1998:3–30.

Miller GA: **The magic number seven plus or minus two.** *Psychol Rev* 1956, **63:**81–97.

Minar N, Burkhart R, Langton C, Askenazi M: *The Swarm Simulation System: A Toolkit for Building Multi-agent Simulations. Working Paper 96-06-042.* Santa Fe: Santa Fe Institute; 1996.

Murphy JT: *Computational social science and high performance computing: A case study of a simple model at large scales.* Santa Fe, NM USA: Proceedings of the 2011 Computational Social Science Society of the Americas Conference; 2011.

Myerson RB: *Game Theory: Analysis of Conflict.* Cambridge, MA USA: Harvard University Press; 1991.

NetLogo: Wilensky 2014 [http://ccl.northwestern.edu/netlogo/]

Niazi M, Hussain A: **A novel agent-based simulation framework for sensing in complex adaptive environments.** *IEEE Sensors J* 2010, **5**(3):1–9.

Nisan N, Roughgarden T, Tardos E, Vazirani W: *Algorithmic Game Theory.* Cambridge, UK: Cambridge University Press; 2007.

North MJ: **A theoretical foundation for computational agent-based modeling.** *2012 World Congress on Social Simulation.* Taipei, Taiwan: National Chengchi University; 2012.

North MJ, Macal CM: *Managing Business Complexity: Discovering Strategic Solutions With Agent-Based Modeling and Simulation.* Oxford, U.K.: Oxford University Press; 2007.

North MJ, Macal CM: **Foundations of and recent advances in artificial life modeling with Repast 3 and Repast Simphony.** In *Artificial Life Models in Software.* 2nd edition. Edited by Adamatzky A, Komosinski M. Heidelberg: Springer; 2009:37–60.

North MJ, Collier NT, Vos RJ: **Experiences creating three implementations of the Repast agent modeling toolkit.** In *ACM Transactions on Modeling and Computer Simulation*, Volume 16. 1st edition. New York: ACM; 2006:1–25.

North MJ, Macal CM, St Aubin J, Thimmapuram P, Bragen M, Hahn J, Karr J, Brigham N, Lacy ME, Hampton D: **Multi-scale agent-based consumer market modeling.** *Complexity* 2010, **15**(5):37–47.

North MJ, Collier NT, Ozik J, Tatara E, Altaweel M, Macal CM, Bragen M, Sydelko P: **Complex adaptive systems modeling with Repast Simphony**. In *Complex Adaptive Systems Modeling*. Heidelberg: Springer; 2013.

Parker J, Epstein JM: **A distributed platform for global-scale agent-based models of disease transmission**. In *ACM Transactions on Modeling and Computer Simulation*, Volume 22. 1st edition. New York: ACM; 2011:1–25.

Parunak HVD, Savit R, Riolo RL: **Agent-based modeling vs. equation-based modeling: a case study and users' guide**. In *Proceedings of Multi-Agent Systems and Agent-Based Simulation*. New York, New York, USA: Springer; 1998:10–25.

Perumulla KS: **Parallel and distributed simulation: traditional techniques and recent advances**. In *Proceedings of the 2006 Winter Simulation Conference*. Edited by Perrone LF, Lawson B, Liu J, Wieland FP. Monterey. California: USA IEEE Press; 2006:84–95.

Rahmandad H, Sterman JD: **Heterogeneity and network structure in the dynamics of diffusion: comparing agent-based and differential equation models.** *Manag Sci* 2008, **54**(5)**:**998–1014.

Repast 3: Accessed 2014 [http://repast.sourceforge.net/repast_3]

Repast for High Performance Computing: Accessed 2014 [http://repast.sourceforge.net/repast_hpc.html]

Repast Simphony: Accessed 2014 [http://repast.sourceforge.net/repast_simphony.html]

Resnick M: *Turtles, Termites, and Traffic Jams: Explorations in Massively Parallel Microworlds*. Cambridge, MA, USA: MIT Press; 1994.

Shang H, Wainer G: **A simulation algorithm for dynamic structure DEVS modeling**. In *Proceedings of the 2006 Winter Simulation Conference*. Edited by Perrone LF, Wieland FP, Liu L, Lawson BG, Nicol DM, Fujimoto RM. Monterey, CA, USA: IEEE Press; 2006:815–822.

Smale S: **Differentiable dynamical systems.** *Bull Am Math Soc* 1967, **73**(6)**:**747–817.

Spivey JM: *The Z Notation: A Reference Manual*. 2nd edition. Oxford, UK: Programming Research Group, University of Oxford; 1992.

StarLogo: Accessed 2014 [http://education.mit.edu/starlogo/]

Swarm: Accessed 2014 [http://www.swarm.org/]

Volterra V: **Variazioni e fluttuazioni del numero d'individui in specie animali conviventi.** *Mem. Acad. Lincei Roma* 1926, **2:**31–113.

Wilensky U: *NetLogo*. Evanston, IL: Center for Connected Learning and Computer-Based Modeling, Northwestern University; 1999.

Wilson WG: **Resolving discrepancies between deterministic population models and individual-based simulations.** *Am Nat* 1998, **151**(2)**:**116–134.

Zeigler BP, Praehofer H, Kim TG: *Theory of Modeling and Simulation: Integrating Discrete Event and Continuous Complex Dynamic Systems*. 2nd edition. San Diego, CA USA: Academic Press; 2000.

6

A moderate role for cognitive models in agent-based modeling of cultural change

Marshall Abrams

Correspondence:
mabrams@uab.edu
Department of Philosophy,
University of Alabama at
Birmingham 900 13th Street South,
HB 414A Birmingham, AL
35294-1260, USA

Abstract

Purpose: Agent-based models are typically "simple-agent" models, in which agents behave according to simple rules, or "complex-agent" models which incorporate complex models of cognitive processes. I argue that there is also an important role for agent-based computer models in which agents incorporate cognitive models of moderate complexity. In particular, I argue that such models have the potential to bring insights from the humanistic study of culture into population-level modeling of cultural change.

Methods: I motivate my proposal in part by describing an agent-based modeling framework, POPCO, in which agents' communication of their simulated beliefs depends on a model of analogy processing implemented by artificial neural networks within each agent. I use POPCO to model a hypothesis about causal relations between cultural patterns proposed by Peggy Sanday.

Results: In model 1, empirical patterns like those reported by Sanday emerge from the influence of analogies on agents' communication with each other. Model 2 extends model 1 by allowing the components of a new analogy to diffuse through the population for reasons unrelated to later effects of the analogy. This illustrates a process by which novel cultural features might arise.

Conclusions: The inclusion of relatively simple cognitive models in agents allows modeling population-level effects of inferential and cultural coherence relations, including symbolic cultural relationships. I argue that such models of moderate complexity can illuminate various causal relationships involving cultural patterns and cognitive processes.

Keywords: Simulation; Culture; Cognition; Analogy; Metaphor; Hermeneutics

Background

Introduction

This essay argues that an underused style of computer modeling—what I call "moderate-complexity" agent-based modeling—has the potential to serve as a source of new insights about processes of cultural change. I'll argue that this style of modeling can be sufficiently fruitful, and sufficiently tractable, to make it worth pursuing. I illustrate this strategy with concrete examples, simulations of processes of cultural change described by anthropologist Peggy Sanday (Sanday PR 1981).

Modeling of cultural change and other social processes has focused primarily on three classes of mathematical and simulation models:

Aggregate models: These are mathematical or computer models which use quantities describing groups of people, institutions, or other broad social factors without representing individual persons by distinct variables (Boyd R and Richerson PJ 1985; Cavalli-Sforza LL and Feldman MW 1981).
Simple-agent models: These models include representations of individual people, but represent individuals' states by a few simple variables. When models of social phenomena are described as complex adaptive systems models, they are typically of this kind.
Complex-agent models: These models represent people by intelligent agents that incorporate complex "heavyweight" (Alam SJ et al. 2010) internal cognitive architectures. This kind of model is common in artificial intelligence research and cognitive science (e.g. (Sun R 2006)).

All three strategies produce useful results. By design, aggregate and simple-agent models ignore most aspects of human thought and interaction, allowing focus on broad patterns within populations. These strategies are not designed to model the effects of more complex, subtle processes within individuals; such processes are the focus of complex-agent models. However, complex-agent models tend to become intractable when more than a handful of individuals are modeled. A fourth approach, though not new, is not common (e.g. (Alam SJ et al. 2010; Bleda M and Shackley S 2012; Reynolds RG 1994; Reynolds RG and Ali M 2008; Thagard P 2000, ch. 7):

Moderate-agent models: In "moderate-complexity-agent", or "moderate-agent" models, agents incorporate abstractions of cognitive processes at a level of complexity intermediate between that of simple-agent and complex-agent models.

The distinction between between moderate-agent models as opposed to simple-agent or complex-agent models is intentionally vague, but I give an illustration below. The difference is partly one of design goals. Like simple-agent models, moderate-agent models abstract from the complexities of human behavior in order to gain insight about patterns in populations. Unlike complex-agent models, moderate-agent models are not intended to capture behavior in a way that might one day approach the sophistication of real humans. Instead, a moderate-agent model trades some of the elegance and tractability that comes from use of simple agents, for the sake of the ability to model *some* effects of the complexities of real human behavior within populations. I argue that moderate-agent models have benefits that justify their exploration, despite challenges they face.

More specifically, I argue that ideas about culture from the humanities and humanistic research traditions within the social sciences—ideas traditionally thought to be resistant to or antithetical to modeling—can be incorporated into agent-based modeling through the use of moderate-agent models. This strategy opens up the possibility of increasing the scope of agent-based modeling to investigating effects of subtleties of human interactions and cultural processes which have rarely, if ever, been addressed by modeling of any kind. If successful, the strategy would give agent-based models broader scope, allowing models of cultural processes with greater realism. This could support more robust interdisciplinary discussion and collaboration with humanistic researchers in areas such as as cultural anthropology, cultural sociology, history, and literature.

In the next section I motivate both the general strategy of moderate-agent modeling of cultural change and my particular modeling framework, POPCO. In the "Methods" section, "POPCO framework" describes this framework. POPCO involves agent-based models at two levels, since it represents cognitive processes of interacting persons (agents) in a population in terms of networks of simple nodes (also agents). "Models of Sanday's hypotheses" illustrates the use of POPCO with models based on Peggy Sanday's (Sanday PR 1981) cross-cultural hypothesis about the origin of relationships between creation stories, sex roles in childrearing, and means of subsistence. The "Results and discussion" section discusses implications of the models I present, and discusses costs and benefits of POPCO and similar modeling strategies. "Conclusions" presents overall conclusions. The appendix provides additional details of the models described in "Models of Sanday's hypothesis".

Motivation

Work by anthropologists and other social scientists illustrates the idea that some elements of a society's culture may "fit together" in a sort of "harmony", or may "cohere" in various respects. For example, Lansing (Lansing JS 2006, 2007) describes a complex water distribution system in one region of Bali. Farmers benefit most, on average, if they coordinate planting and irrigation according to specific schedules. These schedules are coordinated by "water temples", which represent gods and spiritual congregations corresponding to physical components of the water distribution system. Lansing argues that the way in which elements of the religious system symbolize corresponding elements of the physical water system plays a crucial role in guiding individuals' participation in the water distribution system. Effective water management practices thus depend on and reinforce a complex set of religious beliefs, which themselves exhibit some internal harmony. Here we see certain sorts of "fit" not just between physical components of the system of water distribution and practices involving its use, but also between these elements and religious values, symbols, and practices. Lansing and colleagues have simulated relationships between water flow, pest prevalence, and rice production, and modeled game-theoretic interactions between managers of components of the system. However, relationships between these elements and the religious aspect of the cultural system have not been modeled, as far as I know. Capturing change involving similarly complex interrelationships between cultural elements in a population of agents would be difficult with existing modeling strategies. Current models of culture typically represent cultural variation with a few simple variables, in aggregate or simple-agent models (e.g. (Acerbi A et al. 2012; Afshar M and Asadpour M 2010; Alexander JM 2007; Boyd R and Richerson PJ 1985, 2005; Cavalli-Sforza LL and Feldman MW 1981; Enquist M et al. 2010; Grim P et al. 2004; Mueller ST et al. 2010; Nakahashi W 2010)), or represent interactions between cultural components without reference to the role of interacting individuals in a society (Dehghani M et al. 2008; Klüver J and Klüver C 2010; Thagard P 2012).

Internal relations of harmony between elements of culture play a large role in humanistic, and particularly so-called "hermeneutic" studies of culture in anthropology and other disciplines (e.g. (Clark S 1999; Geertz C 1973; González RJ 2001; Lienhardt G 1961; Tilley C 2000)). These approaches often treat culture as composed of a complex fabric of symbolic relationships involving language, arts, behavior, and various relations to the physical world, and exhibiting relations of harmony and tension. Authors in such traditions often argue that culture cannot be studied scientifically. Such authors sometimes argue that explanations involving culture depend on interpretive relationships between elements of language or other cultural elements, rather than causal relationships. There are also authors who argue that though some aspects of culture can be studied scientifically, other aspects are beyond the reach of scientific methods, including modeling (cf. (Fracchia J and Lewontin RC 1999; Peña A 1999)).

I disagree, but I note that it is scholars working in hermeneutical traditions who attempt to study the most subtle of cultural relationships in depth. Even when hermeneutical research does not meet standards of scientific evidence, I believe that it often generates plausible, potentially important insights about human behavior and cognition in particular contexts. Hermeneutical research in anthropology, sociology, history, literature, and other humanistic disciplines matters, because it often focuses on real aspects of human life, which may be ignored if we restrict our vision to phenomena that are most easily studied with current scientific methods. It's thus worth attempting to stretch the boundaries of scientific methods available to us, in order to attempt to bring more of the insights that hermeneutical research provides into scientific research (as social science research has done in the past, e.g. (Atran S and Medin D 2008; Brown MJ and Feldman MW 2009; Dressler WW et al. 2007; Kashima Y 2000; Romney AK et al. 1986; Schultz E 2009)).

I am proposing a somewhat new strategy for modeling culture so as to incorporate more insights from hermeneutical studies into scientific approaches. Claims about interpretive, symbolic, or other supposedly non-causal relationships between cultural elements can in many cases be treated as hypotheses about relationships mediated by cognitive processes. There is no reason, in principle, that we can't model such causal processes in ABMs in order to help understand patterns of cultural change. Other authors have made related points about modeling relationships between cognitive processes and culture (Bleda M and Shackley S 2012; De Block A and Cuypers SE 2012; Kahan DM 2012; Klüver J and Klüver C 2010; Mantzavinos C 2005; Slingerland E 2008; Thagard P 2012) but there have been few attempts to incorporate such ideas into modeling cultural processes in populations of individuals. Moderate-agent strategies are needed to try to model more of the complexity of intracultural interaction characteristic of real human culture. My goal here is to illustrate one such strategy.

The idea that there are metaphorical relationships between cultural elements plays a large role in hermeneutical understandings of culture. For example, (Bird-David N 1990) argues that in some hunter-gatherer societies, a variety of behaviors and attitudes derive from viewing the forest as peoples' parent (cf. (Tilley C 2000, p. 50)), and (Geertz C 1973) argues that betting decisions in Balinese cockfighting events metaphorically represented alliances and antagonisms between groups and individuals. Tilley C (2000) surveys a wide variety of cases in which anthropologists have argued that metaphorical relationships involving natural objects and artifacts play an important role in culture in particular societies. Outside of the study of culture as such, the idea that metaphor can have an important influence on thought has been advocated by Lakoff and his collaborators (e.g. (Lakoff G and Johnson M 2003; Lakoff G 2002)).

Gibbs RW Jr and Colson HL (2012) surveys related behavioral research and evidence from analysis of speech data.

Experiments by Thibodeau and Boroditsky (Thibodeau PH and Boroditsky L 2011) provide evidence for the influence of metaphor in culturally influenced thought processes. For example, in one experiment, American participants were randomly assigned to read either "Crime is a beast ravaging the city of Addison" or "Crime is a virus ravaging the city of Addison." In either condition, the initial sentence was followed by an identical paragraph describing an increase in crime in the fictitious city of Addison. There was a statistically significant effect of the initial sentence on participants' recommended solutions to the crime problem: Those who read the sentence containing "virus" recommended reform and prevention measures relatively more often than those who read the "beast" sentence. Participants who saw the "beast" sentence were relatively more likely to recommend capture and punishment solutions. Apparently, the initial metaphor influenced participants to favor crime prevention measures analogous to those appropriate for preventing the spread of viruses, or analogous to those appropriate for preventing attacks by beasts. Since particular beliefs about viruses and beasts, and attitudes about crime, are specific to some cultures and not others, Thibodeau and Boroditsky's results are evidence of an intracultural influence of metaphor.

It's plausible, though, that the cognitive processes involved in generating and understanding metaphors are often the same as those involved in processing analogies (Gentner D et al. 2001; Holyoak KJ and Thagard P 1995). Some authors explicitly argue for the importance of analogy in relationships between cultural elements (Dehghani M et al. 2009; Hofstadter DR and Sander E 2013; Holyoak KJ and Thagard P 1995; Thagard P 2012). The idea is that analogy might influence culture because it makes certain ideas "feel right" or seem more plausible as a result of analogical relations to propositions the person already accepts. Analogy would therefore provide a subtle influence on culture, making analogically related patterns of thought more likely. This role for analogies is partially supported by existing claims about the role of analogy in defeasible inferences in general problem solving (Forbus KD 2001; Holyoak KJ and Thagard P 1995), science (Bartha PFA 2010; Hesse MB 1966), and law (Picinali F 2011).

There are a number of computer models of analogy processing, with a great deal of associated experimental research validating some of the assumptions of these models (Gentner D et al. 2001; Gentner D and Forbus KD 2011; Holyoak KJ and Thagard P 1995). Neurologically plausible models of analogy processing have been developed (Eliasmith C and Thagard P 2001; Knowlton BJ et al. 2012), and there is research on neural correlates of analogy and metaphor processing (e.g. (Bassok M et al. 2012; Chettih S et al. 2012; Green AE et al. 2012; Knowlton BJ et al. 2012; Maguire MJ et al. 2012; Prat CS et al. 2012)). I attempt to capture interesting relationships between cultural elements, in part, by building Holyoak and Thagard's (Holyoak KJ and Thagard P 1989) ACME model of analogy processing into agents. ACME takes structured proposition inputs, and generates neural networks which identify analogies through the interaction of simple nodes on a network. ACME is a good choice here because it incorporates core assumptions common to most analogy processing models—and not much more—and because it's a relatively simple model whose outputs and operation are easy to investigate with graphical display tools.

My strategy is agent-based at two levels: At the top level, agents represent persons who communicate with each other, process analogies internally, and try to believe and disbelieve propositions in a way that is coherent. Within each person, though, these cognitive processes are modeled by two neural networks, whose nodes are very simple agents.

Methods

POPCO framework

POPCO overview

A POPCO simulation typically begins with the following steps (Figure 1), labeled in this section by top-level source code function names. Details are provided in later sections.

1. A population of POPCO agents, or "persons", is created as specified by the modeler. Each person includes its own list of representations of propositions in predicate/argument form (e.g. "`hunts (man animal)`").
2. `init-pop`: Each person's propositional representations ("propositions" to be brief) are processed by POPCO in order to generate two artificial neural networks in each agent:

    - Nodes in the *belief network* represent degrees of belief/significance for propositions. Weighted links represent inferential relationships, i.e. tendencies to give beliefs similar or dissimilar degrees of belief.
    - Nodes in the *analogy network* represent plausibility of components of an analogy, which are possible pairings of propositions, predicates, or predicates' arguments. Weighted links represent coherence between parts of an analogy, and competition between different analogy components.

    The two networks within each person are not connected by node-node links, but activations in the analogy network affect link weights in the proposition network.
3. The main loop is entered. This performs the following steps, in order:

    (a) `settle-nets`: Neural networks are partially settled.

    (b) `choose-conversers`: Pairs of persons ("speakers" and "listeners") are constructed.

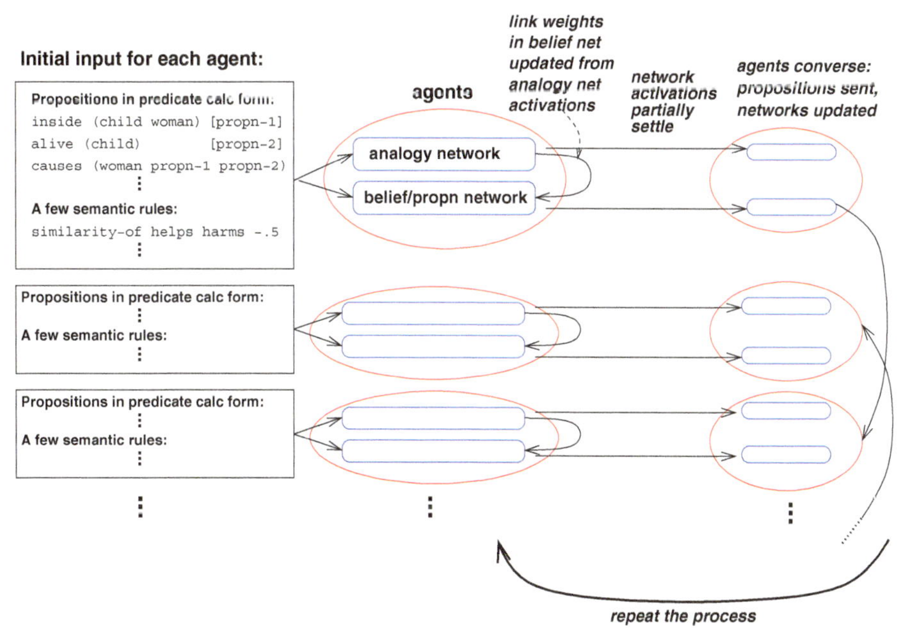

Figure 1 High-level causal structure and sequence of events for POPCO. Ellipses: persons. Rounded rectangles: neural networks. Arrows summarize causal interactions.

(c) `choose-utterances`: For each speaker and corresponding listener, at most one of the speaker's beliefs is chosen to be communicated.

(d) `transmit-utterances`: Utterances are transmitted from each speaker to each listener. This typically causes a persistent influence on the activation of the corresponding belief node in the listener. If the belief is new for the listener, it is added to the belief network.

(e) `transmit-environments`: For each person, perceptual "utterances" are transmitted from a personal "environment" to the person according to specifications by the modeler. These represent the influence of the external world on the mind.

(f) `update-analogy-nets`: If a proposition added in previous steps was new to a listener, new nodes representing possible pairings between the proposition, its components, and appropriate other propositions and components are added to the analogy network.

(g) `update-proposition-nets`: The weight of the link between each pair of belief network nodes, representing propositions P_1 and P_2, is set as a function of the activation of the analogy network node that represents a possible analogical mapping of P_1 and P_2.

The main loop starts over at step 3a, and continues until the modeler causes it to stop.

All activation values and link weights described below are Common Lisp long-floats in POPCO. In Steel Bank Common Lisp (SBCL), the implementation in which I usually run POPCO, long-floats are mapped to double-floats, as allowed by the ANSI Common Lisp standard (American National Standards Institute 1996). Double-floats have a minimum precision of 50 bits, with an 8-bit minimum exponent precision. In practice this means that activation values and weights are specified to 16 or 17 decimal places.

Concepts and initialization

The belief network and the analogy network are constraint-satisfaction networks, in that they attempt to determine an approximate solution to competing constraints. Constraints are represented by links and their weights. Both networks are constructed and updated from propositional inputs provided by the modeler.

Proposition input syntax The initial set of possible beliefs for an agent is specified by storing, in each agent, a set of very simple predicate calculus representations of propositions which the agent might "believe". Proposition inputs have a predicate followed by zero or more arguments in parentheses, and an arbitrary proposition name which can be used to refer to the proposition as a whole. Table 1 lists English sentences, with examples of stylized propositional representations for them on the right. The entry in the first line on the right says that `man`, i.e. males in a particular society, often hunt animals. The "*`propn-1a`*" on the far right is the proposition name. These names allow one to specify higher-order propositions such as the third one, which says that the fact that males hunt animals sometimes causes animals to harm males. The proposition names are also used as parts of node names in belief and analogy networks.

Table 1 POPCO propositional representations for English sentences

Meaning	POPCO representation
Males hunt animals.	`hunts (man animal)` *propn-1a*
Animals harm males.	`harms (animal man)` *propn-2a*
Men hunting sometimes causes harm from animals.	`causes (`*propn-1a propn-2a*`)` *propn-3a*

Analogy network concepts When specifying a POPCO model, the modeler divides the propositional representations for each belief into two sets ("analog sets"). The ACME module (Holyoak KJ and Thagard P 1989), written primarily by Paul Thagard as an implementation of Holyoak and Thagard's model of analogy processing, processes the propositional representations to generate the analogy network.

The core intuition underlying many theories of analogy processing, including Holyoak and Thagard's, is that to form an analogy is to construct a mapping between two sets of propositions, along with their components. Something is perceived as a good analogy if satisfies certain constraints well—if it does this in a way that takes in as much relevant information about the two domains as possible, and avoids conflicting mappings as much as possible:

1. If two propositions are mapped, it's preferable that their components be mapped as well.
2. It's preferable that two elements not be mapped to one element.
3. It's preferable to map concepts which have more rather than less semantic similarity.
4. Mapping more propositions is better than mapping fewer propositions.
5. Mapping causal claims and other higher-order propositions—i.e. those whose arguments are propositions—is more important than mapping other propositions.

To model these ideas, given two analog sets, ACME generates all possible mappings between propositions, predicates, and arguments which preserve syntactic structure and argument types. (ACME has no built-in knowledge of any natural language.) Each possible mapping between two elements is represented by ACME as single node in the analogy network. Links between nodes represent positive and negative constraints. These are designed to enforce the rules of thumb given above as well as possible. The network settling process should then give high activations to nodes representing mappings which compose a good analogy, and low or negative activations to other nodes.

For example, in 1991, before the U.S's entry into the first Gulf War, people working with President George H.W. Bush promoted the idea that Iraq in 1991 was analogous to Germany before World War II. This analogy was used to argue that the U.S. should go to war with Iraq. According to (Holyoak KJ and Thagard P 1995; Spellman BA and Holyoak KJ 1992) to see Iraq in 1991 as analogous to a situation around the time of World War II is a mostly unconscious process, which involves constructing a series of mappings between objects, predicates, and propositions.

Although two pairs of propositions are not enough to capture the idea of an analogy, they allow a simple illustration. Consider the following propositions.

1. Analog set 1:
    (a) Saddam Hussein is the president of Iraq.
    (b) Iraq invaded Kuwait.

2. Analog set 2:

   (a) Adolph Hitler was the führer of Germany.
   (b) Germany occupied Austria.

A reasonable way to understand the proposed analogy between Iraq in 1991 and Germany in 1939 would pair proposition 1a with proposition 2a, and 1b with 2b. Saddam Hussein would then be paired with Hitler, Iraq with Germany, being president with being führer, and so on. As noted above, though, ACME's procedure is to generate nodes for all possible mappings that are syntactically appropriate, and then determine which mappings are correct through the process of settling the analogy network. I summarize the possible mappings that ACME would construct involving elements of the four propositions in Figure 2. Representations of propositions are given in the top and left margins. The existence of a box at the intersection of a row and a column means that ACME will construct a network node representing a possible mapping between those propositions. When ACME creates a node for each of these mappings, it also creates links which capture the constraints mentioned above. For example, the node representing the mapping

fuhrer-of (Adolph Germany) ↔ president-of (Saddam Iraq)

will receive positively weighted links to the nodes representing these mappings:

fuhrer-of ↔ president-of
Saddam ↔ Adolph
Germany ↔ Iraq.

Each of these nodes will get positively weighted links to each other, as well. This collection of nodes represents a potentially coherent set of mappings. The same point applies to the mappings associated with each of the other three boxes. Further, any pair of nodes which map the same element to different elements is an incoherent pair of mappings. ACME will thus generate a negatively weighted link between such nodes. For example, ACME will generate a negatively weighted link between the two nodes representing these mappings:

fuhrer-of (Adolph Germany) ↔ president-of (Saddam Iraq)
occupy (Germany Austria) ↔ president-of (Saddam Iraq).

Figure 3 shows the network that would be generated by ACME for the four propositions given above.

	president-of (Saddam Iraq)	invade (Iraq Kuwait)
fuhrer-of (Adolph Germany)	fuhrer-of ↔ president-of Adolph ↔ Saddam Germany ↔ Iraq	fuhrer-of ↔ invade Adolph ↔ Iraq Germany ↔ Kuwait
occupy (Germany Austria)	occupy ↔ president-of Germany ↔ Saddam Austria ↔ Iraq	occupy ↔ invade Germany ↔ Iraq Austria ↔ Kuwait

Figure 2 Mapping relations in simplified Iraq-Germany analogy.

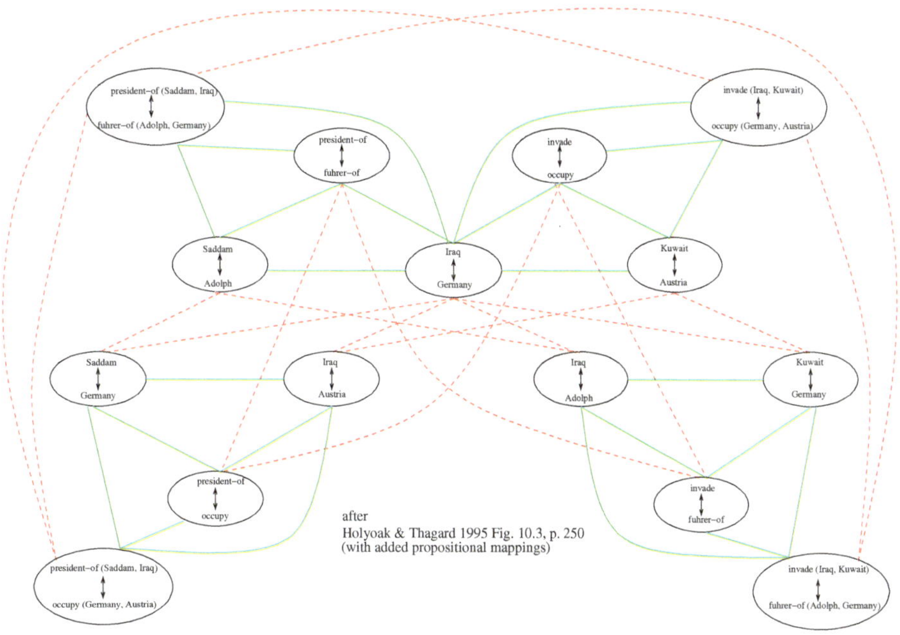

Figure 3 Analogy network generated from propositions 11a, 11b, 22a, and 22b. Ovals represent nodes, with represented mappings listed inside each oval. Solid green lines: positively weighted links; red dashed lines: negatively weighted links.

Note that since ACME doesn't have any built-in knowledge about the meanings of words, it's up to the modeler to supply abstract information about semantics to ACME. This can be done in three ways:

1. Through the choice of how to structure representations of propositions.
2. Through the choice of predicates: If identical predicates are used in the analog sets, and ACME ends up creating a node representing a mapping for these two predicates, this node is always given a little bit of extra positive activation, 0.1.
3. Through explicit specification of semantic similarity between predicates: The modeler can add code that specifies that two predicates are similar to any particular degree, represented by values from -1 to 1. A positive value makes the two nodes more likely to have similar activations; a negative value makes them more likely to have opposite activations. Larger absolute values increase these likelihoods.

Analogy network initialization Here is a more precise description of POPCO's algorithm for generating an analogy network from two analog sets containing proposition representations in predicate/argument form. It is almost the same as the algorithm given by (Holyoak KJ and Thagard P 1989).

1. Within each analog set, divide the propositions into recursively defined equivalence classes of propositions with the same type:
    (a) Propositions are of the same type if they have arguments of the same types in the same order (and therefore the same number of arguments).
    (b) Arguments to a proposition are of the same type if they are both simple names for things (not necessarily identical), or if they are both

propositions of the same type. (POPCO implements the intended behavior of the ACME model rather than a looser rule implemented in (Holyoak KJ and Thagard P 1989). (Example: The first two sentences in Table 1 are equivalent to each other; the third is equivalent to neither. More complicated examples can be found in the appendix. For example, proposition #15 there is not equivalent to proposition #22, despite superficial syntactic similarity. These propositions' first arguments, which are propositions, do not themselves have the same number of arguments.)

2. For each proposition which is a member of the corresponding equivalence classes from the two analog sets, make a "map node", a network node which represents a possible mapping between those two propositions. Do the same for corresponding predicates from each such proposition, and for corresponding arguments, i.e. arguments in the same places in the two propositions' argument lists. Example: These two propositions:
   `helps (woman child) p-Woman-Helps-Child`
   `helps (s-god human) os-God-Helps-Human`
   would result in four map nodes, one for each of these pairs:

p-Woman-Helps-Child	os-God-Helps-Human
helps	helps
woman	s-god
child	human

   Map nodes receive an initial activation value of 0.01.
3. For each proposition map node and its corresponding predicate and argument map nodes, create positive links of weight 0.1 between each pair of nodes. If such a link already exists, add 0.1 to its weight, until an upper limit of 0.5 is reached. (This restriction on analogy network link weights helps avoid large cyclic fluctuations in activation values.)
4. Create negatively weighted links, with weight -0.2, between each pair of distinct map nodes which share a proposition, predicate, or simple argument as one component of the mapping. If such a link already exists, do nothing.
5. If any map nodes for predicates concern identical predicates (i.e. found in propositions in distinct analog sets), create a link from the node *special* (which always has an activation of .99). The weight on this link is 0.1.
6. If the modeler has specified that other mapped predicates have semantic similarity, create a link from *special* to the map node, with the weight specified by the modeler.

Belief network concepts and initialization POPCO is based on the assumption that analogies can influence relationships between beliefs. For example, someone who believed that it was appropriate for the United States to go to war with Germany in World War II, and believed that Iraq in 1991 is analogous to Germany in 1941, might have been more likely to believe that it would be appropriate for the U.S. to go to war with Iraq in 1991. On the other hand, someone who thought that Iraq in 1991 was analogous to North Vietnam in the 1960s, rather than Germany in 1941, might therefore reject the suggestion that the United States should go to ware with Iraq (cf. (Holyoak KJ and Thagard P 1995)).

Nodes in each person's *belief network* represent belief in either a particular proposition or its negation. Activation values of nodes represent degrees of belief, with .99 representing full belief in the proposition associated with the node, and -.99 representing full belief in its negation. Belief nodes are created for each person as specified by the modeler during initialization. Nodes are initialized with activation value 0.0.

Positively weighted links between pairs of nodes represent tendencies to give each node similar degrees of belief; negatively weighted links represent tendencies to give each node opposite degrees of belief. These links are typically created and modified in response to the state of the analogy network (see below), but can also be specified directly by the modeler. There is also a special node, labeled *salient*, which always has the maximum activation, .99. Links from *salient* to belief nodes represent influences external to the person, and result either from explicit specification by the modeler, or from conversational input from other persons in the population. (In practice, the modeler must specify that some nodes in some persons are "perceived"—i.e. that they will have links to the *salient* node; otherwise all belief network activations will remain at 0.0.)

More precisely, weights on links between belief nodes are a function of activations of proposition map nodes in the analogy net, and occasionally, explicit specifications by the modeler, according to the following rules:

1. If there is a map node in the analogy network representing a possible mapping between proposition P_1 and proposition P_2, set the weight of the link between the nodes representing belief in/against P_1 and P_2 to
   - 0.2 times that map node activation, if the activation is > 0, or
   - 0.025 times the map node activation, if the activation is ≤ 0.
2. If the modeler has explicitly specified a weight for a link between P_1 and P_2 (using the function `semantic-iff`), then there will be a link between P_1 and P_2 with that weight, or with the sum of this weight and the weight generated by the previous step (as long as the sum does not exceed the minimum and maximum weights of -1 and 1).

The reason for the difference in the strength of map nodes' effects on negative and positive links is the following. For each successful analogical mapping between two propositions (represented by a map node with high activation), the analogy network will typically contain several unsuccessful mappings involving the same two propositions (represented by map nodes with negative activations). Many mappings are syntactically possible, after all, but no more than one proposition-proposition mapping should be successful. However, it seems reasonable to assume that disanalogies typically have less effect on our thought processes than analogies. If the resulting single positive link and several negative weights to a given belief node were of similar strength, the effects of the negative links would usually overwhelm the effect of the positive link: The disanalogies would have the primary effect. So it's important that the effect of negative map node activations be significantly smaller than the effect of positive activations. However, the precise value of this ratio doesn't seem to make a big difference to POPCO's qualitative results. Setting the multiplier for negative weights (0.025) to 1/8 the size of the multiplier for positive weights (0.2) works well in practice, and reflects the typical number of competing negative links per positive link.

Initialization sequence (init-pop) The full initialization sequence goes like this:

1. Create each person, storing specifications of proposition inputs, analog sets, a list of which propositions should be "perceived" initially, and additional semantic rules for propositions and predicates.
2. For each person, create analog sets and process semantic directives.
3. For each person, create its belief nodes, with activations set to 0.0.
4. For each person, create an environment (a simplified person) containing those propositions (beliefs) which will be perceived.
5. For each person, create the analogy network: Create map nodes and links as described above.
6. For each person, create links between belief nodes, with weights determined as described above.

POPCO main loop

Each time through the main loop, the following operations take place in sequence.

settle-nets Each person's analogy network and belief network undergo 5 iterations of settling. The settling process sets new activations for each network node as follows:

First, new activations for each of a person's nodes are calculated; then each of the nodes' current activations are set to the value of the new activation. Thus node activations are effectively updated in parallel. Following (Holyoak KJ and Thagard P 1989), the new activation a_i' of a node i with current activation a_i is given by a rule based on (Grossberg S 1978). For node i, the sum p_i of inputs from nodes j with positively weighted links w_{ij} to i is:

$$p_i = \sum_j w_{ij} \max(0, a_j) \quad \text{for } w_{ij} > 0 .$$

Similarly, the sum n_i of inputs from nodes j with negatively weighted links w_{ij} to i is:

$$n_i = \sum_j w_{ij} \max(0, a_j) \quad \text{for } w_{ij} > 0 .$$

Note that negative activations do not affect neighboring nodes. The effects of p_i and n_i are scaled by the difference of a_i from the extreme values -.99 and .99. The effect of the previous activation a_i is scaled by a decay value, .1.

$$s_i = a_i * .1 + .99\, a_i + (.99 - a_i)\, p_i + (a_i - .99)\, n_i$$

The new activation a_i' of node i is then s_i, or the maximum or minimum activation values if s_i exceeds them:

$$a_i' = \min(.99,\ \max(-.99,\ s_i))$$

POPCO networks usually settle into a stable state in which activations fluctuate very little, after about 30–200 settling iterations, i.e. 6–40 passes through the main loop. (Since the course of a real person's thought processes needn't come to a stable resolution before being interrupted by input from others, POPCO networks shouldn't have to settle before communication modifies the networks.) The pattern of activations in each network then represents a (perhaps locally) optimal configuration that balances activations in response to weights between connected nodes. A "good" network configuration is one in which pairs of positively-linked nodes usually have activations with the same sign, while negatively-linked nodes usually have activations with opposite signs.

choose-conversers POPCO agents communicate by sending propositions from one agent to another; this is an abstract representation of speech. POPCO creates a list of "converser pairs", each consisting of a speaker person and a listener person. Each person is chosen as a speaker in one converser pair, along with a randomly chosen listener, each time that `choose-conversers` is run. More specifically, for each person, POPCO creates a list of the $n - 1$ other persons, in random order, and selects the first person in the list to be the speaker's listener. The randomized list is created using a Knuth shuffle algorithm. The resulting list of converser pairs is then passed to `choose-utterances`.

choose-utterances For each speaker in a converser pair, POPCO tries to select a belief to be communicated to the pair's listener: Beliefs P_i are randomly selected as candidates for communication in a way that depends on their activation values a_i: For each belief, POPCO generates a new random number $r \in [0, 1)$ from a uniform distribution. If $r < |a_i|$, belief P_i is included in a list of candidates for communication. One belief P_i is randomly chosen from this list. This belief is attached to the converser pair to create a "conversation". The list of all such conversations is passed to `transmit-utterances`.

transmit-utterances For each conversation, information about the speaker's belief is transmitted to the conversation's listener:

1. If the belief communicated is new to the listener:
    (a) A belief node is added to the listener's belief network with initial activation 0.0.
    (b) The original predicate/argument representation of the proposition is stored for processing later by `update-analogy-nets`.
2. If the modeler specified a semantic-iff rule (§1) between this belief and an existing belief in the listener, a link will be created between nodes for these beliefs, with the weight specified by the modeler.
3. If there is not yet a link between the the *salient* node and this belief's node, such a link is created.
4. The weight on the link to *salient* in the listener is set to 0.05 × sign-of(speaker's belief activation), or this value is added to the existing link weight. If the result exceeds the extreme values of -1 or 1, the weight is set to the nearest extreme.

The purpose of the additional weight on the *salient* link is to capture the idea that what's said to someone tends to increase the proposition's salience for the listener, and increases his/her confidence in its truth. Since *salient* always has an activation of .99, a persistent weight on a link from *salient* to a belief node produces a persistent influence on the belief's activation. Note that in the real world, people usually don't convey information about their degrees of belief. Thus the effect of a POPCO utterance on a listener's belief activation depends only on whether the speaker says that the proposition is true (when the speaker's belief activation is positive) or says that its negation is true (when the speaker's belief activation is negative). Incorporation of the new belief's proposition into the analogy network, and subsequent updating of belief node link weights in response, occurs in later processing steps (but before the next round of settling).

transmit-environments As mentioned above, some persons have an "environment" which contains propositions that are repeatedly communicated to the person, as if they were perceived in the person's environment. This communication process differs from that between persons in that:

1. Every "belief" in a person's environment is communicated to the person.
2. The weight added to the link from *salient* is 1.0 rather than 0.05.

update-analogy-nets For each person, if a proposition added in previous steps was new to a listener, then there will be not yet be any nodes in the analogy network corresponding to the original predicate/argument representation of the proposition. This representation, which was stored in one of the preceding two steps, is now processed to add nodes to the analogy network. The procedure for doing this is specified by the analogy network algorithm described above under "Analogy network initialization".

update-proposition-nets For each person, weights of links between belief nodes in the belief network are updated in response to the state of the analogy network, as specified in above under "Belief network concepts and initialization".

Models of Sanday's hypotheses

Sanday's empirical and causal claims

Using an anthropological dataset, the Standard Cross-Cultural Sample, (Sanday PR 1981) reported pairwise correlations across societies between properties in adjoining columns of Table 2, i.e.:

- Between dependence on large game hunting, and greater degrees of male dominance over women.
- Between male dominance, and fathers' emotional distance from children and/or lack of involvement in childrearing.
- Between fathers' emotional distance and lack of involvement in childrearing, and societal emphases on stories about human origins in which humans are created magically by a male or animal creator from a distant place—as opposed to natural creation by a female or couple from nearby.

Sanday proposed an explanation of the correlations in terms of cognitive tendencies linking characteristics in the four dimensions, with the (leftmost) subsistence variable viewed as imposed by the environment rather than influenced by any of the other three. Though Sanday's (Sanday PR 1981, ch. 3) most explicit statement claims that the direction

Table 2 Summary of correlations between cultural variants from (Sanday PR 1981, ch. 3)

Main subsistence	sex roles	fathering	creation stories
large game hunting	men dominant	distant from child	far male/animal creates magically
intermediate/mixed	intermediate		couple creates by natural process
gathering or fishing	men less dominant	close to child	nearby female creates naturally

Correlated variants are listed on the same line.

of causation goes only from left to right in Table 2, other remarks show that she thought there were feedback relations between most of the four variables.

For example, Sanday argued that the danger and struggle for intermittent reward required by large game hunting encourages a view of the power of nature as something apart from humans, which must be mastered by males:

> Consider, for example, a society that relies on large game for food and, perhaps, clothing or items for domestic use. In these cases the hunter and the hunted are engaged in a game of skill in which both have the power to outwit the other .... In such instances power does not come readily to humans; power must be acquired, controlled, and manipulated for human purposes. (Sanday PR 1981 p.65)

I understand Sanday as arguing that, partly because such situations encourage emotional distance from the environment, men are then less likely to be intimately involved with raising young children. Large game hunting also encourages focus on stories in which creation of humans is the result of a magical process controlled by a male, animal, or abstract being in the sky, rather than a more intimate, natural process involving females:

> When large game are hunted, ... males engage in an activity whose outcome is unpredicatable [sic] and entails danger. ... the psychological energy expended in this effort ... is not directed inwardly toward nurturing children or family but toward acquiring and using powers beyond man's dominance. The major source of power is perceived as residing in a supreme being who resides in the sky or in animals. (Sanday PR 1981, pp. 65f)

Some of Sanday's remarks also suggest that creation stories can affect men's involvement with children:

> The nurturant father, on the other hand, is buttressed by the female creator. In these cases, the reproductive functions of women are celebrated both in myth and behavior. In the absence of the female creator, fathers are involved with children in infancy and early childhood either as disciplinarians or not at all. When their major role is to discipline and control, fathers are not unlike supreme beings. They are distant, controlling figures who are removed from biological processes. (Sanday PR 1981, p. 64)

These remarks can reasonably interpreted as proposing that cognitive processes are responsible for the cross-cultural correlations Sanday reported, even if she does not clearly spell out relevant cognitive mechanisms. Sanday also discusses possible social processes by which a culture might be transformed from one that doesn't fit the patterns she describes to one that does, but does not propose precise models.

Various passages in Sanday's book (Sanday PR 1981) are suggestive of the idea that analogical relationships between patterns of thought in the four domains (Table 2) encourage the correlations she reported. Thus I propose that if the causal relations between cultural patterns exist, they are partially mediated by tacit analogical inferences. By simulating a population of communicating persons in which such inferences are available, we can explore models which might provide "how possibly" (Brandon RN 1990; Grim P et al. 2011) explanations of correlations like those that Sanday reported.

These would be "generative" explanations in Epstein's (Epstein JM 2006) sense, not just in showing how the correlations Sanday describes arise from a collection of agents representing persons, but also in showing how they arise from relatively simple networks of "agents" inside persons. Such explanations would provide mechanistic accounts of population processes—processes that depend on hypotheses about symbolic cultural relationships.

More specifically, the simulations described below will illustrate the possibility of providing answers for two questions:

1. What might cause correlations between stories about human origins and concrete behavioral styles concerning male childrearing roles and means of subsistence?

The first simulation (model 1) will illustrate the possibility that these correlations are at least partly due to the influence of analogies, reinforced by communication between members of a society.

2. How might particular origin stories come to be entertained in the first place?

The second simulation (model 2) will illustrate the possibility that one part of an answer is that propositions that are not part of a coherent origin story spread through the population for reasons unrelated to origin stories, childrearing behaviors, or means of subsistence. However, when these propositions come to be entertained together, analogical resonance between concrete behaviors and these propositions can emphasize them in a way that suggests a coherent origin story.

Core features of the simulations

Persons in the models 1 and 2 have:

1. Propositions about current human interactions:
   - (a) Propositions concerning parenting and childbirth. (I call these "parenting" propositions.)
   - (b) Propositions concerning large-game hunting ("hunting" propositions).
2. Propositions about human origins:
   - (a) Propositions characterizing a creator who is from the earth, is female, is nurturing, and created humans from inside her body ("earth origin" propositions about an "earth-based" god).
   - (b) Propositions characterizing a creator who is male, comes from the sky, is both helpful and harsh, and created humans magically ("sky origin" propositions about a "sky-based" god).

The 37 propositional inputs I use to represent these four domains (Appendix) constitute an attempt to provide a simple and somewhat abstract representation of some of the core ideas in Sanday's hypotheses. Within each person, POPCO has the opportunity to map any proposition in set group 1 above to any proposition in set group 2. However, as we'll see, POPCO agents will able to construct two distinct sets of analogical relationships: those relating hunting to sky origin, and those relating parenting to earth origin.

Model 1: What causes the correlations?

1. What might cause correlations between stories about human origins and concrete behavioral styles concerning male childrearing roles and means of subsistence?

Model 1 implements a population which is small enough (100 individuals) so that everyone talks to everyone else. Each person entertains all 37 propositions about parenting, hunting, and both sky-based and earth-based gods. The model assumes that hunting plays a large role in this population: Each hunting proposition is made salient for each person. That is, in the belief network, hunting propositions have links of weight 1 to the *salient* node. All other propositions begin with an activation of 0.0, representing neither belief in the proposition nor in its negation. This last assumption is unrealistic, but allows a simple demonstration of the fact that salient hunting propositions can produce a tendency to encourage sky origin propositions. From this starting state, the model runs for 2000 time steps (iterations of the main loop). Since there is stochastic variation in who talks to whom and in whether a person says anything in a given time step, I ran the same model 50 times with a new random seed each time. The behavior of the model was qualitatively identical in all runs, so I'll simply describe a single run.

Figure 4 displays averages of activations of beliefs in each of four domains: parenting (P), hunting (H), earth origin (OE), and sky origin (OS). Each line in the plot displays the average activation for one person over the first 1500 time steps. (There was little change

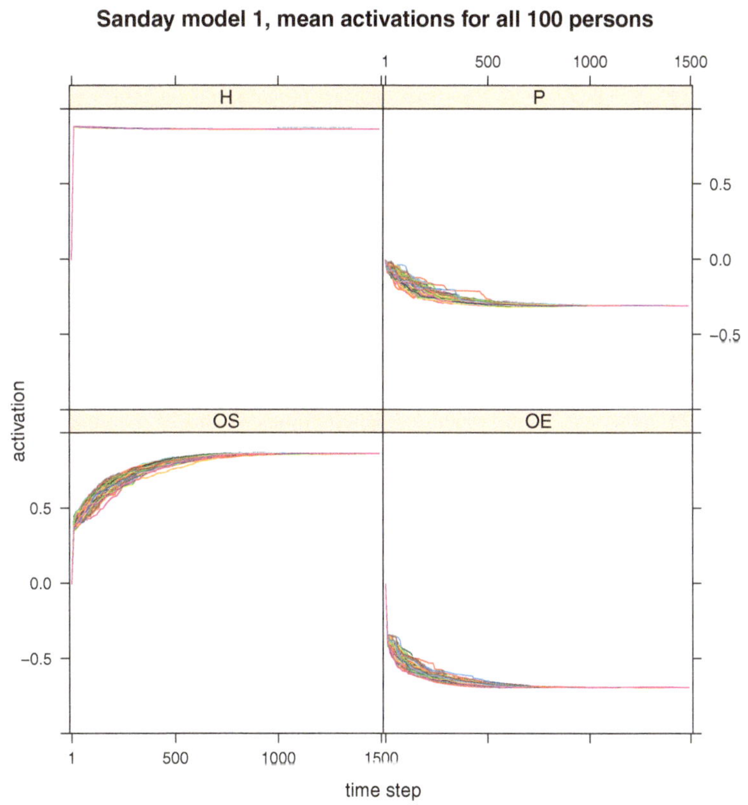

Figure 4 Model 1: Averages of each person's belief activations in the labeled domain, with hunting propositions salient during the run. H: hunting propositions, P: parenting propositions, OS: sky origin propositions, OE: earth origin propositions. One proposition (`o-Human-Alive`), common to both OS and OE, is not included since it doesn't belong exclusively to either domain. Points are plotted every 10 time steps, i.e. at 1, 11, 21, etc.

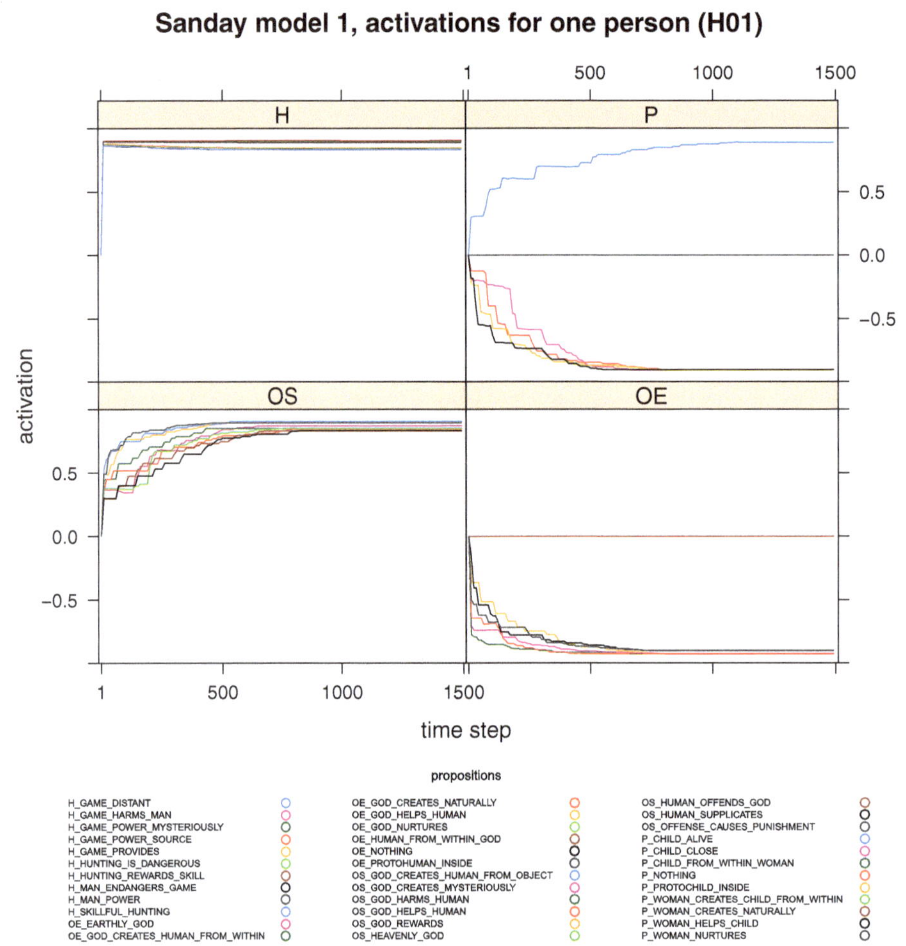

Figure 5 Model 1: Each line represents changing activation of one belief in a single person (H01) from the run summarized in Figure 4. One proposition (`o-Human-Alive`), common to both OS and OE, is not plotted. Points are plotted every 10 time steps.

during the last 500 time steps.) Figure 5 shows activations for individual propositions in a single person over the same time period. All 100 persons exhibited a very similar pattern, in each of the 50 runs.

Figures 4 and 5 show that sky origin beliefs, initially with activations 0.0, immediately acquire positive activations, and stabilize at values near 1 by step 1000. The initial rise in these activations is due to the fact that their pattern of relationships parallels relationships between hunting propositions. POPCO constructs an analogy between hunting propositions and sky origin propositions and as a result, creates positively weighted links between those hunting propositions and sky origin propositions that play similar roles in the two analog sets. Then since hunting propositions have high positive activations, positive activation is transmitted to sky origin propositions.

However, the gradual increase in activations of sky origin propositions after the first few steps is largely due to conversation. Once sky origin propositions acquire positive activations, they thereby acquire a significant probability of being communicated to others. When such a belief is communicated, it causes an increase in the activation of the same belief in the listener. It's this mutual reinforcement due to agreement and conversation

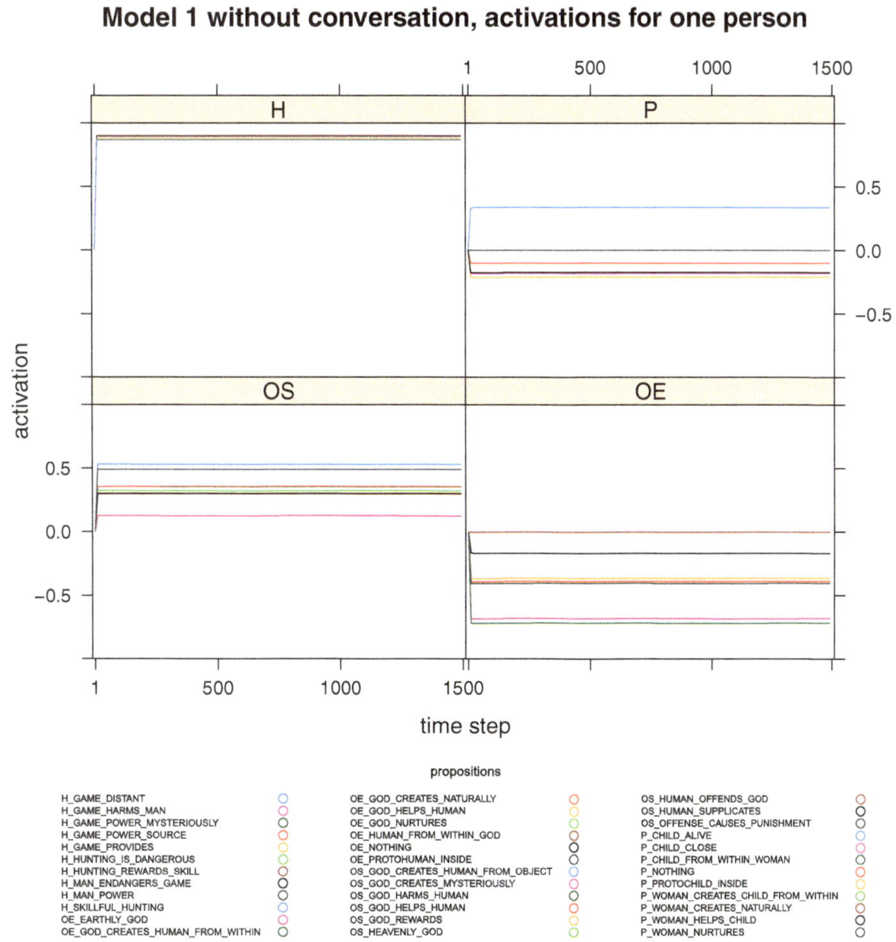

Figure 6 Belief activations for one person, with no conversation. As in Figure 5, one proposition is excluded, and points are plotted every 10 time steps.

that carries sky origin proposition activations to near 1 by tick 1000. The effect of conversation can be seen by comparing Figure 5 with Figure 6, which plots data from a similar model in which no conversation takes place.

Sky origin propositions compete with earth origin propositions to be mapped to hunting and parenting propositions. The analogy network will include a negative link between (a) a node representing a possible mapping of a sky origin proposition to a hunting or parenting proposition *P*, and (b) a node representing a possible mapping between an earth origin proposition and *P* (§1). Thus when, for example, a hunting/sky origin map node acquires a high activation because of its role in the hunting/sky origin analogy, this tends to push down the activation of a competing hunting/earth origin node. If this second node acquires a negative activation, it will usually result in a negatively weighted link between nodes representing the hunting proposition and the earth origin proposition. But the hunting proposition always has a high activation in the (conversational) model described above, so there will be a tendency for the earth origin proposition to receive negative activation. It's because of such effects that most of the earth origin propositions end up having negative activations. Many parenting propositions have negative activations in the model because of similar effects involving possible mappings between parenting propositions and sky origin propositions, as well as positive links between parenting propositions and earth origin propositions. Of course the full story involves effects of many nodes and links at once. (One parenting proposition, `p-Child-Alive`, has high activation. This is due in part to the fact that this proposition gets mapped to `o-Human-Alive`, an origin proposition that plays roles in both sky origin and earth origin analogies. `o-Human-Alive` is also mapped by a hunting proposition, `h-Man-Power`, that has positive activation.)

Though the principles governing the generation of the analogy and belief networks are relatively simple, the generated networks are often complex. Figures 7, 8, and 9 show the analogy network (bottom) and belief network (top) at time steps 0, 10, and 100 in the person whose activations were plotted in Figure 5.

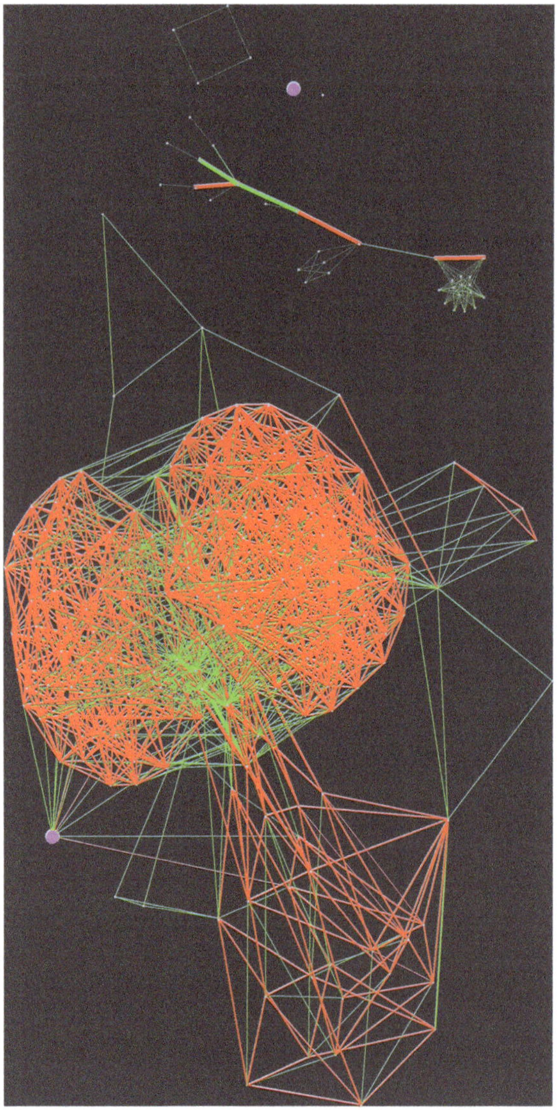

Figure 7 Model 1: Networks in one person at time step 0. Bottom: analogy network; top: belief network. Green links are have positive weight; red links have negative weight.

Figure 8 Model 1: Networks in same person at time step 10. Bottom: analogy network; top: belief network. Thickness of lines indicates absolute value of link weight. Colored triangles show relations between proposition map nodes and belief network links.

(I also performed 50 runs of a model in which parenting propositions rather than hunting propositions were made salient. Again, all runs were qualitatively similar to each other. All runs were like those in model 1 described above, but with behaviors switched to different proposition domains: Earth origin rather than sky origin propositions roughly converged to a high activation. Sky origin and hunting propositions acquired relatively low activations in contrast to their high activations in the original model. Figure 10 shows a typical run of this model.)

What model 1 illustrates is that analogies between beliefs derived from the experience of large game hunting could increase the plausibility of stories which see humans as being created by a physically and emotionally distant, magical being. The model illustrates the possibility that once such an set of origin beliefs became widespread, mutual

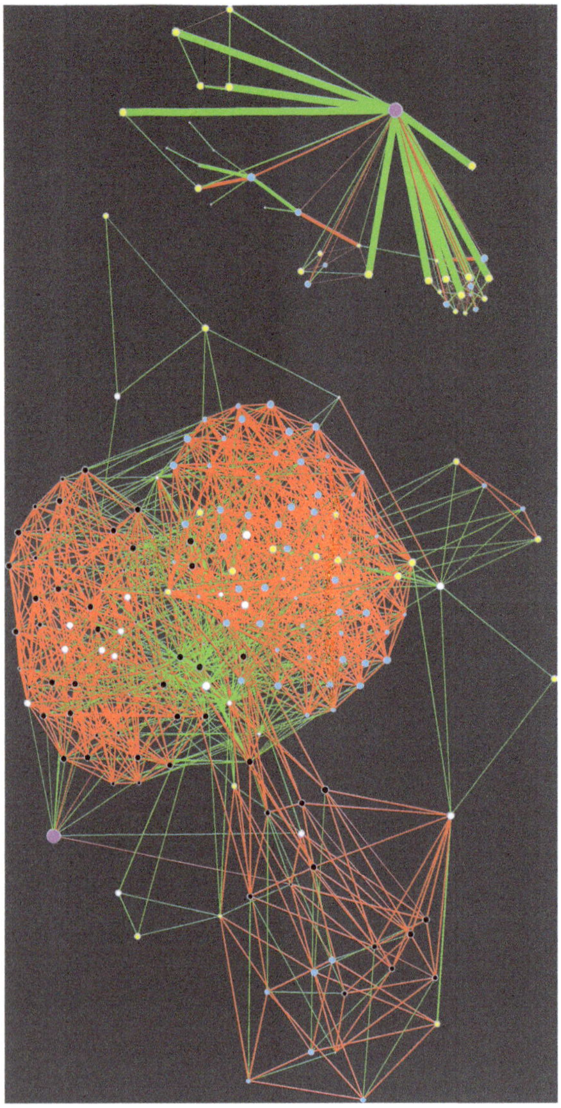

Figure 9 Model 1: Network in same person at time step 100. Bottom: analogy network; top: belief network. White and yellow: positive activation; black and blue: negative activation. Absolute value of activation: size of circle. White/black are used only for proposition map nodes.

reinforcement through social interaction could increase the strength of such beliefs. The model thus suggests a possible mechanism for generating phenomena like those described by Sanday.

Model 2: Where do origin stories come from?

2. How might particular origin stories come to be entertained in the first place?

A story about human origins is a relatively complex whole, including a number of beliefs. How does such a set of beliefs come to be entertained in the first place? Where do those thoughts come from? One possibility is that each component of an origin story originally arises in different persons for reasons that have nothing to do with the possible role of such propositions in an origin story. Such propositions might not have

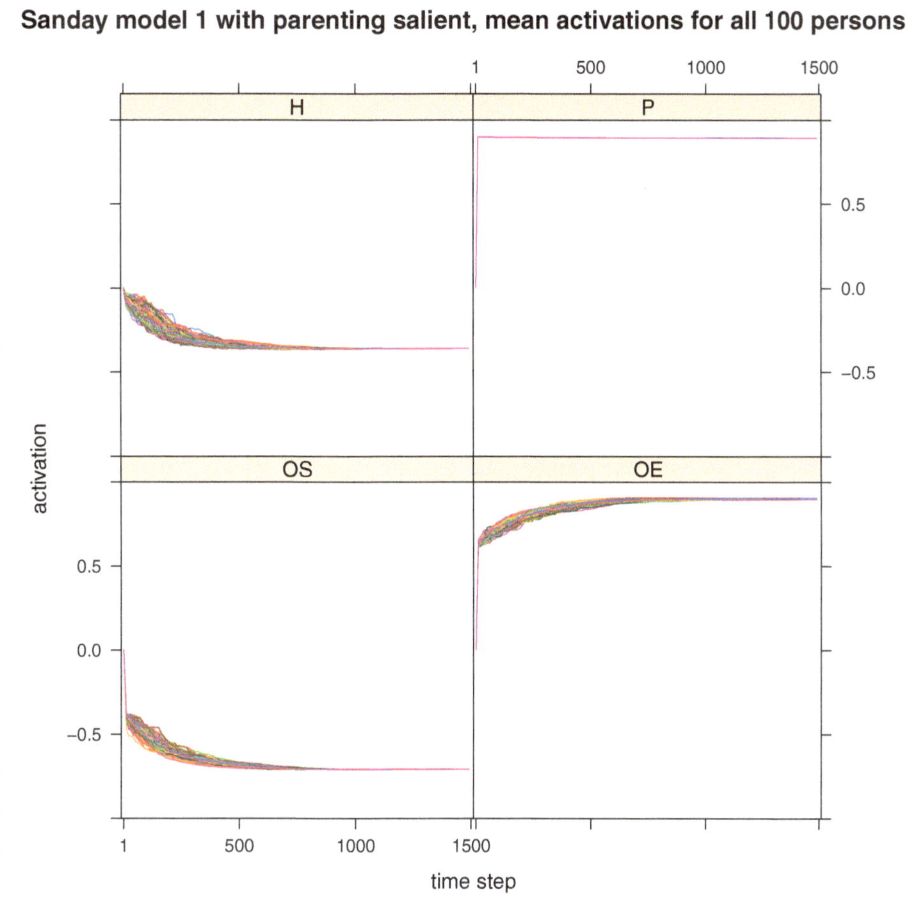

Figure 10 Model 1 with parenting rather than hunting propositions salient during the run. Other specifications are the same as for Figure 4.

great significance, but may nevertheless get mentioned from time to time. After most of the propositions have diffused through the population, a change in the physical or social environment might lead to patterns of belief that resonate analogically with the origin story.

We can model this kind of process by scattering beliefs which might play a role in an earth origin story across different members of the population. In particular, we can alter model 1 by starting with 92 persons who have only hunting propositions, parenting propositions, and sky origin propositions. To this we add eight persons, each of which has all of the propositions just mentioned, plus a single, unique earth origin proposition. We then allow the new model to run for 1000 time steps in the same way as the old model. Although, initially, none of the earth origin propositions in the eight individuals make up a full earth origin analog set, they do end up getting activations other than zero, due to mappings in which they participate imperfectly. Earth origin propositions thus occasionally get mentioned in conversation, and gradually spread through the population. As it turns out, most members of the population acquire most earth origin propositions by time step 1000. (I ran the model 50 times with different random seeds. In at least 27 runs, all 100 members of the population had all eight of the earth origin propositions by time step 1000. In no run were more than 99 of the 800 possible instances of earth origin propositions absent.) Assume, further, that at time step 1000, the physical and/or social environment changes. Intimate parenting becomes more important: We link parenting beliefs to *salient* with weight 1. Hunting become undesirable: We link hunting beliefs to *salient* with weight -1.

Behavior in all runs was qualitatively similar. Figure 11 shows one run (in which all persons had all earth origin propositions by step 1000). We see that once parenting beliefs become salient at step 1000, earth origin beliefs immediately come to have high average activation, as a result of the analogy with parenting beliefs. These activations are subsequently increased by conversation. The negative salience applied to hunting propositions at step 1000 counteracts the effects of earlier positive salience. The analogy between hunting propositions and sky origin propositions, as well as negative links involving earth origin propositions, then gives the sky origin propositions negative average activations. (Models which are the same except that there are no earth origin propositions present anywhere in the population produce similar patterns for sky origin propositions, but these average activations are not as low.)

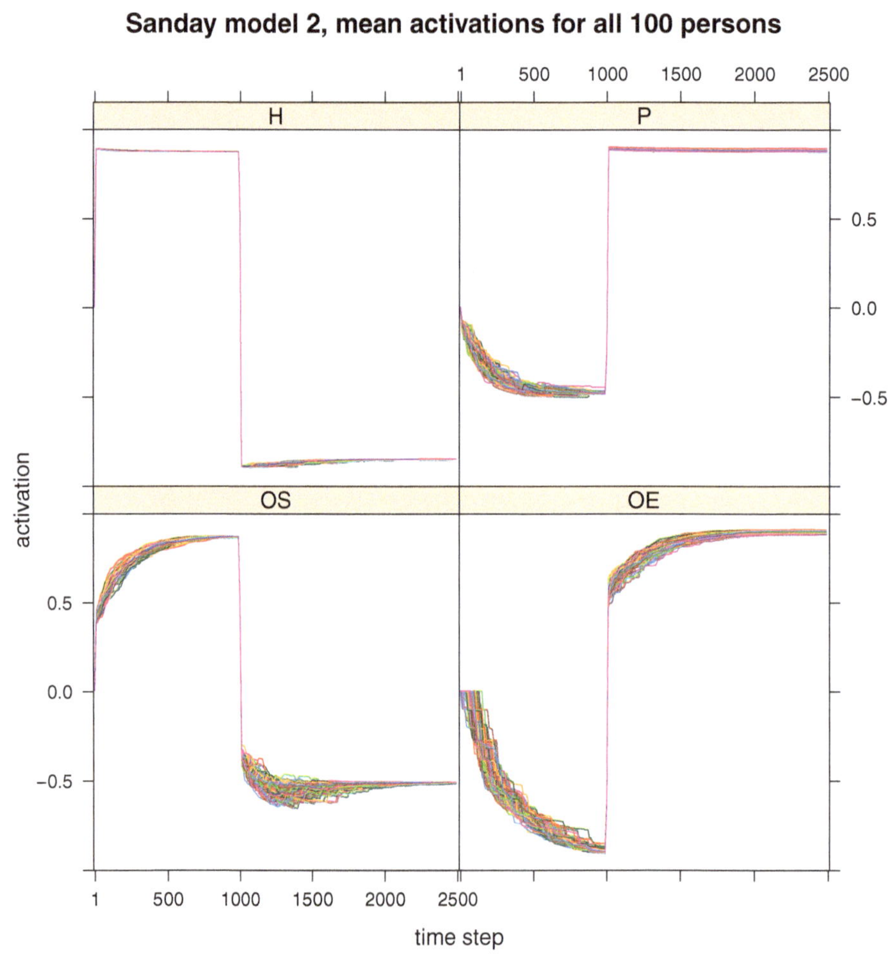

Figure 11 Model 2: Averages of each person's belief activations in the labeled domain. H: hunting propositions, P: parenting propositions, OS: sky origin propositions, OE: earth origin propositions. One proposition (`o-Human-Alive`), common to both OS and OE, is not plotted. Points are plotted every 10 time steps, i.e. at steps 1, 11, 21, etc.

Model 2 illustrates the possibility that an origin story might might come to play a significant role in a culture through a two-part process. First, the components of the story incrementally filter through a population, but play no very special role as a whole in patterns of thought. Later, a cultural change in another domain makes an analogy to this "neutral" set of thoughts salient, thus bringing the origin story into prominence in the culture.

Results and discussion

Models 1 and 2 provide simple illustrations of ways that patterns of culture reported by Sanday might come about. The models provide simple "how possibly" explanations (Brandon RN 1990; Grim P et al. 2011) or "generative" explanations (Epstein JM 2006). What is significant is not that the patterns can be generated by a computer model, but that they can be generated by particular kinds of processes (Epstein JM 2006, pp. 28f)—in this case, processes of analogy construction. Several specific points are worth noting.

First, POPCO is able to illustrate a process by which tendencies to believe cultural elements in different domains are influenced by modeled analogical relationships. Further, we saw that the relatively mild effect of analogy can be reinforced by communication within a population. The idea that the modeling of processes of analogy construction might be useful for understanding cultural relationships is not new (Dehghani M et al. 2009; Holyoak KJ 1982; Holyoak KJ and Thagard P 1995; Thagard P 2012, 1992), but there has been little work on modeling the role of analogical relationships on population-level cultural change. (Thagard (Thagard P 2000, Ch. 7) presents agent-based models of consensus-building in scientific communities, but his focus in these models was not on analogical relationships.)

Second, the simulations illustrated a way of modeling the influence of relations of coherence between elements of a culture on cultural dynamics. Such relations have not played a significant role in modeling of cultural change within populations. In the simulations described above, beliefs changed in frequency in response to inferential relationships connecting them. In most cases these inferential relationships were derived from analogical relations, which themselves are the result of a kind of evaluation of coherence by the analogy network (Thagard P 2000). Also note that many existing models of cultural change are derived from models in population genetics, and the roles that coherence plays in POPCO models suggest properties that are important in biological evolution—epistasis, pleiotropy, and genetic distance.

Third, it's significant that POPCO can model processes by which different ideas diffuse through a population randomly, and then come together to create a new cognitive structure which later becomes important. POPCO should be able to model cases in which propositions spread for one reason—say, because they cohere with propositions X—but later come to influence propositions Y for other reasons (cf. (Gould SJ and Vrba ES 1982; Hurst LD and Peck JR 1996; Wimsatt WC 1999; Wimsatt WC and Griesemer JR 2007)).

Fourth, since the effect of one belief on analogical relationships depends on the presence of other, related beliefs, POPCO illustrates one way to model the idea that elements of real culture often interact in a non-additive manner (Wimsatt WC and Griesemer JR 2007). POPCO also provides a way to model some kinds of cultural "scaffolding" in

Wimsatt and Griesemer's (Wimsatt WC and Griesemer JR 2007) sense, and "attractors" in Sperber's (Sperber D 1996) sense: The effect of an utterance on the internal state of a listener in POPCO depends on the listener's other background beliefs and experiences.

The reader will have noticed that in the simulations described above, members of a population ultimately converged to the same beliefs. Large scale societies are nothing like this, of course, and anthropological studies of small, self-contained societies don't always show uniformity of opinion. However, convergence to consensus is what we should usually expect in models in which everyone influences everyone else. Lehrer K and Wagner C (1981) derive this conclusion from basic facts about matrices for a specific model of communication, but the point seems more general. Alexander JM (2007) presents a number of models involving imitation of neighbors' behavior in games, in which maintenance of diversity requires social network structures other than full communication between all agents. Morris S (2000) (cf. (Vega-Redondo F 2007)) showed that for players which imitate the best responses in simple multi-player coordination games on a network, the spread of a response depends on ratios between certain node degrees. In more recent investigations I have been exploring effects of social network structure in POPCO models, and have found that POPCO models can maintain diversity of opinion on a structured network. (POPCO's social network functions were added by Kristen Hammack.) Finally, note that interactions between opinions in real societies are much more complex than in any model, and are influenced by changes in the physical environment as well. It may be that such interactions can by themselves prevent convergence of opinion in real societies, even if everyone were influenced by everyone. Indeed, (Mueller ST et al. 2010) have illustrated the possibility of maintaining distinct clusters of beliefs in a population because of dependence of some beliefs on others, rather than because of network structure. POPCO may able to implement similar patterns, but this is a topic for future investigation.

Given the complexity of a POPCO simulation, the following related questions express natural worries:

- Can't *any* result be gerrymandered using a POPCO simulation? There seem to be too many degrees of freedom.
- Doesn't the complexity of POPCO models obscure the patterns that are in fact responsible for the simulation results? If you get an interesting result using POPCO, how will you know which features of the model were responsible for it?

In trying to construct the models described above, I did make a number of somewhat arbitrary choices, as the Appendix illustrates. On the other hand, you can't just put *anything* into a POPCO model—not if the goal is to try to capture preexisting ideas about how certain complex cultural variations interact with each other. My experience so far is that it's in fact quite difficult to capture ideas like Sanday's in a POPCO model. I think that the goal of trying to model a certain kind of cognitive process which plausibly might operate in the world places severe constraints on modeling with POPCO.

POPCO agents do lack the transparency of agents in simple-agent models, and this is a drawback. In order to understand why POPCO agents behave as they do, one sometimes has to investigate their internal processes. POPCO is designed to make this process as easy as possible, though. A person's neural networks can be investigated using interactive network display tools such as GUESS, and I have also built a tool in NetLogo (Wilensky U 1999) for quick observation of certain patterns. Further, Common Lisp, in which POPCO is implemented, makes it easy to interactively query any aspect of a simulation's state at any time.

Finally, investigation of what's responsible for a result in POPCO may require varying parameters and constructing alternative models, as in other contexts (cf. Figure 6). In the case of POPCO simulations, sometimes this means going outside of POPCO. For example, I've found it helpful, in investigating effects of network structure in POPCO, to build related simple-agent models in NetLogo.

Conclusions

I've described the POPCO agent-based framework for simulating cultural variation in populations, and presented models illustrating some of its capabilities. POPCO is unusual in providing a relatively simple framework for studying cultural variation, which at the same time allows changes in culture to depend on analogical and other coherence relationships between beliefs. This allows us to model the effects on cultural dynamics of some of the symbolic relationships thought to be important in hermeneutical studies of culture.

Sanday PR (1981) used statistical methods and proposed hypotheses relating cultural variants that were explicitly causal. Sometimes researchers in hermeneutical traditions consider such strategies inappropriate for the study of culture. Nevertheless, because of the kinds of cultural variants that Sanday discussed, the complex and subtle nature of their relationships, and the authors upon whose work she drew to construct her hypotheses, I view Sanday's research in (Sanday PR 1981) as substantially indebted to hermeneutical research traditions. (The quotations in section "Sanday's empirical and causal claims" hint at this dimension of her work.) The models I describe above attempt to capture some of the subtle relationships Sanday proposed, albeit in an abstract form. These models thus represent initial steps in modeling effects of such complex and subtle cultural phenomena in populations. I believe that the models described above suggest some of the potential of moderate-agent modeling for investigating relationships between cultural phenomena. I suggest that moderate-agent strategies, though currently rare, have the potential to help bridge gaps between insights from humanistic research and scientific methods.

Appendix: Propositional specifications for Sanday models

Table 3 lists the propositional representations used as inputs to POPCO in the Sanday simulations. Figure 12 gives Lisp source code input to POPCO to specify additional semantic relationships, which go beyond those captured by analogical relationships discovered by POPCO. Proposition prefixes such as "oe-", "os-", "h-", and "p-" mainly serve to remind the modeler about how propositions are grouped, although they are used in some routines used to display data. The prefixes "s-" and "e-" in "s-god" (sky-based god) and "e-god" (earth-based god) cause POPCO to treat s-god and e-god as distinct objects, rather than one object with the properties of both sorts of creator. Note that although the correlations that Sanday found involved male parenting, her explanation of these correlations had more to do with female parenting.

Table 3 Propositional inputs for models 1 and 2

Predicate	Arguments	Proposition name	Intended meaning
Parenting:			
1. alive	(child)	*p-Child-Alive*	A child is alive.
2. intimate-agent	(woman, child)	*p-Child-Close*	Women, children are emotionally intimate.
3. inside	(child, woman)	*p-Protochild-Inside*	A child (fetus) is initially inside a woman.
4. process-from-to	(*p-Protochild-Inside, p-Child-Alive*)	*p-Child-From-Within-Woman*	There is a process that leads from (3) to (1).
5. creates	(woman, *p-Child-From-Within-Woman*)	*p-Woman-Creates-Child-From-Within*	Woman are the cause of the process in (4).
6. natural-process	(*p-Woman-Creates-Child-From-Within*)	*p-Woman-Creates-Naturally*	The preceding process is natural.
7. helps	(woman, child)	*p-Woman-Helps-Child*	Women help (nurture, etc.) their children.
8. causes	(nothing, *p-Woman-Helps-Child*)	*p-Woman-Nurtures*	Nothing causes women to nurture children.
9. nothing	(nothing)	*p-Nothing*	(Has no real meaning, but is useful.)
Hunting:			
10. feels-power	(man)	*h-Man-Power*	Men feel powerful, able to control nature.
11. power-source	(game, *h-Man-Power*)	*h-Game-Power-Source*	Game is a source of this power.
12. mysterious-process	(*h-Game-Power-Source*)	*h-Game-Power-Mysteriously*	Game being source of power is mysterious.
13. hunts-endangers	(man, game)	*h-Man-Endangers-Game*	Men hunting is dangerous to game.
14. harms	(game, man)	*h-Game-Harms-Man*	Game sometimes harms men.
15. causes	(*h-Man-Endangers-Game, h-Game-Harms-Man*)	*h-Hunting-Is-Dangerous*	(13) is a cause of (14)
16. hunts-skillfully	(man, game)	*h-Skillful-Hunting*	Hunting involves skill.
17. helps	(game, man)	*h-Game-Provides*	Game helps men (by providing food, etc.).
18. causes	(*h-Skillful-Hunting, h-Game-Provides*)	*h-Hunting-Rewards-Skill*	Skill in hunting causes game's benefit.
19. distant-agent	(game, man)	*h-Game-Distant*	Game and men are emotionally distant.
Both origin domains:			
20. alive	(human)	*o-Human-Alive*	Humans are alive.

Table 3 Propositional inputs for models 1 and 2 *(Continued)*

Earth origin:			
21. inside	(human, e-god)	*oe-Protohuman-Inside*	Human(s) began inside e-god.
22. process-from-to	(*oe-Protohuman-Inside, o-Human-Alive*)	*oe-Human-From-Within-God*	There's a process leading from (21) to (20).
23. causes	(e-god, *oe-Human-From-Within-God*)	*oe-God-Creates-Human-From-Within*	An e-god causes this process.
24. natural-process	(*oe-God-Creates-Human-From-Within*)	*oe-God-Creates-Naturally*	An e-god doing so is a natural process.
25. helps	(e-god, human)	*oe-God-Helps-Human*	An e-god helps humans, is nurturing, etc.
26. causes	(nothing, *oe-God-Helps-Human*)	*oe-God-Nurtures*	Nothing causes an e-god to be nurturing.
27. close	(e-god, human)	*oe-Earthly-God*	e-god is physically close to humans.
28. nothing	(nothing) [Has no real meaning, but is useful.]	*oe-Nothing*	(Has no real meaning, but is useful.)
Sky origin:			
29. creates	(s-god, *o-Human-Alive*)	*os-God-Creates-Human*	An s-god causes (20).
30. mysterious-process	(*os-God-Creates-Human*)	*os-God-Creates-Mysteriously*	The process in (29) is a mysterious process.
31. offends	(human, s-god)	*os-Human-Offends-God*	Humans offend s-god (sometimes).
32. harms	(s-god, human)	*os-God-Harms-Human*	s-god harms humans (sometimes).
33. causes	(*os-Human-Offends-God, os-God-Harms-Human*)	*os-Offense-Causes-Punishment*	Offending s-god causes punishment.
34. supplicates	(human, s-god)	*os-Human-Supplicates*	Humans supplicate, pray, etc. to s-god.
35. helps	(s-god, human)	*os-God-Helps-Human*	s-god helps humans (sometimes).
36. causes	(*os-Human-Supplicates, os-God-Helps-Human*)	*os-God-Rewards*	Supplicating is what causes s-god to help.
37. distant	(s-god, human)	*os-Heavenly-God*	s-god is physically distant from humans.

```
; Semantic similarities between predicates, with negative values for antonymic
; relationships.  Note: *ident-weight* is a number which represents the semantic
; similarity between identical predicates in ACME.  Asterisks appearing just
; after a left parenthesis are multiplication operators; others are parts of names.
(similar 'helps 'harms (* -1 *ident-weight*))
(similar 'feels-power 'alive (* -.75 *ident-weight*))
(similar 'woman 'human (* .5 *ident-weight*))
(similar 'man 'human (* .5 *ident-weight*))
(similar 'distant 'distant-agent (* .5 *ident-weight*))
(similar 'close 'intimate-agent (* .5 *ident-weight*))
(similar 'offends 'harms (* .5 *ident-weight*))
; "Soft-iff" semantic-like relationships between propositions:
(semantic-iff 'oe-Earthly-God 'os-Heavenly-God -.5)
(semantic-iff 'oe-God-Creates-Naturally 'os-God-Creates-Mysteriously -.5)
(semantic-iff 'os-Heavenly-God 'os-God-Creates-Mysteriously .1)
(semantic-iff 'oe-God-Creates-Naturally 'oe-God-Creates-Human-From-Within .5)
(semantic-iff 'oe-God-Nurtures 'oe-God-Creates-Human-From-Within .1)
(semantic-iff 'oe-Human-From-Within-God 'oe-God-Creates-Human-From-Within .5)
(semantic-iff 'oe-God-Creates-Human-From-Within 'os-God-Creates-Human-From-Object -.5)
```

Figure 12 Input specifications for additional semantical relationships.

Competing interests

The author declares that there are no competing interests.

Author's contributions

All writing, program design, simulation runs, and analysis were performed by Marshall Abrams. As noted in the text, POPCO includes the ACME component, which was written primarily by Paul Thagard. Kristen Hammack added social network functionality to POPCO and modified procedures for communication between persons. (Social network functionality was not used in the simulations reported here, but Hammack's communication code was used.) All other coding of POPCO, and all coding for particular simulations were performed by Marshall Abrams.

Acknowledgements

I'm grateful for very helpful feedback from four anonymous reviewers, and patience and advice from Muaz Niazi. Paul Thagard made source code for ACME/COHERE developed by him along with others freely available. Kristen Hammack implemented POPCO's social network functionality (not used here), modified its communication processes, and did other work with POPCO. Hammack's work was funded by a Dean's Grant from the UAB College of Arts and Sciences and the UAB Office of Research and Economic Development. I'm grateful to the organizers, mentors, and participants in the 2011/2012 NEH Summer Institute on Computer Simulations in the Humanities, which provided a fertile environment for the initial development of this project and enriched my understanding of philosophical and methodological issues concerning complex adaptive systems. Thanks particularly to J. McKenzie Alexander, Ted Carmichael, and Chuck Turnitsa. For discussion and feedback on work leading up to this paper, I'm also grateful to Aaron Bramson, Alan Sprague, Andrew Keitt, Anthony Bowman, Armin Schultz, Bill Wimsatt, Brenda Smith, Chris Kyle, Daniel Singer, David Henderson, David Morrow, Emily Schultz, Graham Sack, Greg Pence, Harold Kincaid, Hilton Root, John T. Murphy, Julia Morley, Karsten Steuber, Marjorie Price, Marvin Croy, Maura Priest, Nicholas Payette, Omar Lizardo, Patrick Grimm, Paul Roth, Peter Gildenhuys, Phillip Honenberger, Rob Boyd, Ron Mallon, Russell Gonnering, Scott Weingart, Sharyn Jones, Steve Crowley, Steve Kimbrough, Ted Benditt, and students in my courses. Though it's not usually mentioned in acknowledgments, I'm among many who are grateful for excellent free software, in particular Steel Bank Common Lisp, GUESS, R, and NetLogo, not to mention LaTeX. Last, but not least, this work was supported in part by the research computing resources acquired and managed by UAB IT Research Computing. Access to UAB's Cheaha computing cluster was invaluable. (Any opinions, findings and conclusions or recommendations expressed in this material are those of the author and do not necessarily reflect the views of the University of Alabama at Birmingham.)

References

Acerbi A, Ghirlanda S, Enquist M: **Old and young individuals' role in cultural change.** *J Artif Soc Soc Simul* 2012, **15**(4):1. [http://jasss.soc.surrey.ac.uk/15/4/1.html]

Afshar M, Asadpour M: **Opinion formation by Informed agents.** *J Artif Soc Soc Simul* 2010, **13**(4):5. [http://jasss.soc.surrey.ac.uk/13/4/5.html]

Alam SJ, Geller A, Meyer R, Werth B: **Modelling contextualized reasoning in complex societies with "Endorsements".** *J Artif Soc Soc Simul* 2010, **13**(4):6. [http://jasss.soc.surrey.ac.uk/13/4/6.html]

Alexander JM: *The Structural Evolution of Morality.* Cambridge, UK: Cambridge University Press; 2007.

American National Standards Institute, Information Technology Industry Council: *Am Nat Stand Inf Tech: programming language — Common LISP.* InterNational Committee for Information Technology Standards 1996. [Web version available at http://www.lispworks.com/documentation/common-lisp.html]

Atran S, Medin D: *The Native Mind and the Cultural Construction of Nature.* Cambridge, Massachusetts: MIT Press; 2008.

Bird-David N: **The giving environment: another perspective on the economic system of gatherer-hunters.** *Curr Anthropology* 1990, **31**(2):189–196.

Bartha PFA: *By Parallel Reasoning: The Construction and Evaluation of Analogical Arguments.* Oxford, UK: Oxford University Press; 2010.

Bassok M, Dunbar KN, Holyoak KJ: **Introduction to the special section on the neural substrate of analogical reasoning and metaphor comprehension.** *J Exp Psych: Learning, Memory, Cogn* 2012, **38**(2):261–263.
Bleda M, Shackley S: **Simulation modelling as a theory building tool: the formation of risk perceptions.** *J Artif Soc Soc Simul* 2012, **15**(2):2. [http://jasss.soc.surrey.ac.uk/15/2/2.html]
Boyd R, Richerson PJ: *Culture and the Evolutionary Process.* Chicago, IL: University of Chicago; 1985.
Boyd R, Richerson PJ: *The Origin and Evolution of Cultures.* Oxford, UK: Oxford University Press; 2005.
Brandon RN: *Adaptation and Environment.* Princeton, New Jersey: Princeton University; 1990.
Brown MJ, Feldman MW: **Sociocultural epistasis and cultural exaptation in footbinding, marriage form, and religious practices in early 20th-Century Taiwan.** *Proc Natl Acad Sci* 2009, **106**(52):22139–22144.
Cavalli-Sforza LL, Feldman MW: *Cultural Transmission and Evolution: A Quantitative Approach.* Princeton, New Jersey: Princeton; 1981.
Clark S: *Thinking With Demons: The Idea of Witchcraft in Early Modern Europe.* Oxford, UK: Oxford University Press; 1999.
Chettih S, Durgin FH, Grodner DJ: **Mixing Metaphors in the Cerebral Hemispheres: What Happens when Careers Collide?** *J Exp Psych: Learning, Memory, and Cogn* 2012, **38**(2):295–311.
De Block A, Cuypers SE: **Why Darwinians should not be afraid of Mary Douglas—and vice versa: the case of disgust.** *Phil Soc Sci* 2012, **42**(4):459–488.
Dehghani M, Tomai E, Forbus K, Klenk M: **An integrated reasoning approach to moral decision-making.** In *Proceedings of the Twenty-Third AAAI Conference on Artificial Intelligence,* Menlo Park, California: AAAI; 2008.
Dehghani M, Sachdeva S, Ekhtiari H, Gentner D, Forbus K: **The role of cultural narratives in moral decision making.** In *Proceedings of the 31th Annual Conference of the Cognitive Science Society.* Austin, Texas: University of Texas; 2009. [http://csjarchive.cogsci.rpi.edu/Proceedings/2009/papers/397/index.html]
Dressler WW, Balieiro MC, Ribeiro RP, dos Santos JE: **A prospective study of cultural consonance and depressive symptoms in urban Brazil.** *Soc Sci & Med* 2007, **65:**2058–2069.
Eliasmith C, Thagard P: **Integrating structure and meaning: a distributed model of analogical mapping.** *Cogn Sci* 2001, **25:**245-286.
Enquist M, Eriksson K, Ghirlanda S: **Critical social learning: a solution to Rogers's paradox of nonadaptive culture.** *Am Anthropol* 2010, **109**(4):727–734.
Epstein JM: *Generative Social Science: Studies in Agent-Based Computational Modeling.* Princeton, New Jersey: Princeton University Press; 2006.
Forbus KD: **Exploring analogy in the large.** Edited by Gentner D, Holyoak K, Kokinov B. Cambridge, MA: MIT Press; 2001:23–58.
Fracchia J, Lewontin RC: **Does culture evolve?** *Hist Theory* 1999, **38**(4):52–78.
Geertz C: *The Interpretation of Cultures: Selected Essays.* Oxford, UK: Oxford University Press; 1973.
Geertz, C: **Deep play: notes on the Balinese cockfight.** In *The Interpretation of Cultures: Selected Essays* Oxford, UK; 1973.
Gentner D, Bowdle B, Wolff P, Boronat C: **Metaphor is like analogy.** In *The analogical mind: Perspective from cognitive science.* Edited by Gentner D, Holyoak KJ, Kokinov BN. Cambridge, MA: MIT Press; 2001:199–253.
Gentner D, Holyoak KJ, Kokinov BN (Eds): *The Analogical Mind: Perspectives from Cognitive Science.* Cambridge, MA: MIT Press; 2001.
Gentner D, Forbus KD: **Computational models of analogy.** *WIREs Cogn Sci* 2011, **2:**266–276.
Gibbs RW Jr, Colson HL: *Interpreting Figurative Meaning.* Cambridge, UK: Cambridge University Press; 2012.
González RJ: *Zapotec Science.* Austin, Texas: University of Texas; 2001.
Gould SJ, Vrba ES: **Exaptation—a missing term in the science of form.** *Paleobiology* 1982, **8:**4–15.
Green AE, Kraemer DJM, Fugelsang JA, Gray JR, Dunbar KN: **Neural correlates of creativity in analogical reasoning.** *J Exp Psych: Learning, Memory, and Cogn* 2012, **38**(2):264–272.
Grim P, Kokalis T, Alai-Tafti A, Kilb N, St Denis P: **Making meaning happen.** *J Exp & Theor Artif Intell* 2004, **16**(4):209–243.
Grim P, Rosenberger R, Rosenfeld A, Anderson B, Eason RE: **How simulations fail.** *Synthese* 2011:1–24. [http://dx.doi.org/10.1007/s11229-011-9976-7]
Grossberg S: **A theory of visual coding, memory, and development.** In *Formal theories of visual perception*. Edited by Leeuwenberg ELJ, Buffar HFJ. Chichester; New York: Wiley; 1978.
Hesse MB: *Models and Analogies in Science.* Notre Dame, Indiana: University of Notre Dame Press; 1966.
Hofstadter DR, Sander E: *Surfaces and Essences: Analogy as the Fuel and Fire of Thinking.* New York: Basic Books; 2013.
Holyoak KJ: **An analogical framework for literary interpretation.** *Poetics* 1982, **11**(2):105–126.
Holyoak KJ, Thagard P: *Mental Leaps: Analogy in Creative Thought.* Cambridge, Massachusetts: MIT Press; 1995.
Holyoak KJ, Thagard P: **Analogical mapping by constraint satisfaction.** *Cogn Sci* 1989, **13:**295–355.
Hurst LD, Peck JR: **Recent advances in understanding of the evolution and maintenance of sex.** *Trends Ecol Evol* 1996, **11**(2):46–52.
Kahan DM: **Cultural cognition as a conception of the cultural theory of risk.** In *Handbook of Risk Theory: Epistemology, Decision Theory, Ethics and Social Implications of Risk.* Edited by Hillerbrand R, Sandin P, Roeser S, Peterson M. Dordrecht, The Netherlands: Springer; 2012:725–760.
Kashima Y: **Maintaining cultural stereotypes in the serial reproduction of narratives.** *Pers Soc Psychol Bull* 2000, **26**(5):594–604.
Klüver J, Klüver C: *Social Understanding: On Hermeneutics, Geometrical Models and Artificial Intelligence.* Dordrecht, The Netherlands: Springer; 2010.
Knowlton B J, Morrison RG, Hummel JE, Holyoak KJ: **A Neurocomputational system for relational reasoning.** *Trends Cogn Sci* 2012, **16**(7):373–381.
Lakoff G: *Moral Politics: How Liberals and Conservatives Think.* 2nd edition. Chicago, Illinois: University of Chicago Press; 2002.
Lakoff G, Johnson M: *Metaphors We Live By.* 2nd edition. Chicago, Illinois: University of Chicago Press; 2003.
Lakoff G: *Moral Politics: How Liberals and Conservatives Think.* 2nd edition: University of Chicago Press; 2002.
Lansing JS: *Perfect Order: Recognizing Complexity in Bali.* Princeton, New Jersey: Princeton University; 2006.
Lansing JS: *Priests and Programmers: Technologies of Power in the Engineered Lanscape of Bali.* Princeton, New Jersey: Princeton University; 2007. [1991]. [First edition 1991. Reprinted, with a new foreward and preface, 2007]

Lehrer K, Wagner C: *Rational Consensus in Science and Society.* Dordrecht, The Netherlands: D. Reidel Publishing; 1981.
Lienhardt G: *Divinity and Experience: The Religion of the Dinka.* Oxford, UK: Oxford University Press; 1961.
Maguire MJ, McClelland MM, Donovan CM, Tillman GD, Krawczyk DC: **Tracking cognitive phases in analogical reasoning with event-related potentials.** *J Exp Psych: Learning, Memory, and Cogn* 2012, **38**(2):273–281.
Mantzavinos C: *Naturalistic Hermeneutics.* Cambridge, UK: Cambridge University Press; 2005. [Translated from German by Darrell Arnold in collaboration with the author]
Morris S: **Contagion.** *Rev Econ Stud* 2000, **67:**57–78.
Mueller ST, Simpkins B, Rasmussen L: **Incorporating representation when modeling cultural dynamics: Analysis of the bounded influence Conjecture.** In *Proceedings of the Workshop on Cognitive Social Sciences: Grounding the Social Sciences in the Cognitive Sciences?* Edited by Sun R. Troy, NY 12180, USA: Rensselaer Polytechnic Institute; 2010:29–34. [Technical Report 2010-RS-0001]
Nakahashi W: **Evolution of learning capacities and learning levels.** *Theor Popul Biol* 2010, **78:**211–214.
Peña A: **On the role of mathematical biology in contemporary historiography.** *Hist Theory* 1999, **38**(4):101–120.
Picinali F: **Structuring inferential reasoning in criminal fact finding: an analogical theory.** *Law, Probability, and Risk* 2011, **11:**197–223.
Prat CS, Mason RA, Just MA: **An fMRI investigation of analogical mapping in metaphor comprehension: the influence of context and individual cognitive capacities on processing demands.** *J Exp Psych: Learning, Memory, and Cogn* 2012, **38**(2):282–294.
Reynolds RG: **An introduction to cultural algorithms.** In *Proceeedings of the Third Annual Conference on Evolutionary Programming.* Edited by Sebald A, Fogel LJ. Singapore: World Scientific Press; 1994:131–139.
Reynolds RG, Ali M: **Computing with the social fabric: the evolution of social intelligence within a cultural Framework.** *IEEE Comput Intell Mag* 2008, **3**(1):18–30.
Romney AK, Weller SC, Batchelder WH: **Culture as consensus: a theory of culture and informant accuracy.** *Am Anthropologist* 1986, **88**(2):313–338. [New Series]
Sanday PR: *Female Power and Male Dominance.* Cambridge, UK: Cambridge University Press; 1981.
Schultz E: **Resolving the anti-antievolutionism dilemma: a brief for relational evolutionary thinking in anthropology.** *Am Anthropologist* 2009, **111**(2):224–237.
Slingerland E: *What Science Offers the Humanities: Integrating Body and Culture.* Cambridge, UK: Cambridge University Press; 2008.
Spellman BA, Holyoak KJ: **If Saddam is Hitler then who is George Bush? analogical mapping between systems of social roles.** *J Pers Soc Psych* 1992, **62**(6):913–933.
Sperber D: *Explaining Culture: A Naturalistic Approach.* Oxford, UK: Blackwell; 1996.
Sun R (Ed): *Cognition and Multi-Agent Interaction.* Cambridge, UK: Cambridge University Press; 2006.
Thagard P: *Coherence in Thought and Action.* Cambridge, Massachusetts: MIT Press; 2000.
Thagard P: **Mapping minds across cultures.** In *Grounding Social Sciences in Cognitive Sciences.* Edited by Sun R. Cambridge, Massachusetts: MIT Press; 2012:35–62.
Thagard P: *Conceptual Revolutions.* Princeton University Press; 1992.
Tilley C: *Metaphor and Material Culture.* Oxford, UK: Wiley-Blackwell; 2000.
Thibodeau PH, Boroditsky L: **Metaphors we think with: the role of metaphor in reasoning.** *PLoS ONE* 2011, **6**(2):e16782. [http://dx.doi.org/10.1371%2Fjournal.pone.0016782]
Wilensky U: **NetLogo** 1999. [http://ccl.northwestern.edu/netlogo] [Computer modeling environment].
Wimsatt WC: **Genes, memes, and cultural heredity.** *Bio & Philo* 1999, **14:**279–310.
Wimsatt WC, Griesemer JR: **Reproducing entrenchments to scaffold culture: the central role of development in cultural evolution.** In *Integrating Evolution and Development.* Edited by Sansom R, Brandon RN. Cambridge, Massachusetts: MIT Press; 2007:227–323.
Vega-Redondo F: *Complex Social Networks.* Cambridge, UK: Cambridge University Press; 2007.

7

Information and phase transitions in socio-economic systems

Terry Bossomaier[1*†], Lionel Barnett[2†] and Michael Harré[3†]

*Correspondence: tbossomaier@csu.edu.au
†Equal contributors
[1]Centre for Research in Complex Systems, Charles Sturt University, Panorama Ave, Bathurst NSW 2795, Australia
Full list of author information is available at the end of the article

Abstract

We examine the role of information-based measures in detecting and analysing phase transitions. We contend that phase transitions have a general character, visible in transitions in systems as diverse as classical flocking models, human expertise, and social networks. Information-based measures such as mutual information and transfer entropy are particularly suited to detecting the change in scale and range of coupling in systems that herald a phase transition in progress, but their use is not necessarily straightforward, possessing difficulties in accurate estimation due to limited sample sizes and the complexities of analysing non-stationary time series. These difficulties are surmountable with careful experimental choices. Their effectiveness in revealing unexpected connections between diverse systems makes them a promising tool for future research.

Keywords: Phase transitions; Mutual information; Transfer entropy; Social networks; Stock markets; Expertise

Review

Diamonds are not a good very long term investment! They are steadily turning into graphite. It will take millions of years, but the most stable form of carbon at room temperature and pressure is graphite. Thus diamonds will undergo a phase transition to graphite, albeit over a very long timescale.

When we normally think of phase transitions we think of the states of matter, ice melting to water or water turning to steam. They are order/disorder transitions. In graphite the carbon atoms are linked together in layers. The layers can slide over one another giving graphite its excellent lubricant properties. In diamond the carbon atoms are linked together in a three dimensional structure with each carbon at the centre of a tetrahedron linked to carbons at all four corners. Thus carbon has to go through a major structural reorganization to change from diamond to graphite – the way the atoms are connected together changes dramatically.

We can easily see the outcomes of a phase transition of diamond to graphite or a solid turning into a liquid. But can we construct measures which go through a minimum or maximum at the transition? This turns out to be a surprisingly difficult question to answer. It gets even more difficult if we look for measures which can apply to systems in general, not just to the physical systems above. Organizations, societies, economies, stock markets all go through radical reorganization, but should these changes be called phase transitions? This paper explores a metric based on information theory (Shannon 1948a) which is empirically a quite general indicator of a phase transition in progress. It then considers a much newer metric which, on examples to date, might be a predictor of impending transitions: the crashing of stock markets is an example of such a transition, early warning of which, would be highly desirable.

In this paper we explore two closely related phase transitions which appear in social economic systems. One is herding behaviour, where individuals behave more and more in the same way. The second is the connectivity avalanche, in which all of the elements of a system are becoming connected to one another.

To begin with we explore theoretical issues. Section "Overview of phase transitions and metrics" discusses the characteristics of phase transitions and the the peak in mutual information that usually accompanies them. It also introduces the idea of transfer entropy, a recent extension to mutual information, which in some cases is known to peak before a transition to synchronous behaviour. With this background, the first example, in section "Mutual information for phase transitions in a simple flocking model", considers a computational example derived from physics. The computation of mutual information from continuous data is tricky, involving difficult decisions on bin sizes and other statistical issues, which are also considered in this section. The next two sections discuss phase transitions in two areas in the social/humanities domain. Section "Phase transitions in socio-economic systems" discusses how peaks in mutual information occur around stock market crashes. Section "Phase transitions in the acquisition of human expertise" discusses the reorganization of strategy in the human brain during the acquisition of expertise. Section "Inferring social networks with transfer entropy" discusses how transfer entropy is calculated in practice using the example of inferring social networks from time series data. Finally we conclude in section "Conclusions" with some opportunities for further work.

Overview of phase transitions and metrics

The simple physical notion of a phase transition, such as ice melting to water, is surprisingly hard to transfer to non-physical systems, such as society and organisations. This section will try to first look at the physical intuition behind the transition and then move on to look at some of the possible metrics.

The essential feature of a phase transition is a structural reordering of some kind, usually an order-disorder change. This usually involves some sort of long range order – everything gets connected so that things can be reconnected in a different way. In physical systems, we can define an *order parameter*. Transitions occur around particular values of the order parameter. We can make this idea more intuitive by looking at random graphs, section "Random graphs and phase transitions".

A peak in *mutual information*, (section "Mutual information" is a widespread measure of a phase transition (Gu et al. 2006; Wilms et al. 2011). Mutual information theory is a very general indicator of critical transition, which incorporates all correlation effects and all nonlinear behaviour as well. Demonstrations for specific systems include the Vicsek model of section "Vicsek model background" and work by Matsuda et al. (1996), who obtained the mutual information for Ising systems, demonstrating that it was a good measure for critical behaviour in such systems that have second order phase transitions. There are, however, two distinct classes of phase transitions that are relevant to this discussion.

Second order phase transitions are indicated by peaks in mutual information and whereas *first order* phase transitions are indicated by discontinuities in the entropy, rather than the mutual information, see (Solé et al. 1996) for a brief discussion of the general differences in such systems. First order phase transitions are caused by discontinuities in the first derivative of a thermodynamic variable rather than the second derivative as is the case with second order phase-transitions, hence their names. For a discussion of these issues in terms of complex systems, and computation at the edge of chaos in particular, see (Langton 1990a).

There are other metrics. Recently, for example, *Fisher Information*, has been used in this context by Prokopenko et al. (2011). Another common indicator is *critical slowing down*, where the time taken to respond to perturbation increases near a transition (Dai et al. 2012), often accompanied by logarithmic oscillations. The latter have been extensively studied in stock markets by Sornette (2001).

Although the transitions referred to so far are of a singular nature, there are other mechanisms. In some cases a system may *flicker*, across the transition (Scheffer et al. 2012), spending increasing amounts of time in the alternative state. Thus flickering can also be an indicator of an impending irreversible system change.

The attraction of mutual information for this paper is its intuitive link to large scale order near the transition and its close relationship to *transfer entropy* (section "Transfer entropy and granger causality"). Although mutual information and related indicators co-occur with phase transitions across many systems, they have two shortcomings in terms of *prediction:* the precise timing depends on many factors and an exact prediction of when a transition will occur is fraught with difficulty; and the sign of a transition is not determined. Work remains to be done on how the peak in mutual information relates to the ordered and dis-ordered phases.

Random graphs and phase transitions

The idea of a random graph, introduced by Erdős and Rényi (1960), is to start with a set of N, nodes and add edges at random. At first the edges create small graph fragments. The total number of possible edges (without duplicates) is of the order of N^2 but for quite a small number of edges, of order N, large components form. Now as adding an edge may join two components together, the total connectivity rises very rapidly until every node in the graph becomes connected by some path to every other node. This rapid rise is referred to as the connectivity avalanche and represents a phase transition.

Before the connectivity avalaanche, very few nodes are connected. Since the connected components are small the path lengths between nodes are also small. But as the components get rapidly bigger during the avalanche, more and more nodes become connected, but there is often only a single long path between them. As more nodes are added after the connectivity avalanche, they effectively provide short cuts and the average path length goes down again. This increase in path length is the analogue of the long range order seen during a phase transition.

As Green (Green and Newth 2005) showed, random graphs underlie many complex systems; thus a rigorous mapping to a random graph, demonstrating an isomorphism, guarantees that the system will have a phase transition. Although Green has done this for some example systems, the underlying principle is generally useful even where an exact isomorphism has not been demonstrated.

Mutual information

Shannon (1948a) defined *mutual information* in the course of obtaining the maximum information which could be transmitted across a channel. It is a functional, mapping from probability distributions of random variables to scalars. In Shannon's formulation, for a random variable X defined on a discrete alphabet with probability mass function $p(x)$, the information, q_i, sometimes called the *surprisal*, obtained from an event x_i is given, in bits, by Eq. 1.

$$q_i = -log_2(p(x_i)) \tag{1}$$

If we average the information according to the probablity of each event occurring, we end up with the Shannon *entropy*, Eq. 2.

$$H(X) \equiv -\sum_x p(x) \log_2 p(x) \tag{2}$$

With some caveats, this has the familiar interpretation as the average length of the shortest description of an observation from that variable, or how 'uncertain' a particular random variable is (Cover and Thomas 2006). *Mutual* information between two random variables, X and Y with joint probability mass function $p(x, y)$ is given

$$I(X; Y) \equiv H(X) + H(Y) - H(X, Y) \tag{3}$$

$$= -\sum_{x,y} p(x, y) log_2 \frac{p(x, y)}{p(x)p(y)} \tag{4}$$

Mutual information is thus a functional of the *joint* distribution of X and Y. It is symmetric in X and Y, and can be given a natural interpretation as

> the reduction in uncertainty in one variable from knowing the other, or the amount of information about X contained in Y.

Derivations and implications of these properties are given elsewhere (Cover and Thomas 2006)[a]. It forms a measure of interdependence of two variables. Unlike correlation, however, it is sensitive to nonlinear interactions, works on general nonparametric inference, and naturally performs well on discrete data. These qualities have led to an interest in the use of mutual information-based measures to automatically detect diverse classes of associations in data sets with few assumptions as to the functional form of the relationship (Reshef et al. 2011; Slonim et al. 2005).

Transfer entropy and granger causality

Mutual information (and its conditional variant), applied to stochastic processes, measure *contemporaneous* statistical dependencies between stochastic dynamical processes evolving in time; that is, given joint processes X_t, Y_t, the mutual information $I(X_t : Y_t)$ *at a given time t* might be read as:

> The amount of information about the present of X resolved by the present of Y.

However, it would seem desirable to have an information-theoretic measure capable of capturing *time-directed* statistical dependencies between processes - *information flow*, if you will. To this end, the most obvious extension to contemporaneous mutual information is *time-delayed* (lagged) mutual information between processes. Thus one might consider, say, $I(X_t : Y_{t-1}, Y_{t-2}, \ldots)$ as a candidate measure for directed information flow from the past[b] of Y to the present of X. This quantity might be read as:

> The amount of information about the present of X resolved by the past of Y.

There is, however, a fundamental shortcoming to lagged mutual information as a measure of directed information flow: it fails to take into account *shared history* between processes. In his seminal paper (Schreiber 2000), Schreiber recognised that this could lead to spurious imputation of directed information flow and introduced a new measure (Kaiser and Schreiber 2002; Schreiber 2000) which explicitly takes account of a "shared past" between processes. Formally, the *transfer entropy* from process Y_t to process X_t may be defined as:

$$T_{Y \to X} \equiv I(X_t : Y_{t-1}, Y_{t-2}, \ldots | X_{t-1}, X_{t-2}, \ldots) \tag{5}$$

Thus in contrast to lagged mutual information, the past of the process X is conditioned out. Eq. (5) might be read as:

> The amount of information about the present of X resolved by the past of Y *given the past of* X.

An elegant, minimal example illustrating the necessity of conditioning out the past of the target process may be found in (Kaiser and Schreiber 2002). Note that the processes X_t, Y_t in (5) may be fully *multivariate*. Furthermore, given a third (possibly multivariate) jointly stochastic process Z_t, say, any common influence of Z_t on X_t and Y_t may be taken into account by conditioning, in addition, on the past of Z:

$$T_{Y \to X|Z} \equiv I(X_t : Y_{t-1}, Y_{t-2}, \ldots | X_{t-1}, X_{t-2}, \ldots, Z_{t-1}, Z_{t-2}, \ldots \tag{6}$$

This quantity is known as *conditional transfer entropy*, and is in particular used to define *pairwise-conditional* information flows

$$T_{X_i \to X_j | X_{[ij]}} \qquad i \neq j \tag{7}$$

between component variables $X_i \to X_j$ of the system, conditioned on the remaining variables $X_{[ij]}$, where the notation denotes omission of the variables X_i, X_j. The set of pairwise-conditional information flows may be viewed as a weighted, directed graph, the i, jth entry quantifying information flow between individual elements X_i, X_j of the system.

Clive Granger won the Nobel Prize in economics for a closely related *parametric* measure in econometric theory (and, more recently, applied extensively to neural time series data (Ding et al. 2006)), the Wiener-Granger causality (Geweke 1982; Granger 1969; Wiener 1956). Here, rather than "information flow", the measure is designed to reflect "causal" influence of one process on another, premised on a notion of causality whereby a causal effect temporally precedes and *helps predict* its influence. The measure is based on linear vector autoregressive (VAR) modelling: suppose that the (again, possibly multivariate) "predictee" process X_t is modelled as a vector linear regression on its own past, as well as on the past of the "predictor" process Y_t:

$$X_t = A_1 X_{t-1} + A_2 X_{t-2} + \ldots + B_1 Y_{t-1} + B_2 Y_{t-2} + \ldots + \varepsilon_t \tag{8}$$

We may consider that, given regression coefficients A_k, B_k, (8) furnishes a *prediction* for the present of the variable X, based on its own past and that of the variable Y. A suitable measure for the magnitude of the *prediction error* is given by the *generalised variance*

(Barett et al. 2010), defined as the determinant $|\text{cov}(\varepsilon_t)|$ of the covariance matrix of the residuals ε_t. Given a realisation of the processes, it may be shown that the generalised variance is proportional to the *likelihood* of the model (8); regression coefficients A_k, B_k may thus be calculated within a maximum likelihood framework so as to minimise the generalised variance[c].

We may now compare the regression (8) with a prediction model for X based only on its own past:

$$X_t = A_1' X_{t-1} + A_2' X_{t-2} + \ldots + \varepsilon_t' \tag{9}$$

We then say that Y *Granger-causes* X iff the full model (8) furnishes a significantly better prediction than the reduced model (9). By "significantly better", we mean that the null hypothesis of zero causality:

$$H_0 : B_1 = B_2 = \ldots = 0 \tag{10}$$

should be rejected at a given significance level. Now the linear regression model (9) is *nested* in (i.e. is a special case of) the model (8), and standard theory (Hamilton 1994) tells us that the appropriate statistical test for the null hypothesis H_0 of (10) is a likelihood-ratio test. The Granger causality statistic is then formally the log-likelihood ratio

$$\mathrm{F}_{Y \to X} \equiv \log \frac{|\text{cov}(\varepsilon_t')|}{|\text{cov}(\varepsilon_t)|} \tag{11}$$

This quantity may be read as (*cf.* transfer entropy):

> The degree to which the past of Y helps predict the present of X *over and above the degree to which X is already predicted by its own past.*

By another classical result (Wilks 1938) the asymptotic distribution of $\mathrm{F}_{Y \to X}$ under the null hypothesis is χ^2 with number of degrees of freedom equal to the difference in the number of parameters between the full and reduced models; this enables significance testing of Granger causality[d]. Parallel to the transfer entropy case, a third jointly distributed process Z_t may be conditioned out by appending its past to both the full and reduced regressions, yielding the *conditional Granger causality* (Geweke 1984) $\mathrm{F}_{Y \to X|Z}$. As in (7), pairwise conditional causalities

$$\mathrm{F}_{X_i \to X_j | X_{[ij]}} \quad i \neq j \tag{12}$$

may be calculated[e].

A powerful feature of Granger causality is that it admits a *spectral decomposition*; that is, Granger-causal influence may be measured at specific frequencies, or in specific frequency bands:

$$\mathrm{F}_{Y \to X} = \frac{1}{2\pi} \int_{-\pi}^{\pi} \mathrm{f}_{Y \to X}(\omega)\, d\omega \tag{13}$$

where $\mathrm{f}_{Y \to X}(\omega)$ is the Granger causality at frequency ω. We refer the reader to (Barnett and Seth 2011; Geweke 1982, 1984) for definitions and details.

For both transfer entropy and Granger causality, the issue of *stationarity* arises. Although formally both quantities are well-defined for *non*stationary processes—the result then depends on the time t—empirically, estimation will generally require stationarity. The exception is where multiple synchronised realisations of the processes are available, but this is rarely the case in practice. Otherwise, nonstationarity must be dealt with by *windowing* time series data; that is, dividing it into approximately stationary segments. Then a trade-off must be made between shorter window length, where estimation suffers through lack of data, and longer window length, where stationarity may be only approximate; in either case there is a risk of spurious inference. Pre-processing (e.g. detrending, differencing, filtering,...) may help improve stationarity.

Both measures, it may be noted, are invariant under a rather broad class of transformations. Barnett et al. (Barnett et al. 2011; Geweke 1982), showed that Granger causality is invariant under arbitrary stable, invertible digital filtering. Transfer entropy is invariant under a still wider class of nonlinear invertible transformations involving lags of the respective time series. In practice, though, even theoretically invariant transformations may impact causal inference (Barnett and Seth 2011).

Regarding the relationship between transfer entropy and Granger causality, it is proven in (Barnett et al. 2009a) that in the case where all processes have a jointly multivariate Gaussian distribution, the measures are entirely equivalent (and that, furthermore, a stationary Gaussian autoregressive process must be linear (Barnett et al. 2010)). Where the measures differ markedly is in the type of data to which they are naturally applicable, and the ease with which empirical estimation may be effected. Granger causality, based as it is on linear regression, is in general not suited to causal inference of *discrete-valued* data. On the other hand, for continuous-valued data, estimation of Granger causality is generally straightforward, as the comprehensive and well-understood machinery of linear regression analysis applies. There are, furthermore, mature and reliable software packages available for Granger casual estimation (Cui et al. 2008; Seth 2010). Further advantages of Granger causality are that (i) insofar as it is model-based with a known likelihood function, standard techniques of *model order estimation* (such as the Akaike or Bayesian Information Criteria (McQuarrie and Tsai 1998)) may be deployed, and (ii) asymptotic distributions for the sample statistic are known. For transfer entropy it is not clear how much history should be included in estimates, nor how the statistic may be easily significance tested, beyond standard non-parametric (but computationally intensive) methods such as permutation testing (Anderson and Robinson 2001; Edgington 1995). However, recent work in this area by Barnett and Bossomaier (2012), goes some way to resolving these issues. If a fairly general parametric model is assumed, then the transfer entropy again becomes a log-likelihood ratio, as for the Granger Causality. Moreover, its distribution is χ^2, enabling straightforward statistical estimates.

It is also known (Kaiser and Schreiber 2002) that, for continuous variables, in-sample estimation of transfer entropy is problematic insofar as (in contrast to mutual information) the sample statistic in general fails to converge under naïve refining of state-space partitioning. Thus more sophisticated but less well-understood (and computationally expensive) techniques such as adaptive partitioning or kernel estimation must be used. Nonetheless, transfer entropy is attractive due to its non-parametric "model agnostic" nature, particularly when interactions are likely to be highly nonlinear and hence unsuited to linear modelling. Table 1 shows a summary comparison of the measures.

A promising application of transfer entropy is in the construction of information-theoretic *complexity* measures. In (Tononi et al. 1994) the authors introduce a "neural complexity" metric $C_N(X)$ designed to capture a notion of network complexity based on *integration/segregation balance* (Barnett et al. 2009b). The idea is that a multi-element dynamical system exhibiting "complex" behaviour will tend to lie somewhere between

Table 1 Comparison between transfer entropy and Granger causality

Feature	Transfer entropy	Granger causality
Parametric	No	Yes: linear VAR model
Predictive	No	Yes
Frequency decomposition	No	Yes
Transformation-invariance	Nonlinear filter	Linear filter
Estimation in sample	Hard	Easy
Model order estimation	Unknown	AIC, BIC, cross-validation...
Statistical distribution	Unknown	Known (asymptotic)
Software implementation	*Ad hoc*	Many available packages

extremes of highly integrated, where every element tends to affect every other, and highly segregated, where elements behave almost independently. In the former case, a system will generally behave chaotically, while in the latter it will tend to decompose into simple independent processes (*cf.* "edge-of-chaos" phenomena (Langton 1990b)). These correspond to the random graph with few edges and the highly connected graph in section "Random graphs and phase transitions". The complex behaviour lies at the phase transition, or connectivity avalanche.

The original Tononi-Sporns-Edelman measure $C_N(X)$ averages mutual information across bipartitions of the system (Figure 1, left-hand figure); it is, however, extremely unwieldy to calculate in practice and, moreover, fails to capture information *flow* as expounded above.

Seth (Seth et al. 2011, 2006) has developed an alternative Granger causality-based measure, *causal density*, which is both more computationally manageable and captures complexity as mediated by *time-directed* influences; it admits a transfer entropy-based analogue:

$$\mathrm{cd}(X) \equiv \frac{1}{n(n-1)} \sum_{i \neq j} \mathrm{T}_{X_i \to X_j | X_{[ij]}} \tag{14}$$

where n is the number of variables; i.e. causal density is the average of the pairwise-conditional information flows (7) (Figure 1, right-hand figure). Again, cd(X) captures integration/segregation balance: for a highly integrated system the measure assumes a low value, since for each pair of variables X_i, X_j much of the information flow $X_i \to X_j$ is already resolved by the remaining variables $X_{[ij]}$, and conditioned out. For a highly segregated system the measure also assumes a low value, since the lack of coupling between variables results in comparatively few significant pairwise information flows.

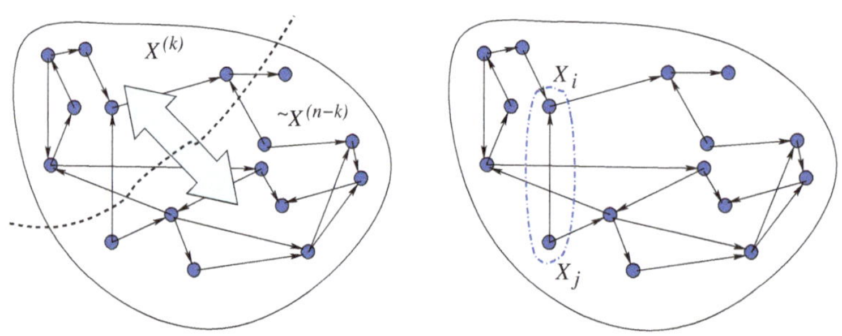

Figure 1 The Tononi-Sporns-Edelman neural complexity measure $C_N(X)$ (left figure) and Seth's causal density cd(X) (right figure).

Computational aspects of mutual information

There are numerical problems associated with the computation of mutual information. Of these there are two very obvious and related issues. The first is a consequence of the naive, i.e. fixed and equidistant, discretisation: some of the discrete bins (i) may contain no elements and therefore have a probability $p_i = 0$ while at the same time some $p_{i,j} \neq 0$ and in these cases the mutual information diverges, therefore both p_i and p_j need to be *absolutely continuous* (Lin 1991) with respect to each other to avoid such issues. Next we discuss how both of these issues can be addressed in practice, the first by defining a form of mutual information that is naturally continuous and therefore does not require an ad hoc discretisation step and the other addresses the case where, in a naturally discrete system numerical artefacts resulting in divergences can be avoided.

The choice of number of bins, bin boundaries and which estimator to use are all the subject of intensive research. Techniques such as adaptive partitioning (Cellucci et al. 2005; Darbellay and Vajda 1999) exist to optimise the bin selection. Bias-corrected estimators from histograms are also numerous, including shrinkage (Hausser and Strimmer 2009), Panzeri-Treves (1996), the Nemenmen-Shafee-Bialek (Nemenman et al. 2002; Wolpert and Wolf 1995) estimator, quadratic extrapolation (Strong et al. 1998).

For this experiment, where large data sets are to be handled, the most computationally rapid choices are paramount. To this end, Wicks et al. (2007) selected equal-width bins, using an intuitive length scale of $2R$ as the width of the bin (with number of bins equal to the $L2R$). Discretising the signal into bins, we calculate MI using an uncorrected *plugin estimator* between (presumed scalar) variables X and Y, based on occupancy values of the joint histogram. The term "plugin estimator" is used in the standard statistical sense: we simply substitute our estimated values for the distribution into the formula for mutual information, rather than construct an estimator by other means such as maximum likelihood.

Cellucci et al. (2005) show that equal bins can give very poor results, particularly when the datasets are small. They propose an adaptive binning algorithm in which the bins for X and Y are sized so that each contains the same expected number of values. The computational overhead of this procedure is at worst $n \log n$, corresponding to sorting each of the X and Y values.

Explicitly, for the histogram matrix Q of bin occupancy counts, we estimate the joint probability mass function p by normalising

$$\hat{p}_{X,Y}(x,y) = \frac{Q_{xy}}{\sum_i \sum_j Q_{ij}} \tag{15}$$

Then marginal probabilities can be estimated

$$\hat{p}_X(x) = \sum_j \hat{p}_{X,Y}(x,j)\hat{p}_Y(x,y) = \sum_i \hat{p}_{X,Y}(i,y) \tag{16}$$

from which the mutual information follows from Eq. 4.

For N_E adaptive bins along each axis the MI estimate simplifies to

$$\hat{I}(X,Y) = \sum_x \sum_y \hat{p}(x,y) \log_2 N_e^2 \hat{p}(x,y) \tag{17}$$

where N_e is the number of bins.

This estimator is subject to finite sample bias (Treves and Panzeri 2008), but for the purely comparative purposes of our experiment is empirically sufficient, and computationally very cheap.

The first is that while mutual information is well defined for continuous systems as shown at least as early as (Shannon 1948a), this is often difficult to implement in practice as it does not give rise to an obvious algorithm that does not first involve an initial discretisation step, thereby losing the information content of possibly important relationships.The second issue is that of systems that are naturally continuous. The continuous analogue of entropy and mutual information given in equations 2 and 3, the derivation of the entropy can be found in (Shannon 1948b):

$$H_{cont}(X) = -\int_X p(x)\log(p(x))dx \tag{18}$$

$$I_{cont}(X;Y) = \int_X\int_Y p(x)p(y)\log\left(\frac{p(x,y)}{p(x)p(y)}\right)dxdy \tag{19}$$

where x and y are continuous variables. Such continuous systems occur in research on channel coding (MacKay 2003), chaotic dynamics (Granger and Lin 1994) and signal detection theory (Parra et al. 1996) amongst many others. A natural first estimate of 19 is to divide the continuous spaces X and Y into fixed bin sizes of δX and δY and uses the count of the occupancy of each bin in order to estimate the probability of bin occupancy. This has been used successfully in many different studies that have discretised continuous variables but it has been pointed out that it is often difficult to estimate the mutual information with such an approach (Kraskov et al. 2004). Other estimators include Gaussian-kernel inference, (Moon et al. 1995), spline estimators (Daub et al. 2004), and k-nearest neighbour estimators (Kraskov et al. 2004).

An alternative that retains much of the continuous properties of $I_{cont}(X;Y)$ is provided in (Kraskov et al. 2004). The key idea is to measure the distance between element e_i and its nearest neighbour e_j in the x-direction and e_k in the y-direction. For every element $e_i, i \in \{1,\ldots,N\}$ located in the continuous space $Z = X \times Y$ a rank ordering of distances between i and all other elements can be constructed based on a distance $d(e_i(x), e_j(x))$ (distance between elements in the x-direction) and $d(e_i(y), e_k(y))$ (distance between elements in the y direction). The number of points that lie within the vertical strip defined by $d(e_i(x), e_j(x))$ are then counted: n_x and similarly for $d(e_i(y), e_k(y))$: n_y. This is called $k = 1$ *clustering* as it is based on the nearest neighbour, in the case of the second nearest neighbour being used to define $d(e_i(x), e_k(x))$ and $d(e_i(y), e_k(y))$ then $k = 2$ etc. For a system with N interacting particles in $Z = X \times Y$ the mutual information is approximated by (this is $I^{(2)}$ in (Kraskov et al. 2004)):

$$I_{cont}(X;Y) \approx \psi(N) + \psi(k) - 1/k - \langle\psi(n_x) + \psi(n_y)\rangle \tag{20}$$

where $\psi(n)$ is the Digamma function: $\psi(n) = \Gamma^{-1}(n)d\Gamma(n)/dn$, $\Gamma(n) = (n-1)!$ (as n is a simple counting function and so always an integer) and $\langle\ldots\rangle$ is the average over all i $\in \{1,\ldots,N\}$ and across all samples.

As a numerical approximation this is an effective method that eliminates many of the difficulties and scales as $O(n)$. Kraskov et al. (2004) have shown that it is an effective estimate of the mutual information between two coupled Gaussians where the exact solution is known as well as for gene expression analysis, independent component analysis and data clustering (Kraskov et al. 2004). However, some data is naturally discrete in nature and the development of accurate measures of mutual information to accommodate this is an important area of recent research.

Kraskov's algorithm described above can be thought of as an adaptive partitioning technique in the sense that the kth nearest neighbour decides the width of a bin counting technique. An alternative is to start by discretising the space $Z = X \times Y$ into a grid and then adapting the bin sizes such that each 'vertical' or 'horizontal' strip in phase space has the same number of elements. The joint probability is then the occupancy of the rectangles formed by the intersection of these equally populated strips in X- and Y-space.

The calculations used in Cellucci et al's work (Cellucci et al. 2005) starts with N, the number of elements in Z-space, and the number of partitions that will be used to divide up the X- and Y-axes, labelled N_e. Each ith partition of the X-axis contains, by definition, a set $S_X(i)$ of sizeN/N_e elements, equally each jth partition of the Y-axis contains a set $S_Y(j)$ of size N/N_e. The joint probability $P_{X,Y}(i,j) = \text{count}(S_Y(j) \cap S_X(i))/N$, i.e. the count of the size of the intersecting set of the X and Y partitions divided by the total number of elements in the system. The mutual information is then:

$$I(X;Y) \approx \sum_{i=1}^{N_e} \sum_{j=1}^{N_e} P_{X,Y}(i,j) \log(N_e^2 P_{X,Y}(i,j)) \tag{21}$$

where the N_e^2 term accounts for the case of statistically independent distributions of $P_X(i)$ and $P_Y(j)$.

Mutual information for phase transitions in a simple flocking model

Herding or flocking behaviour has been the subject of many studies and it serves here as an illustration of the computation of mutual information and its attendant difficulties. Despite the apparent simplicity of mutual information measures, there is no one simple way which works in general for every data set. In the fields of bioinformatics and neuroinformatics in particular, much research has been done on the estimating of entropies from samples. Dozens of estimators and many code-bases (Daub et al. 2004; Hausser and Strimmer 2009; Ince et al. 2009; Slonim et al. 2005) are available for the task.

For observations generated by a simple *parametric model,* the mutual information functional may sometimes be calculated analytically from the underlying distribution. However, in the case that we wish to use a non-parametric model for our distributions, the procedure is rather more complicated.

Vicsek model background

To eliminate the complexities of real-world data, we use a synthetic data set, the well-studied model of self-propelled particles (SPP) of Vicsek (Vicsek et al. 1995). The SPP model is one of many accepted to undergo a phase transition with varying noise as a control parameter.

The SPP algorithm is a special case of Reynold's "Boids" flocking algorithm, (Reynolds 1987), remarkable for the small set of rules required to produce its rich behaviour. It has the virtues of trivial implementation, topologically simple configuration space, as there are no repulsive forces, and a small number of control parameters. Moreover, there is no known closed-form analytic relationship between system order and control parameters, much as with many experimental data sets. Details of the model and analysis of its phases of behaviour are extensively studied elsewhere (Aldana et al. 2007; Chaté et al. 2008; Czirók and Vicsek 2000; Grégoire and Chaté 2004; Wicks et al. 2007).

The SPP process is given as:

> The only rule of the model is: at each time step a given particle driven with a constant absolute velocity assumes the average direction of motion of the particles in its neighborhood of radius r with some random perturbation added (Vicsek et al. 1995).

Vicsek's original specification is for an ensemble of particles moving in two spatial dimensions on the unit square with periodic (i.e. toroidal) boundary conditions. The model admits straightforward generalisation to other spatial dimensions, and alternate interaction topologies, but it is this original configuration that is used here. The system is parametrised by the temperature-analog noise parameter η, a fixed particle absolute speed v_0, a number N of particles, an interaction radius, R, and a system side-length L. Particles travel at constant speed with random changes in direction as specified by the noise parameter. When two particles come within the interaction radius, their directions of movement are moved closer together. An *order parameter*, the magnitude of mean velocity (or "mean transport"), reflects the phase transition (Figure 2 with the side-length is set to 1 and particle interaction radius R to $1/L$).

Initial conditions assign to all particles positions uniformly at random on the unit square and velocities with a uniformly distributed angle. The simulation is run until the transient behaviours have died out, at which time the system reflects a degree of order dependent upon the noise – see Figure 3.

For different values of the noise control parameter, the system exhibits qualitatively different behaviour: for low noise parameters, the system exhibits spontaneous symmetry breaking, with particles tending to align in one specific direction. At high values of the noise parameter the particles approximate a random walk (and for maximal noise is precisly equal to a random walk). In between the system exhibitions transient regularity, and "clumpy" clustering.

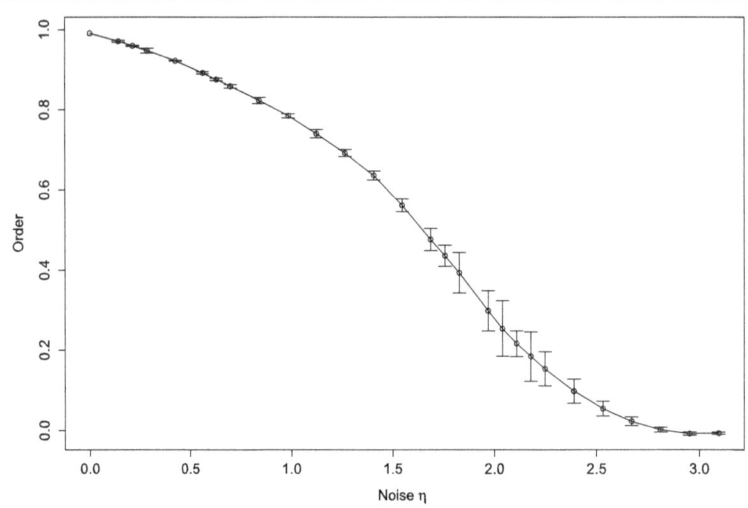

Figure 2 Order parameter versus noise in the Vicsek SPP model. Parameters: $v_0 = 0.00265$, $R = 1/L = 0.0177$, 1000 particles. Fitted line reflects a locally linear nonparametric estimate of order, error bars reflect conditional MSE estimates.

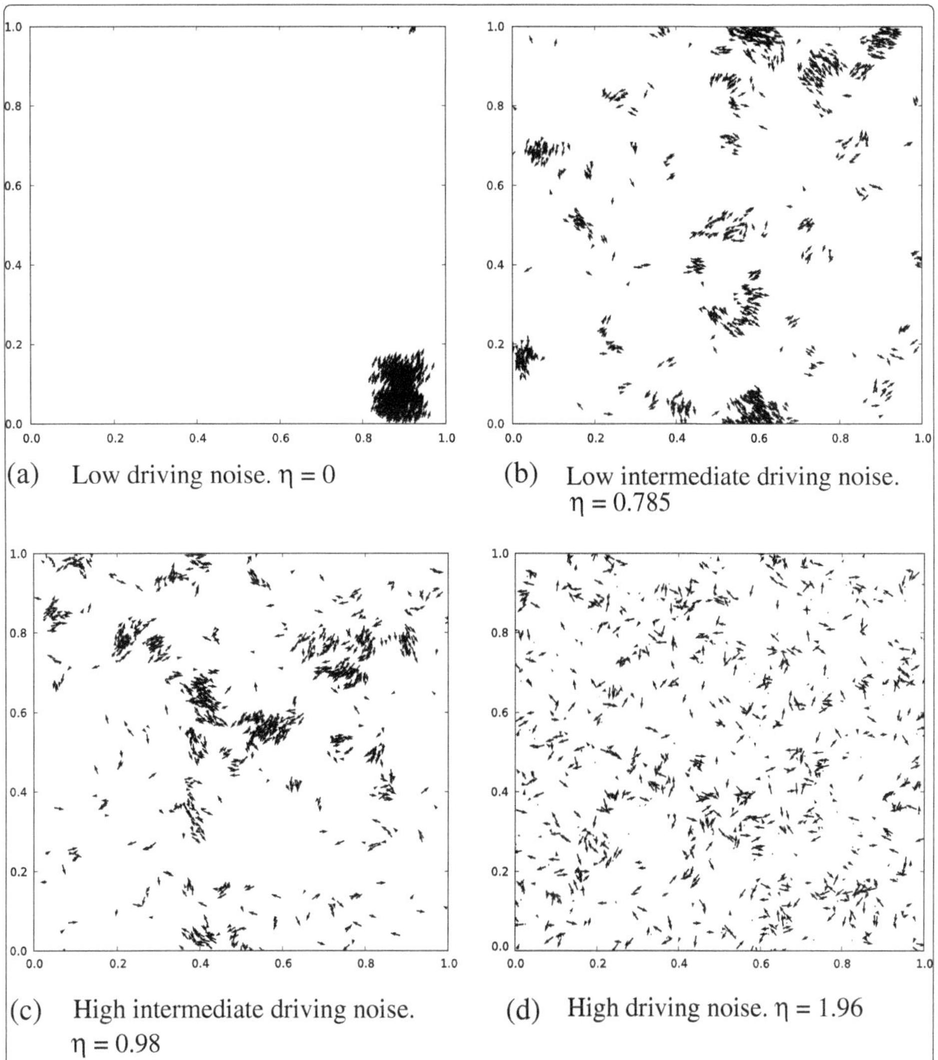

Figure 3 One time step of the two dimensional Vicsek model, with various driving noise levels. Parameters: $v_0 = 0.00265$, $R = 1/L = 0.0177$, 1000 particles. (**a**) Low driving noise. $\eta = 0$. (**b**) Low intermediate driving noise.$\eta = 0.785$. (**c**) High intermediate driving noise. $\eta = 0.98$. (**d**) High driving noise. $\eta = 1.96$.

Wicks et al. (2007) then demonstrated under some simplifying assumptions that there was also a peak in mutual information around the same noise values as for the phase-transition. This is shown in Figure 4 reproduced from their paper.

This model, therefore, serves as a good illustration of the subtleties of calculating mutual information. It also serves as a possible heuristic for the economic phase transitions of section "Overview of phase transitions and metrics" whereby market traders can self-organise to 'flock' in their trading behaviour.

Phase transitions in socio-economic systems

The stock market is one complex system we would all like to understand and predict. Unfortunately it goes through bubbles and crashes, anticipated sometimes, but rarely precisely predicted. These phenomena *seem* like phase transitions: the market undergoes a

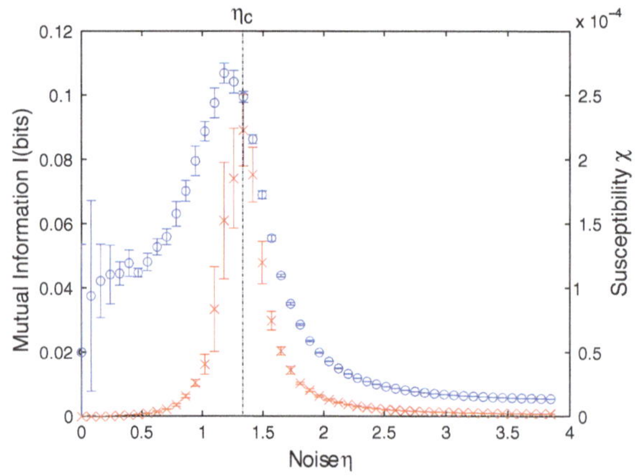

Figure 4 Co-occurrence of mutual information peak and the magnetic susceptibility phase transition as modelled by Vicsek. The mutual information, I (circles) peaks at approximately the same point as the susceptibility, χ (crosses) and with the critical noise $\eta_C = 1.33$ marked. The system parameters are: interaction radius, $R = 1$, time step $\delta t = 1$; particle speed $v_0 = 0.15$; and particle density $\rho = 0.31$ per unit area with arbitrary length units. 3000 particles were used. Reprinted from Wicks et al. (2007) Figure 5 with permission.

radical reorganisation. Harré and Bossomaier (2009) showed that they are indeed phase transitions, exhibiting a peak in mutual information.

They calculated mutual information between pairs of equities in five different industry and commercial sectors. Figure 5 shows the results for a period of over a decade.

For each equity the *daily deltas* were computed, the summary of the day's trading which indicates if the stock fell or rose overall. The distributions of these were then used to calculate the mutual information between stocks (Harré and Bossomaier 2009).

The vertical red band in late 2002 shows a peak in maximum mutual information for almost all equities. This corresponds to a significant correction in the Dow Jones index, but unlike other notable crashes does not have a name. In the bottom right hand corner

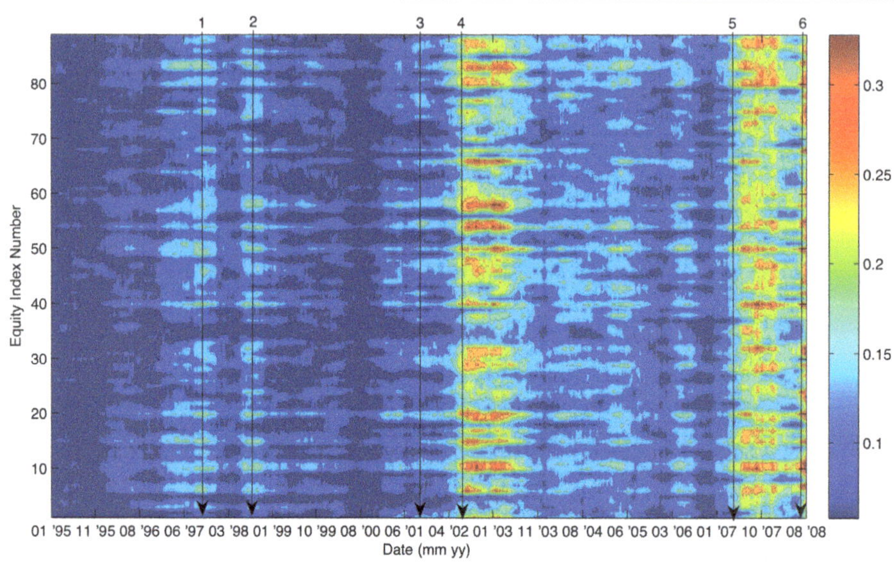

Figure 5 Mutual Information for a set of equities over a twelve year period. Reprinted from Europhysics Letters (Harré and Bossomaier 2009).

there is another red patch. The lowest numbered equities are the financial stocks. This fledgling tradition in late 2005 is undoubtedly linked to the subprime meltdown.

These are empirical results and as such there is no explicit order parameter. However, the Vicsek self-propelled particle model of section "Mutual information for phase transitions in a simple flocking model" provides an econophysical analogy of a stock market with two equities. If we consider each particle to be a trader and the axial position to be the perceived instantaneous investment value of a stock, then any trader has a view at any given time of the value of each stock. Her velocity is indicative of the rate of change of perception of value of each stock, and thus the trade likelihood. Since in most cases stock perceptions are cyclical, periodic boundary conditions are not too implausible.

Just as a change in order parameter can elicit behaviour change from the SPP model, so can an endogenous change in market optimism induce variation in stock market perceptions across all traders. As they approach the phase transition from the random side, stock movements become increasingly synchronised. In the ordered phase all traders are moving in the same direction and the market is surging or crashing.

The Vicsek phase transition is visible in the average *velocity* of the particles (traders). But empirically we observe the mutual information peak in the equity *prices*. We would expect this though. In the random phase there are no links between the trades of different equities and their prices are uncorrelated. In the ordered phase their prices changes are rigidly locked, but the entropy in their pricing is now very low. Thus the mutual information must peak somewhere in the intermediate state. Note that, in this interpretation, the phase transition would begin as the crash or bubble is forming: it is not the bubble or crash itself, but the need for sufficiently wide time windows makes this distinction moot.

The theoretical underpinnings of bifurcations and phase transitions in economics and finance have been around for many years. In the 1970's the mathematical framework of 'catastrophe theory' (Rosser 2007; Zeeman 1980) became a popular field of research as it provided one of the first formalisations of macro-economics that included a notion of both an equilibrium and non-linear state transitions (Zeeman 1974). This formalism provided a parsimonious description of bull and bear markets and market crash dynamics based on bullish bubbles using a small number of macro-economic parameters such as the proportion of chartists (traders who base their strategies on historical prices) and the proportion of fundamentalists (traders who base their strategies on the underlying business).

Such theoretical considerations have played an important role in socio-economic systems, but it was not until the onset of massive databases and high performance computing that it became possible to empirically study the 'microscopic' dynamics of the relationships between equities. Recent work has shown that there is an order parameter in financial markets (Plerou et al. 2003) (section "Overview of phase transitions and metrics"). This order parameter measures the the net demand: before a phase transition, net market demand is zero; after the phase transition (when the market is no longer in equilibrium), the net demand either favours buying or selling.

Such *hysteresis* effects have been the basis of recent work in macro-economic models (Wolpert et al. 2012) as well. In this work a control parameter, mean tax rate, is varied in order to move from a low growth to a high growth game-theoretic equilibrium. Interestingly, the model applies to varying the parameter by either the market (free-market model) or a centralised government (socialist model).

Phase transitions in the acquisition of human expertise

Thomas Kuhn in his famous book, *The Structure of Scientific Revolutions* (Kuhn 1962) discussed the idea of paradigm shifts in science or human knowledge, where everything is reorganised. Relativity and quantum mechanics were major paradigm shifts of the twentieth century. Much earlier Copernicus' idea, that planets travel around the sun, rather than everything around the earth, was a dramatic shift in thinking about the solar system.

Such shifts seem to occur in human thinking, where we learn to join the dots in different ways. Yet it is difficult to find ways to measure such changes. Since expertise requires a long time to develop, at least 10,000 hours according to Eriksson (1998), or the acquisition of 50,000 or more "chunks" according to Simon (Simon and Chase 1973), now thought to be as many as 300,000 (Gobet and Simon 2000). Thus any measurements on a single individual would have to take place over a long period.

Harré et al. found a solution to this using online data mining. Where decisions are recorded online, they can be analysed in large numbers, providing a quite different way of inferring shifts in human thinking and knowledge organisation. To do this they used the game of Go. This game is extraordinarily simple in structure and rules, but is as demanding for human players as chess. Moreover, the best computer programs do not come close to human experts at the time of writing in early 2012.

Figure 6 shows a sample Go board. Black and white stones are placed on the intersections, or *points* of a 19x19 grid. Black begins anywhere in the board, but typically in one of the corners. Stones do not move. They are removed from the board when they die. Stones require a contact along the grid, either directly, or via other stones of the same kind, with an unoccupied point, or *liberty*. This simple setup defines one of the oldest and most challenging of all board games.

Such developmental transitions have been implicated in the activation networks used to model human cognition as well as artificial intelligence systems (Shrager et al. 1987).

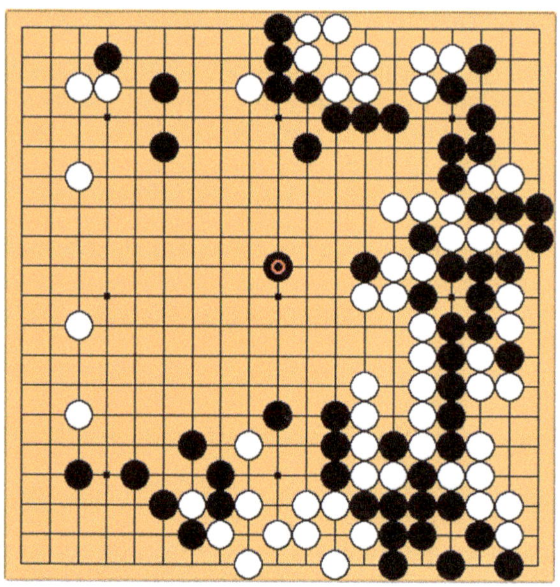

Figure 6 Illustrative Go Board (taken from Sensei's Go Library.

These models, using activation networks, emphasise the role of network topology in how information is accessed, implying that as topologies of associative networks change via learning there is the possibility of a network-wide phase transition occurring. Since this earlier work, many other theoretical models have argued that cognitive structures undergo significant reorganisation at intermediate stages of development. This has included models for epileptic fits (Percha et al. 2005) and clustering algorithms used to learn complex data relationships (Bakker and Heskes 2003).

Finding such transitions in cognitive data is more difficult though, although there has been some evidence of non-linear changes through work on the 'inverted-U' effect in novice-skilled-expert comparative studies (Rikers 2000). This effect is based on the observation that while increasing skill increases the quality of decisions (in medical practitioners, for example), other factors, such as the recall of individual case information and the ability to elaborate more extensively on such cases, peaks for the 'skilled' subjects but were equally low for both the 'expert' and and the 'novice' groups. Such inverted-U observations have been made in chess (Gobet 1997), meditation practitioners (Brefczynski-Lewis et al. 2007) and emotional responses to art (Silva 2005) This work implies an intermediate point where cognitive organisation changes significantly, but as many studies only have a small number of skill levels, i.e. three: novice-skilled-expert, dramatic changes in a dependent variable such as depth of search or recall is often difficult to observe.

The use of entropy as an implicator of phase transitions in cognitive studies has also had some success in recent studies. The developmental transition of generalising a mechanical manipulation into a mathematical insight of the underlying logic, an 'a-ha!' moment has recently been reported using entropy and based on notions of self-organising criticality (Stephen et al. 2009). In this direction, some of the most exciting work has been carried out in transfer entropy (Dixon et al. 2010) applied to self-organising criticality and how it is the entropy that drives massive transitional changes in cognitive structures.

Finally, in a more basic experimental paradigm, Dutilh et al. (2010). have used the speed-accuracy trade-off and Catastrophe Theory in simple decision making to postulate that even some of our most primitive decision making processes might implicate phase transition-like behaviour. (See section "Overview of phase transitions and metrics").

To find phase transitions in the development of expertise we use the same metric, a peak in mutual information. The order parameter is the level of expertise. In Go this is measured in Dan ranks, up to 8 Dan Amateur and 9 Dan professional. Up to 1 Dan Amateur has a separate set of ranks, 26 kyu, with 26 being the weakest and 1 the strongest.

For each rank Harré et al. studied a game tree – every possible move that can happen in a small region (7×7). The moves within the region are taken from actual games, thus they do not have to be sequential: a player may play somewhere else and come back to the region of analysis later.

We need three probability distributions to calculate the mutual information. Firstly there is the joint probability, $p(q, m)$, where q is a position and m is a move made at that position. Then we need the marginal probabilities, $p(q)$ and $p(m)$ of the position and move occurring respectively. For a 7×7 region there are approximately 3^7 possible positions. Some of these will be illegal, but, more importantly, many of them will never occur in actual play. Thus the analysis is tractable.

Figure 7 shows the mutual information as a function of rank. The MI peaks at around the transition from amateur to professional, agreeing with other evidence that a radical shift in strategic thinking occurs at this juncture.

Inferring social networks with transfer entropy

Social networks have rapidly become one of the dominant features of today's world. Understanding when such networks undergo phase transitions can be very important, such as in modelling the spread of disease or public opinion at election time. Numerous tools have been developed to measure information flowing between agents in a network, such as the analysis of email traffice (Kossinets and Watts 2006). But in some cases no direct measures of connectivity are availabe.

We encountered such a situation in customer records in the financial industry. Hence we developed techniques based on mutual information and transfer entropy of investment time series. This section discusses this methodology.

In (Bossomaier et al. 2010) a detailed data set of 42 million records describing the investment profiles of 1.5 million customers over a 24 month period was analysed with the aim of understanding the social networks among clients. In that study, pairwise (unconditional) mutual information between investment histories—lagged and unlagged—was calculated with the aim of identifying relationships between investment behaviour patterns that could be ascribed to social interaction.

Given that lagged mutual information is likely to be a misleading indicator of time-directed information flow (see Section "Transfer entropy and granger causality"), the study was recently repeated using transfer entropy. The exercise highlighted several features and difficulties with the practicalities of estimating and significance testing transfer

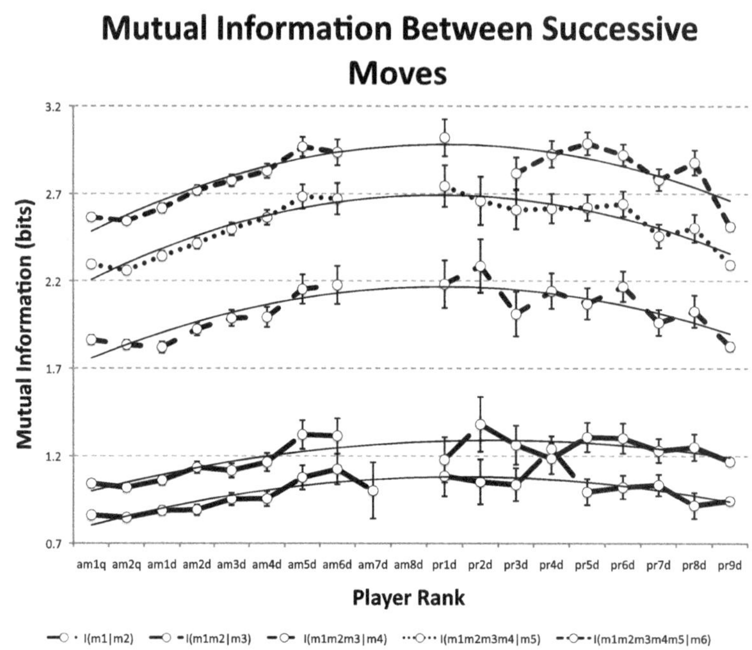

Figure 7 Mutual information between the next move in a given position and the position itself. The missing data points in the centre of plot is due to unreliable estimates, see (Harré et al. 2011) for details. Reprinted from (Harré et al. 2011).

entropy in large datasets. Critically, while a large number of investment history time series were available, they were of short length; in practice only about 20 months' data was generally available per client record. This was largely due to the significant fraction of *missing data*, necessitating a principled approach to the handling of missing data in statistical estimation. Initially investment histories with more than 4 months' missing data were excluded; all subsequent analysis was performed on a per-product basis. It was found that (again due largely to the short length of histories) there were many duplicate time series within a product. These are statistically indistinguishable, so only *unique* time series were analysed, each corresponding to a specific group of customers within the given product. The final number of unique histories was of the order of 500–5000 per investment product.

As in the original study, it was deemed that the actual magnitude of monthly investments was not relevant; monthly investment states were thus classified simply into '+' (account credited), '-' (account debited) or '0' (no change). This discretisation of investment states (a practice commonplace in econometric analysis) also makes estimation of transfer entropies less problematic (*cf.* Section "Transfer entropy and granger causality"). The stance taken on remaining missing data was that it should be "conditioned out"; that is, statistics were estimated *conditional on all relevant investment states being valid (non-missing)*. Furthermore, as an attempt to control for common influences on all customers (within the given population of investment histories), transfer entropy was conditioned on a *consensus sequence*, C_t obtained by taking the most prevalent valid state in the population at each time step. Thus conditional transfer entropy was calculated as

$$\mathrm{T}_{Y\to X|C} \equiv \mathrm{I}(X_t : Y_{t-1}|X_{t-1}, C_{t-1} \text{and } X_t \neq *,\ X_{t-1} \neq *,\ Y_{t-1} \neq *) \tag{22}$$

where '$*$' denotes missing data. Due again to short history length, only one lag of history was taken into consideration. In sample, (22) was calculated by estimating the requisite probabilities as frequencies.

In order to derive networks of causal information flow, we calculated pairwise transfer entropies conditioned on the appropriate consensus sequence between all pairs of time-series X_i, X_j in the selected populations. These were then tested for statistical significance (see below) and significant information flows presented as directed graphs, weighted by the actual value of the transfer entropy statistic (Figure 8).

Note, however, that due to the large number of time series it was not possible to calculate pairwise-conditional statistics [Section "Transfer entropy and granger causality", eq. (7)]. Thus if there is e.g. significant information flow from $X_i \to X_j$ and also from $X_j \to X_k$ then it is likely that the information flow $X_i \to X_k$ will appear as significant too, even if there is no *direct* information flow from X_k to X_i; i.e. the apparent information flow $X_i \to X_k$ is intermediated by X_j.

Pairwise information statistics were permutation tested to establish significance, against the null hypothesis of zero information flow. For each statistic, 1000 random permutations of the time series of the causal variable were generated (missing data is held in-place during permutation) to disrupt any possible causal effects. The unpermuted statistic was then tested against the resulting surrogate null distribution to derive a p-value, representing the probability that a value at least as large as the statistic being tested might be obtained by chance under the null hypothesis. A subtlety is that, while the number of (unique) values for a permuted statistic can be quite small—many different

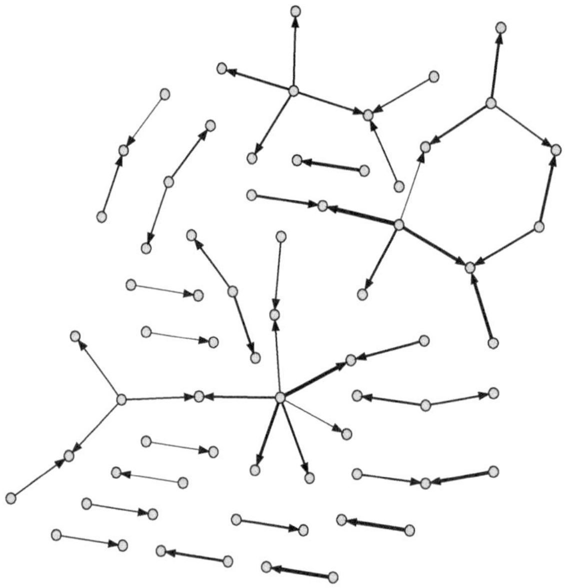

Figure 8 Sample social network inferred from statistically significant pairwise transfer entropies between client investment histories for a single investment product, conditioned on a consensus sequence (see text for details). The thickness of the arrows corresponds to the magnitude of information flow. After pre-processing there were 1429 unique investment histories for this product. The resulting network features 68 of these histories linked by 50 directed edges.

permuted sequences will in general give rise to the same sample statistic—some values are very rare, and will frequently not be discovered by any of the 1000 permutations. It is thus possible that a p-value can come out as zero; this is patently unsatisfactory, as it would reject the null hypothesis at *any* significance level, thus giving rise to Type I errors (false positives). To address this issue, we use the fact that the *maximum* sample information flow statistic will be obtained when the (lagged) causal sequence is *identical* to the causee sequence. In general there will be only one possible permutation with this property, which thus occurs with probability[f]

$$p_{Imax} \equiv \frac{n_0!\ n_+!\ n_-!}{(n_0 + n_+ + n_-)!} \tag{23}$$

where n_0, n_+ and n_- are the number of $0, +$ and $-$ states respectively in the investment history sequence being tested. The following procedure was implemented to mitigate the effects of spurious zero p-values: if a p-value was empirically calculated to be zero—i.e. the test statistic was larger than all permutation statistics—the resulting p-value was set to p_{Imax} rather than zero. This does not preclude the possibility that the "true" p-value is actually larger than p_{Imax}, but can at least be expected to reduce substantially the number of false-positives that might otherwise arise due to zero p-values.

A further issue to be addressed is that we are performing *multiple* hypothesis tests (Miller 1981); i.e. one for each pairwise information flow statistic within the population under consideration. Under this scenario, at any given significance level, α, we would expect approximately $\alpha \times$ (number of pairwise statistics) Type I errors (false positives). There are various established approaches to controlling for this effect, which generally require a stronger level of evidence for rejection of null hypotheses. Unfortunately, in our situation where the number of simultaneous hypothesis tests may be very large [it will be

Table 2 Number of significant information flow sample statistics by product at significance level $\alpha = 0.01$, with per-time-series family-wise error rate correction (see text)

Product	LMI uncond	LMI cond	TE uncond	TE cond
1	29449	2971	7540	1986
2	739	109	179	75
3	191	94	112	50
4	112	116	90	31
5	12	19	30	15
6	38	27	25	15
7	60	6	20	8
8	7	10	6	4
9	52	7	18	3

LMI = Lagged mutual information, TE = Transfer entropy; conditional/unconditional refers to conditioning on the population consensus sequence.

$n(n-1)$ where n is the number of unique time series in the test population] common approaches such as controlling for *false discovery rate* or *family-wise error rate* (Benjamin and Hochberg 1995) are highly likely to yield *no* significant statistics at acceptable significance levels; essentially, Type I errors will have been traded off for an unacceptably high Type II (false negative) error rate. As a compromise, for each time-series X_j in the given population, we estimated significance for information flow statistics by controlling the family-wise error rate among *all time series that might potentially exhibit a significant causal influence on series* X_j - i.e. for all statistics $X_i \to X_j$. Family-wise error rates were calculated at a significance level $\alpha = 0.01$ using the well-known *Bonferroni correction* (Miller 1981).

Numbers of significant information flows under this scheme are displayed in Table 2. We see that conditioning out the common influence represented by the consensus sequence usually (but not always) reduces the number of significant statistics. Transfer entropy generally yields fewer significant information flows; this is consistent with the known deficiencies of lagged mutual information as a measure of directed information flow, whereby spurious information flows may be reported due to shared historical influences (Section "Transfer entropy and granger causality").

Worthwhile future work on this study would include comparison of the *causal density* dynamic complexity measure (Section "Transfer entropy and granger causality") between investor networks; however, this presents a technical challenge with regard to the difficulties mentioned above in obtaining true pairwise-conditional transfer entropies.

Conclusions

Mutual information is a useful indicator of phase transitions. It peaks in the same region as other indicators, such as the magnetic transition in the Vicsek model. We have shown that the calculation of mutual information is fraught with difficulty, but it can be used with care to find phase transitions in socio-economic and cognitive systems.

Transfer entropy, a conditioned extension of lagged mutual information, is closely related to mutual information and is a powerful new technique. It can be used to infer causal information flows within complex systems (Lizier 2008) and holds out the possibility of being able to predict phase transitions before they occur. Of particular interest for future study would be to investigate the behaviour of the information-theoretic dynamical complexity measures described in Section "Transfer entropy and granger causality" with regard to phase transitions and the application to socio-economic systems from organisational change to the onset of recessions.

Endnotes

[a]These informational properties can be extended, with some qualifications, to continuous random variables as well — see, for example, (Gray 1991) — but as random variable considered herein are either discrete or discretised before analysis, these extensions will not be discussed here. Generalisations of mutual information to more than two variables also exist.

[b]Here, and below, we leave ambiguous the number of lags to be included in expressions of this type; in principle one might include the *entire* past of a process, while for empirical estimation (or on domain-specific grounds) lags may be truncated at some finite number.

[c]It is a standard result in time series analysis that maximum likelihood estimation of the regression parameters in (8) is equivalent to minimisation of the "total variance" $\text{trace}(\text{cov}(\varepsilon_t))$, e.g. by a standard ordinary least squares (OLS) (Hamilton 1994). Other equivalent approaches are by solution of the Yule-Walker equations for the regression, e.g. via the LWR algorithm or one of its variants (Morettin 1984).

[d]In the case of a *univariate* predictee variable X, the Granger statistic is sometimes defined as the R^2-like statistic $\exp(\mathrm{F}_{Y\to X}) - 1$, which has an asymptotic F- rather than a χ^2 null distribution (Hamilton 1994).

[e]In fact this is probably the most commonly encountered variant of Granger causality, at least in the neuroscience literature; confusingly, it is frequently this quantity that is referred to as "multivariate" (conditional) Granger causality, as opposed to the case $\mathrm{F}_{Y\to X|Z}$ where the individual variables X, Y, Z are themselves multivariate.

[f]This will not be precisely the case if e.g. there number of $+$ states and $-$ states in the sequence is the same; in this case the permutation derived by swapping $+$ and $-$ states will yield an additional maximum information sequence. We do not believe this affects significance test results unduly.

Competing interests

The authors declare that they have no competing interests.

Author's contributions

All authors read and approved the final manuscript.

Acknowledgements

This work was supported by US Air Force grant AOARD 104116, Australian Research Council Grants LP0453657 and DP0881829. Lionel Barnett is supported by the Dr Mortimer and Theresa Sackler Foundation. Dan Mackinlay produced the Vicsek diagrams and contributed helpful discussions.

Author details

[1]Centre for Research in Complex Systems, Charles Sturt University, Panorama Ave, Bathurst NSW 2795, Australia. [2]Department of Informatics, University of Sussex, Falmer, Brighton BN1 9QH, UK. [3]Department of Civil Engineering, University of Sydney, Sydney, NSW, Australia.

References

Aldana, M, Dossetti V, Huepe C, Kenkre VM, Larralde H: **Phase transitions in systems of self-propelled agents and related network models.** *Phys Rev Lett* 2007, **98:**095702.

Anderson, MJ, Robinson J: **Permutation tests for linear models.** *Aust NZ J Stat* 2001, **43:**75–88.

Bakker, B, Heskes T: **Clustering ensembles of neural network models.** *Neural Netw* 2003, **16**(2):261–269.

Barnett, L, Bossomaier T: **Transfer entropy as a log-likelihood ratio.** *Phys Rev Lett* 2012, **109:**138105.

Barnett, L, Seth AK: **Behaviour of Granger causality under filtering Theoretical invariance and practical application.** *J Neurosci Methods* 2011, **201**(2):404–419.

Barnett, L, Barrett AB, Seth AK: **Granger causality and transfer entropy are equivalent for Gaussian variables.** *Phys Rev Lett* 2009a, **103**(23):238701.

Barnett, L, Buckley CL, Bullock S: **Neural complexity and structural connectivity.** *Phys Rev E* 2009b, **79**(5):51914.

Barrett, AB, Barnett L, Seth AK: **Multivariate Granger causality and generalized variance.** *Phys Rev E* 2010, **81**(4):41907.

Benjamini, Y, Hochberg Y: **Controlling the false discovery rate: A, practical and powerful approach to multiple testing.** *J Roy Stat Soc B* 1995, **57:**289–300.

Bossomaier, T, Standish RK, Harré M: **Simulation of trust in client-wealth management adviser relationships.** *Int J Simul Process Model* 2010, **6:**40–49.

Brefczynski-Lewis, J, Lutz A, Schaefer H, Levinson D, Davidson R: **Neural correlates of attentional expertise in long-term meditation practitioners.** *Proc Natl Acad Sci* 2007, **104**(27):11483.

Cellucci, CJ, Albano AM, Rapp PE: **Statistical validation of mutual information calculations: Comparison of alternative numerical algorithms.** *Phys Rev E* 2005, **71**(6):66208.

Chaté H, Ginelli F, Grégoire G, Raynaud F: **Collective motion of self-propelled particles interacting without cohesion.** *Phys Rev E* 2008, **77:**046113.

Cover, TM, Thomas JA: *Elements of Information Theory.* Wiley Series in Telecommunications and Signal Processing. New York: Wiley-Interscience; 2006.

Cui, J, Lei X, Bressler SL, Ding M, Liang H: **BSMART: a Matlab/C toolbox for analysis of multichannel neural time series.** *Neural Netw, Special Issue Neuroinformatics* 2008, **21:**1094–1104.

Czirók, A, Vicsek T: **Collective behavior of interacting self-propelled particles.** *Physica A: Stat Theor Phys* 2000, **281:**17–29.

Darbellay, GA, Vajda I: **Estimation of the information by an adaptive partitioning of the observation space.** *IEEE Trans Inf Theory* 1999, **45:**1315–1321. [http://ieeexplore.ieee.org/xpls/abs_all.jsp?arnumber=761290]

Dai, L, Vorselen D, Korolev KS, Gore J: **Generic indicators for loss of resilience before a tipping point leading to population collapse.** *Science* 2012, **336**(6085):1175–1177.

Daub, CO, Steuer R, Selbig J, Kloska S: **Estimating mutual information using B-spline functions - an improved similarity measure for analysing gene expression data.** *BMC Bioinformatics* 2004, **5:**118.

Ding, M, Chen Y, Bressler S: **Granger causality: Basic theory and application to neuroscience.** In *Handbook of Time Series Analysis*. Edited by Schelter, S, Winterhalder M, Timmer J. Wienheim: Wiley; 2006:438–460.

Dutilh, G, Wagenmakers E, Visser I, Van Der Maas H: **A phase transition model for the speed-accuracy trade-off in response time experiments.** *Cogn Sci* 2010.

Dixon, J, Stephen D, Boncoddo R, Anastas J: **The self-organization of cognitive structure.** *Psychol Learn Motiv* 2010, **52:**343–384.

Edgington, ES: *Randomization Tests.* New York: Marcel Dekker; 1995.

Erdős, P, Rényi A: *On the Evolution of Random Graphs.* Budapest: The Hungarian Academy of Sciences; 1960.

Ericsson, K: **The scientific study of expert levels of performance general implications for optimal learning and creativity 1.** *High Ability Studies* 1998, **9:**75–100.

Geweke, J: **Measurement of linear dependence and feedback between multiple time series.** *J Am Stat Assoc* 1982, **77**(378):304–313.

Geweke, J: **Measures of conditional linear dependence and feedback between time series.** *J Am Stat Assoc* 1984, **79**(388):907–915.

Gobet, F: **A pattern-recognition theory of search in expert problem solving.** *Thinking Reasoning* 1997, **3**(4):291–313.

Gobet, F, Simon H: **Five seconds or sixty? Presentation time in expert memory.** *Cogn Sci* 2000, **24:**651–682.

Granger, CWJ: **Investigating causal relations by econometric models and cross-spectral methods.** *Econometrica* 1969, **37:**424–438.

Granger, C, Lin J: **Using the mutual information coefficient to identify lags in nonlinear models.** *J Time Ser Anal* 1994, **15**(4):371–384.

Gray, RM: *Entropy and Information Theory.* New York: Springer-Verlag; 1991. [http://ee.stanford.edu/~gray/it.html]

Green, D, Newth D: **Towards a theory of everything: grand challenges in complexity and informatics.** *Complexity Int* 2005, **8:**. [http://www.complexity.org.au/ci/vol08/index.html]

Grégoire, G, Chaté H: **Onset of collective and cohesive motion.** *Phys Rev Lett* 2004, **92**(2):025702.

Gu, SJJ, Sun CPP, Lin HQQ: **Universal role of correlation entropy in critical phenomena.** *J Phys A: Math Gen* 2006. [http://arxiv.org/pdf/quant-ph/0605164]

Hamilton, JD: *Time Series Analysis.* Princeton: Princeton University Press; 1994.

Hausser, J, Strimmer K: **Entropy inference and the James-Stein estimator with application to nonlinear gene association networks.** *J Mach Learn Res* 2009, **10:**1469–1484.

Harré, M, Bossomaier T: **Phase-transition – behaviour of information measures in financial markets.** *Europhysics Lett ERA A* 2009, **87:**18009.

Harré, M, Bossomaier T, Gillett A, Snyder A: **The aggregate complexity of decisions in the game of go.** *Eur Phys J B ERA A* 2011, **80:**555–563.

Ince, RA, Petersen RS, Swan DC, Panzeri S: **Python for information theoretic analysis of neural data.** *Front Neuroinformatics* 2009, **3:** [http://www.hubmed.org/display.cgi?uids=19242557]

Kaiser, A, Schreiber T: **Information transfer in continuous processes.** *Physica D* 2002, **166:**43–62.

Kraskov, A, Stögbauer H, Grassberger P: **Estimating mutual information.** *Phys Rev E* 2004, **69:**066138–066153.

Kossinets, G, Watts DJ: **Empirical analysis of an evolving social network.** *Science* 2006, **311**(5757):88–90.

Kuhn, T: *The Structure of, Scientific Revolutions*: University of Chicago Press; 1962.

Langton, C: **Computation at the edge of chaos: Phase transitions and emergent computation.** *Physica D: Nonlinear Phenomena* 1990a, **42:**12–37.

Langton, CG: **Computation at the edge of chaos.** *Physica D* 1990b, **42:**12–37.

Lin, J: **Divergence measures based on the Shannon entropy.** *Inf Theory, IEEE Trans* 1991, **37:**145–151.

Lizier, J, Prokopenko M, Zomaya A: **A framework for the local information dynamics of distirbuted computation in complex systems.** *Phys Rev E* 2008, **77:**026110.

MacKay, D: *Information Theory, Inference, and, Learning Algorithms.* Cambridge: Cambridge University Press; 2003.

Matsuda, H, Kudo K, Nakamura R, Yamakawa O, Murata T: **Mutual information of I sing systems.** *Int J Theor Phys* 1996, **35**(4):839–845.

McQuarrie, ADR, Tsai CL: *Regression and, Time Series Model Selection.* Singapore: World Scientific Publishing; 1998.

Miller, RG: *Simultaneous Statistical, Inference.* New York: Springer-Verlag; 1981.

Moon, YI, Rajagopalan B, Lall U: **Estimation of mutual information using kernel density estimators.** *Phys Rev E* 1995, **52:**2318–2321.

Morettin, PA: **The Levinson algorithm and its applications in time series analysis.** *Int Stat Rev* 1984, **52:**83–92.

Nemenman, I, Shafee F, Bialek W: **Entropy and inference, revisited.** In *Advances in Neural Information Processing Systems 14 Volume 14.* Edited by Dietterich, TG, Becker S, Ghahramani Z, Cambridge, The MIT Press; 2002.

Parra, L, Deco G, Miesbach S: **Statistical independence and novelty detection with information preserving nonlinear maps.** *Neural Comput* 1996, **8**(2):260–269.

Panzeri, S, Treves A: **Analytical estimates of limited sampling biases in different information measures.** *Netw Comput Neural Syst* 1996, **7:**87–107. [http://www.ingentaconnect.com/content/apl/network/1996/00000007/00000001/art00006]

Percha, B, Dzakpasu R, Zochowski M, Parent J: **Transition from local to global phase synchrony in small world neural network and its possible implications for epilepsy.** *Phys Rev E* 2005, **72**(3):031909.

Plerou, V, Gopikrishnan P, Stanley H: **Two-phase behaviour of financial markets.** *Nature* 2003, **421**(6919).

Prokopenko, M, Lizier JT, Obst O, Wang XR: **Relating fisher information to order parameters.** *Phys Rev E* 2011, **84**(4):041116. [http://lizier.me/joseph/publications/2011-Prokopenko-RelatingFisherInfoToOrderParams.pdf]

Reshef, DN, Reshef YA, Finucane HK, Grossman SR, McVean G, Turnbaugh PJ, Lander ES, Mitzenmacher M, Sabeti PC: **Detecting novel associations in large data sets.** *Science* 2011, **334**(6062):1518–1524.

Reynolds, CW: **Flocks, herds and schools: A distributed behavioral model.** In *SIGGRAPH '87 Proceedings of the 14th Annual Conference on Computer Graphics and Interactive Techniques, Volume 21.* New York: ACM; 1987:25–34.

Rikers, R, Schmidt H, Boshuizen H: **Knowledge encapsulation and the intermediate effect.** *Contemp Educ Psychol* 2000, **25**(2):150–166.

Rosser Jr, J: **The rise and fall of catastrophe theory applications in economics: Was the baby thrown out with the bathwater?.** *J Econ Dyn Control* 2007, **31**(10):3255–3280.

Scheffer, M, Carpenter SR, Lenton TM, Bascompte J, Brock W, Dakos V, van de Koppel J, van de Leemput IA, Levin SA, van Nes EH, Pascual M, Vandermeer J: **Anticipating critical transitions.** *Science* 2012, **338**(6105):344–348.

Schreiber, T: **Measuring information transfer.** *Phys Rev Lett* 2000, **85**(2):461–464.

Seth, AK: **A MATLAB toolbox for Granger causal connectivity analysis.** *J Neurosci Methods* 2010, **186:**262–273.

Seth, AK, Barrett AB, Barnett L: **Causal density and integrated information as measures of conscious level.** *Phil Trans R Soc A* 2011, **369:**3748–3767.

Seth, AK, Izhikevich E, Reeke GN, Edelman GM: **Theories and measures of consciousness: an extended framework.** *Proc Natl Acad Sci USA* 2006, **103**(28):10799–10804.

Shannon, CE: **A mathematical theory of communication.** *Bell Syst Tech J* 1948a, **27:**379–423.

Shannon, C: **The bell system technical journal 27.** *Math Theory Commun* 1948b:379–423.

Shrager, J, Hogg T, Huberman B: **Observation of phase transitions in spreading activation networks.** *Science* 1987, **236**(4805):1092.

Silvia, P: **Emotional responses to art: From collation and arousal to cognition and emotion.** *Rev Gen Psychol* 2005, **9**(4):342.

Simon, H, Chase W: **Skill in Chess: Experiments with chess-playing tasks and computer simulation of skilled performance throw light on some human perceptual and memory processes.** *Am Sci* 1973, **61**(4):394–403.

Slonim, N, Atwal GS, Tkačik G, Bialek W: **Information-based clustering.** *Proc Natl Acad Sci* 2005, **102:**18297–18302. [http://www.hubmed.org/display.cgi?uids=16352721]

Solé R, Manrubia S, Luque B, Delgado J, Bascompte J: **Phase transitions and complex systems.** *Complexity* 1996, **1**(4):13–26.

Sornette, D, Johansen A: **Significance of log-periodic precursors to financial crashes.** *eprint arXiv:cond-mat/0106520* 2001.

Stephen, D, Boncoddo R, Magnuson J, Dixon J: **The dynamics of insight Mathematical discovery as a phase transition.** *Mem Cogn* 2009, **37**(8):1132–1149.

Strong, SP, Koberle R, de Ruyter van Steveninck RR, Bialek W: **Entropy and information in neural spike trains.** *Phys Rev Lett* 1998, **80:**197–200. [http://arxiv.org/abs/cond-mat/9603127]

Tononi, G, Sporns O, Edelman GM: **A measure for brain complexity Relating functional segregation and integration in the nervous system.** *Proc Natl Acad Sci USA* 1994, **91:**5033–5037.

Treves, A, Panzeri S: **The upward bias in measures of information derived from limited data samples.** *Neural Comput* 2008, **7**(2):399–407.

Vicsek, T, Czirók A, Ben-Jacob E, Cohen I, Shochet O: **Novel type of phase transition in a system of self-driven particles.** *Phys Rev Lett* 1995, **75:**1226–1229. [http://arxiv.org/abs/cond-mat/0611743]

Wicks, RT, Chapman SC, Dendy R: **Mutual information as a tool for identifying phase transitions in dynamical complex systems with limited data.** *Phys Rev E* 2007, **75:** [http://arxiv.org/pdf/physics/0612198]

Wiener, N: **The theory of prediction.** In *Modern Mathematics for Engineers.* Edited by Beckenbach, EF. New York: McGraw Hill; 1956:165–190.

Wilks, SS: **The large-sample distribution of the likelihood ratio for testing composite hypotheses.** *Ann Math Stat* 1938, **6:**60–62.

Wilms, J, Troyer M, Verstraete F: **Mutual information in classical spin models.** *J Stat Mech: Theory Exp* 2011, **2011:**P10011. [http://arxiv.org/abs/1011.4421]

Wolpert, DH, Wolf DR: **Estimating functions of probability distributions from a finite set of samples.** *Phys Rev E* 1995, **52**(6):6841–6854. [http://arxiv.org/abs/comp-gas/9403001]
Wolpert, DH, Harré M, Olbrich E, Bertschinger N, Jost J: **Hysteresis effects of changing the parameters of noncooperative games.** *Phys Rev E* 2012, **85:**036102. [http://link.aps.org/doi/10.1103/PhysRevE.85.036102]
Zeeman, E: **On the unstable behaviour of stock exchanges.** *J Math Econ* 1974, **1:**39–49.
Zeeman, E: *Catastrophe Theory*: Reading: Addison-Wesley Pub. Co; 1980.

8

Dynamic agent composition for large-scale agent-based models

Fanny Boulaire[1*], Mark Utting[2,3] and Robin Drogemuller[1]

* Correspondence: Fanny.Boulaire@qut.edu.au
[1]Queensland University of Technology, Gardens Point – P Block level 8, Brisbane Qld 4000, Australia
Full list of author information is available at the end of the article

Abstract

Purpose: This paper describes *dynamic agent composition*, used to support the development of flexible and extensible large-scale agent-based models (ABMs). This approach was motivated by a need to extend and modify, with ease, an ABM with an underlying networked structure as more information becomes available. Flexibility was also sought after so that simulations are set up with ease, without the need to program.

Methods: The dynamic agent composition approach consists in having agents, whose implementation has been broken into atomic units, come together at runtime to form the complex system representation on which simulations are run. These components capture information at a fine level of detail and provide a vast range of combinations and options for a modeller to create ABMs.

Results: A description of the dynamic agent composition is given in this paper, as well as details about its implementation within MODAM (*MOD*ular *A*gent-based *M*odel), a software framework which is applied to the planning of the electricity distribution network. Illustrations of the implementation of the dynamic agent composition are consequently given for that domain throughout the paper. It is however expected that this approach will be beneficial to other problem domains, especially those with a networked structure, such as water or gas networks.

Conclusions: Dynamic agent composition has many advantages over the way agent-based models are traditionally built for the users, the developers, as well as for agent-based modelling as a scientific approach. Developers can extend the model without the need to access or modify previously written code; they can develop groups of entities independently and add them to those already defined to extend the model. Users can mix-and-match already implemented components to form large-scales ABMs, allowing them to quickly setup simulations and easily compare scenarios without the need to program. The dynamic agent composition provides a natural simulation space over which ABMs of networked structures are represented, facilitating their implementation; and verification and validation of models is facilitated by quickly setting up alternative simulations.

Keywords: Agent-based model; Dynamic composition; Large-scale; Electricity distribution network

Background

Agent-based modelling (ABM) has been used successfully over the last decade to model different aspects of the electricity sector. Its use was initially mainly for the analysis of power market design for large-scale electricity systems when deregulation happened (North et al. 2002; Batten et al. 2006; Weidlich 2008). These models aimed at investigating the interactions between the physical infrastructure at the transmission level (high voltage networks), and the economic behaviour of market participants to help engineer markets in the electricity sector. The application of ABM to the electricity distribution network (medium and low voltage networks) is not as widespread as that of the transmission networks, but is becoming more studied, especially as new technologies (rooftop solar panels, batteries, ...) are appearing on the market and transforming the way electricity is consumed, produced and traded (Cai et al. 2011; Institute for Energy and Transport 2014).

Agent-based modelling has seen a rise in popularity for its capacity to provide some insight as to how a system responds to changes from the entities' responses and interactions and the environment, by capturing information at a fine level of detail over space and time using simple rules (Klügl and Bazzan 2012; Macal et al. 2006). It is therefore particularly suited to model the electricity grid which is currently going through a phase of transformation. The way the grid is going to be used is changing, with consumers now also becoming producers and installing new technologies that are changing the flows of electricity on the networks. Communication between the different network assets will become more prominent, impacting further its management but also providing many opportunities. Information about where, how and when these changes are going to happen is important as averages are not sufficient to inform planners appropriately.

Within this context, we have developed a modelling and simulation (M&S) application to answer questions relating to the planning of the future grid and to assess the impact of the integration of decentralised generators (DGs) on a distribution grid owned by Ergon Energy (Ergon Energy 2013). This M&S application supports the network planning process by providing an understanding of the evolution of the network in terms of load and voltages, over space and time, and finding the most economical solution in terms of network upgrades. The full platform uses two modelling techniques: agent-based modelling (Castiglione 2006) and particle swarm optimisation (PSO) (del Valle et al. 2008). The ABM approach was chosen for a few reasons. One is its capacity to capture the information at a fine level of detail both geographically and over time, allowing customers behaviours in relation to their usage of new technologies to be represented and linked to the network structure with accuracy. Another one is that in this application context, the past is no predictor of the future, because very little is known about the impact of the large-scale integration of DGs on distribution networks. By capturing the functioning of the different entities and their relationships to one another, insight into what might happen using simple rules can be gained. In our M&S application, ABM is used to run a large number of scenarios of possible futures, and its output (load duration curves at any node on the network) is passed to the PSO module. This module evaluates which network assets can be installed or upgraded to ensure safe, reliable and economical delivery of electricity. Details on the overall M&S application can be found in (Boulaire et al. 2012b).

This paper focuses solely on the ABM part of the M&S application; more specifically on the technical aspect in building a large-scale agent-based model in regards to the requirements of the project. The following requirements were defined, where the model needed to:

a) Take into account the physical characteristics of the network (assets and their connections) as well as the different actors and the environment, influencing the flows on the network;
b) Be able to represent the evolution of the system over many years (long-term planning) but with a fine level of detail that captures how the different elements operate over half-hourly periods;
c) Deal with large and varied datasets coming from corporate databases to populate the model - in terms of configuration of the network and characteristics of its elements, and also allow for different data types for the output of the simulation;
d) Be able to evolve along with the changes in the network and the consumers market
    a. With the addition of new technologies over time, as they became available;
    b. and different ways in using them, e.g. comparing behavioural or policy impact depending on how the technology is used;
e) Allow creating various scenarios with ease so that simulations can be set up on a daily basis by power engineers, who are not programmers.

Further to these model requirements, goals were identified for its implementation:

- An independent author (a developer that is adding new agents) does not need to modify previously written code;
- The models need to be *assembled* following a "*code-free*" approach, where
    - A model user does not need to read or modify the Java code
        - Behaviours of agents can be added/changed without coding – both at the beginning when setting up the model, and during the run,
        - A user can try different models without going into the code, but simply by combining different aspects of the model
    - There is no need for recompiling when adding agents to the model

Two early implementations of our M&S application, based on Repast (Argonne National Laboratory 2014) and MASON (Luke et al. 2005), exposed various shortfalls with building large-scale models using existing model building approaches in regards to our needs. These are summarised in Table 1. The limitations, requirements and goals mentioned above, led to the development of MODAM (MODular Agent-based Model), a framework that builds flexible and extensible large-scale ABMs, using *dynamic agent composition*. This approach consists in having the agents built at runtime by bringing together the physical representation of the elements (assets) and their different behaviours that will specify their actions.

The solutions implemented in MODAM, in response to the shortfalls identified are also given in Table 1, and are discussed further in the paper.

Table 1 Shortfalls of existing model building approaches and software systems, and MODAM solutions

Shortfalls of existing model building approaches and software systems	MODAM solutions
• Central simulation class is responsible for	• Decentralised factories create the physical properties of an agent (assets) and its behavioural properties (behaviours) separately
a. Instantiating all agents;	
b. Defining relationships between instantiated agents;	
c. Reading the data used to build agents and relationships (if the model is data-driven);	a. Several asset factories create assets and the relationships between them;
d. Complex option handling done as centralised code.	b. Several behaviour factories attach behaviours to assets;
	c. Each asset factory can be independently parameterised with data providers;
	d. Each factory handles its own options.
• Each agent is a single class.	• Separation of assets and behaviours allows
a. Variation of behaviour requires many subclasses;	a. Mix-and-match construction of agents at runtime;
b. Each agent is created independently of others.	b. Gathering certain entities in groups.
• Non deterministic order of agents' execution	• Deterministic simulation runs for each random seed (reproducibility of results)

The MODAM framework enforces the separation of assets and behaviours, and combines this key idea with several other techniques (data providers, factories, interfaces, channels (runtime data attributes), use of reflection by a module manager) to fully support the dynamic composition of agents. Because we are building large-scale ABMs where thousands of agents are represented and for which we have sufficiently accurate knowledge, data is used to populate the model, defining the way the agents are in relation to one another as well as their properties. Assets and behaviours are created within components according to their type, and data providers are called on to populate the individuals' entities, with the model coming together at runtime. This facilitates setting up large-scale simulations and has the additional property of not requiring the user to program. This approach builds on the vision set by (Hamill 2010) of having a library of building blocks for agent-based models. These building blocks would capture a specific environment, or agent, or group of agents and by bringing them together a modeller can set up a simulation more easily, which is especially interesting for the non-programmer. This approach, *dynamic agent composition*, is the key contribution of this paper.

This paper describes this approach, to building flexible and extensible large-scale ABMs. First, the dynamic agent composition is motivated in Section 3 using an example of the implementation of an electric vehicle agent. An overview of the approach is then given in Section 4, followed in Section 5 by more details describing the asset and behaviour models, and a description of how these elements come together at runtime to create a simulation. Section 6 discusses the challenges and the benefits in using this method. Finally, our work is put in relation to other work in the domain of agent-based modelling, and composition.

References to applications of the electricity sector are made throughout this paper to illustrate the use of the *dynamic agent composition* in a concrete manner. More details of the electricity models and simulation results can be found in (Boulaire et al. 2013a; Boulaire et al. 2012a; Boulaire et al. 2012b).

Motivating example of a dynamic agent composition

Before formalising what is meant by *dynamic agent composition*, this section motivates this approach by introducing a simple example of an electric vehicle (EV) agent. This EV agent is to be used within the context of planning the electricity grid, to understand how the increasing number of vehicles and the way they are used will impact the current infrastructure.

An EV is a mobile agent, which is not bound by its geographic characteristics. It can however have properties of location that will indicate where it connects to the grid to charge. It can be considered as a mobile battery, which can be limited to recharge only, but is also able to discharge to support the premise consumption it is attached to, if needed. It has a state of charge which is the amount of energy that is left in the battery, and from which charging requirements are calculated. Therefore it has similar properties to a battery with a few additional ones.

Following from this, an EV has at least two main properties, from the network viewpoint, that influence its behaviour: driving characteristics, and a charging regime. These need to be accounted for when implementing the rules within the EV agent. Figure 1 gives a schematic representation of the composition of an agent, which is made of an electric vehicle asset to which two behaviours are added. The asset includes all the physical and data attributes of the vehicle, such as location, battery capacity, state of charge, maximum charge rate.

While these two behaviours need to be described within the rules, each of them can have multiple implementations. For example, we can think of two implementations for the driving characteristics:

- The location parameter is informed by whether or not the vehicle is at a premise, using a Boolean variable that is set to true for the timesteps after which the EV has reached the premise. This time of arrival can be randomly chosen from a probability distribution curve derived from typical home arrival times. Similarly, its charging state is randomly set from probability distribution curves of typical vehicle trips;
- The location parameter is informed by a traffic simulation system that knows where and when the vehicle has arrived. Its charging state is calculated from the traffic simulation system that knows the exact trip the EV did for the day.

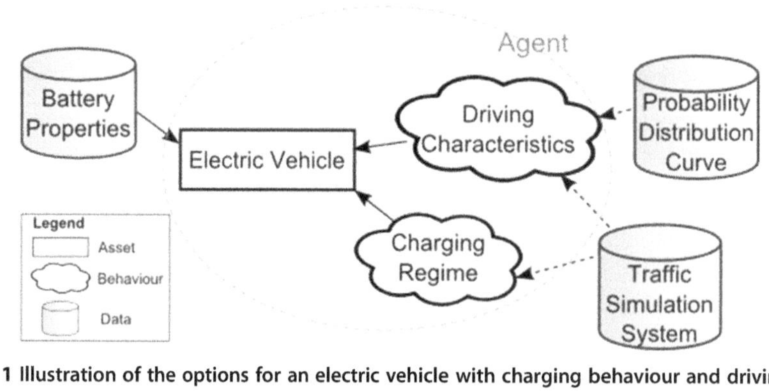

Figure 1 Illustration of the options for an electric vehicle with charging behaviour and driving behaviours.

Similarly, we can also have two implementations for the charging regime:

- The EV can charge at any time of the day, as soon as the signal is set for it (e.g. as soon as the vehicle arrives at the premise);
- The EV charges only during set periods of time, which can be defined by a policy setting (set) or informed from communication with a central controller (dynamic).

For each behaviour type, only two options are presented here, but we can imagine many more alternative behaviours.

If the behaviours and the static information were implemented in one same agent class, these alternative behaviours could be implemented by subclassing an electric vehicle agent class for example, or by calling behaviour objects defined in other classes. In this example, this would result in four agents that the user could choose from, which is equivalent to our implementation. However, when increasing the number of behaviours' options, using the dynamic agent composition would result in lesser implementation needs compared to subclassing existing agents implemented in one class.

Having that flexibility in combining the behaviours is important. Comparison of the impact of behaviours of agents is then facilitated by simply swapping a behaviour with another one.

Further, if a behaviour can be used by two asset types, this separation avoids writing extra code and enables reuse. This is the case for the charging regime in our example, which can be used not only to describe charging characteristics of an electric vehicle but also of a battery, whether it is privately owned or is grid operated. This property of reusability is important as it reduces the development time, and the risk of possible implementation mistakes.

In addition to these properties of flexibility and reusability, having the asset and the behaviours separated also offers ease in extending the model. Indeed, there is no need to modify the code of an agent if a new behaviour type is to be added, even if that behaviour is defined by an independent author. This is quite interesting, especially when more information becomes available as the project evolves, or when information/data changes, e.g. due to new applications.

Finally, data is also used to populate the agents' properties, both for assets and behaviours, which offers additional flexibility in the definition of the agents. For example, projections of EV uptake can be used to create x EV assets in year 1 of the simulation, x' in year 2, etc., and additional data to specify the properties of these assets. Behaviours can be associated to these assets following different percentages of expected behavioural profiles of their users passed in the data (e.g. y% of the population are expected to charge at anytime, and y' % at a set time). These behaviour percentages can also be varied to see the impact an incentive might have on the overall network (e.g. y' could be increased to find the percentage at which users should be encouraged to sign up to a tariff incentive, that would benefit the grid).

These different elements are brought together at runtime, creating agents and the agent-based model through a linking mechanism. This dynamic composition of the agents offers great flexibility and extensibility of the ABM, and means that a modeller does not necessarily need to program; they can just combine assets, behaviours and data to create an ABM. Details are given in the following sections.

Method

Overview of the dynamic agent composition

This section defines the dynamic agent composition and the different types of compositions that are possible, enabling flexibility, extensibility and reusability in the definition of an agent. Then an overview of how a large-scale ABM is built, using this composition, is given.

Definition of dynamic agent composition

We define *dynamic agent composition* as the process of bringing together at runtime the following distinct entities:

- An asset - the physical properties of an agent (static information);
- One or many behaviours - the rules the asset is subject to, to make its decisions (dynamic information);
- Data

where an agent is defined as:

$$Agent = Asset + Behaviours$$

and the data is used to populate either or both the asset and the behaviour attributes. Data is not a requirement, but offers greater flexibility and facilitates the creation of large-scale models. Combinations of these three entities are then held in a component, or module, for their implementation.

Figure 2 shows graphically this dynamic agent composition.

Here, when we mention one asset or one behaviour, we mean one class of asset and behaviour rather than an instance of an asset or behaviour. An asset class will typically have attributes and *setter/getter* methods to populate and access their values, while a behaviour class will contain the rules, held in *start(), step(), stop()* methods.

Data can be used not only to populate the asset and behaviour attributes, but also to determine how many of these are to be created, as well as what their relationship to one another is. Consequently, data reading is not happening at the individual agent level but at a higher level, the factory level, where data is used to populate the individual agents. More detail is given in section 5.

This type of breakdown of the agent into asset and behaviour is especially suited to our domain application, where the physical structure of the distribution network, which is quite static, is to be represented along with the way electricity is consumed or flows over it, which is dynamic. While changes in the infrastructure can happen with upgrades and extension of the network, these are quite slow compared to the dynamic behaviour of

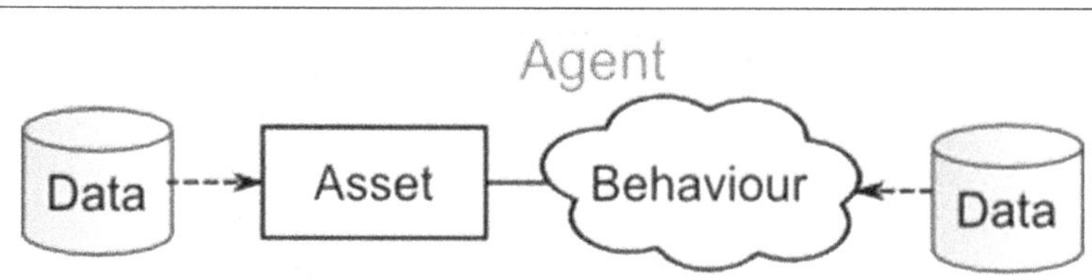

Figure 2 Dynamic agent composition.

the usage of electricity which is the model's state variable, whether it is described as a load, a voltage or a current in the simulations. The assets are likely to be used differently for reasons independent of the asset characteristics, although still within their properties' limits, such as when new policies are implemented that affect user behaviours. Being able to easily change the behaviour facilitates their quick assessment on the system as a whole.

Agent composition types

Many types of agent compositions can be done, which shows the properties of extensibility, flexibility and reusability in building an agent-based model. Figure 3 illustrates many of these compositions.

The base case when extending an agent-based model consists of creating a new agent, which means creating a new asset and its associated behaviours. With the agent composition, an agent-based model can be extended by simply defining new behaviours and applying them to an existing asset, increasing the number of available agent types. This is illustrated with Behaviour B1 for example, which also has alternative implementations (B1', and many up to B1^n).

In addition to extensibility, this example illustrates flexibility, as it is possible to choose any of the available behaviours for a new agent type.

Reusability is illustrated with behaviour B2 which can be used by both Asset A1 and Asset A2. In this case, the two assets have very distinct properties; however, one of their distinct features is that they have in common a set of rules to describe an aspect of their behaviour. This could be for example the case for batteries and electric vehicles, which could both be using the same rules for charging control algorithms.

In these three cases of agent composition, an independent author does not need to modify previously written code. New classes can be created, implementing interfaces to

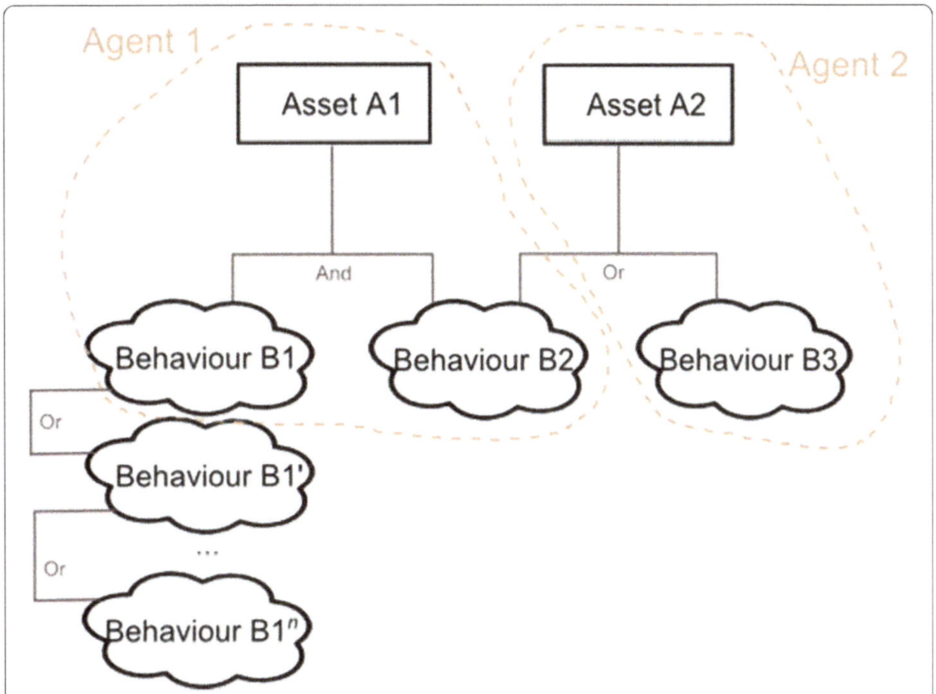

Figure 3 Example of the properties of extensibility, flexibility and reusability when creating agents.

specify whether they are assets or behaviours, and these can then be called at simulation setup to compose the required agent.

Figure 3 shows that an agent is composed of an asset and one or many behaviours. The opposite is also possible, where a behaviour can operate over one or many assets, and update their state at once, within one timestep. Such an example can be found when implementing a global voltage analysis algorithm (load flow (Morton 2003)) which runs over a group of assets, and updates the assets' voltage at each timestep. This is such that the group of assets over which the analysis is happening needs to be balanced, and therefore have a central place of calculation, considering all the assets at once.

Building the agent-based model

When defining an agent-based model, many agents are created and put in relation to one another to form the system over which they will evolve. Using dynamic agent composition, bringing the different assets, behaviours and data together will define a model. Figure 4 illustrates building blocks, or modules, holding these three entities (assets, behaviours and data) where MODAM is the framework that glues them together to form an agent-based model. These modules might contain information relating to many asset or behaviour types at once, or individual ones, depending on the needs. While assets and behaviours are defined individually in their own class, the modules enable them to be grouped together to form sub-systems.

In Figure 4, we can see four modules that contain information relating to assets describing the network: the network assets module (which contains lines, buses, transformers, switches, etc.), solar photovoltaic (PV), battery and EV assets modules. Four modules are also available for the description of the behaviours, where some can contain many behaviours (e.g. the network behaviours), and others share their behaviours across assets (e.g. the battery behaviour for EVs and batteries). Finally, many datasets are represented that inform the assets or the behaviours, and that can be interchanged or used in combination depending on the need of the model or the

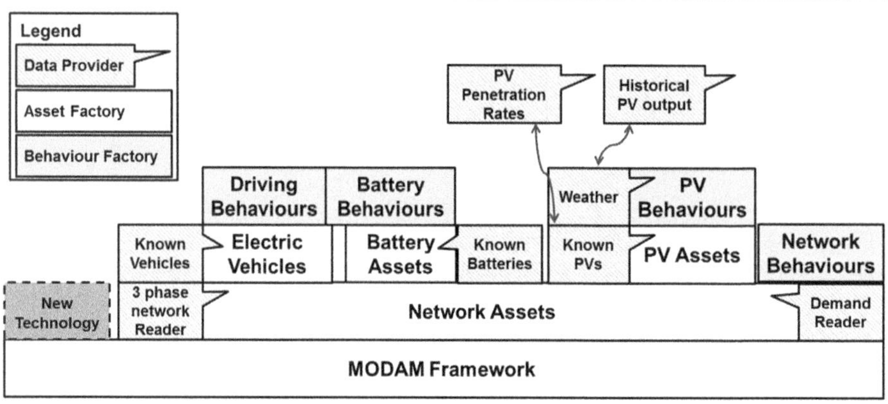

Figure 4 The MODAM framework is the foundation of the ABM model; it connects the different parts of the model. Here a network model describing the network elements and their behaviour is defined. Photovoltaic (PV), electric vehicle (EV) and battery modules have been added to understand how they can support or hinder the functioning of the network. This can be extended in many ways – for example, a 'New Technology' model could be defined and added to the tool to represent any new technology that might impact the network.

availability of the information. For example, for the solar PVs, the assets can be described using known information about their location and their characteristics (that can be used to initialise simulations from the end of the observation period), or using solar PV penetration rates (when making predictions about the future of expected placement and rating of the panels). Similarly, datasets can be used to inform the behaviours, as is the case for solar PV behaviours, which can use a weather model to calculate the output of the solar panels at individual locations taking into account the passage of clouds, or using historical data of solar PV output.

Having described the theoretical framework of the dynamic agent composition, the following section describes how it was implemented.

Results

Implementation of the dynamic agent composition

Our M&S application was developed using the Eclipse IDE (The Eclipse Foundation 2012) which is a widely-used open source platform made of a base workspace and an extensible plug-in system for customizing the environment. Using Eclipse on top of OSGi (Open Service Gateway initiative) and Eclipse plugins, which have strong support for modularity, supported our requirement of a flexible and extensible model environment, and was a natural fit to the definition of our components, or modules, which can each be implemented within their own plugin.

This section describes how our agents are created, using different approaches to the implementation of the assets and behaviours models, and how the data is used to populate them. But first, a UML diagram of the main interfaces used in the MODAM framework is presented in Figure 5; this diagram will be used to support our explanations throughout this section. As a first introduction, an interface named *IABMState* is at the centre of the framework and holds all the elements for a simulation. It sets a scheduler (*IScheduler*), so has access to the simulation time (*ISimTime*), and has access to assets (*IAsset*) which may have one or many behaviours (*IBehaviour*) associated with them. These assets and behaviours are created by factories (*IAssetFactory* and *IBehaviourFactory*) which are populated by data providers (*IDataProvider*). Each asset has a demand object (*IDemand*) that contains an extensible set of named values that can be set; these are the state variables of the simulation. Finally, the scheduler schedules the behaviours which update the demand at each time step during the simulation. More details for each of these entities are given throughout this section.

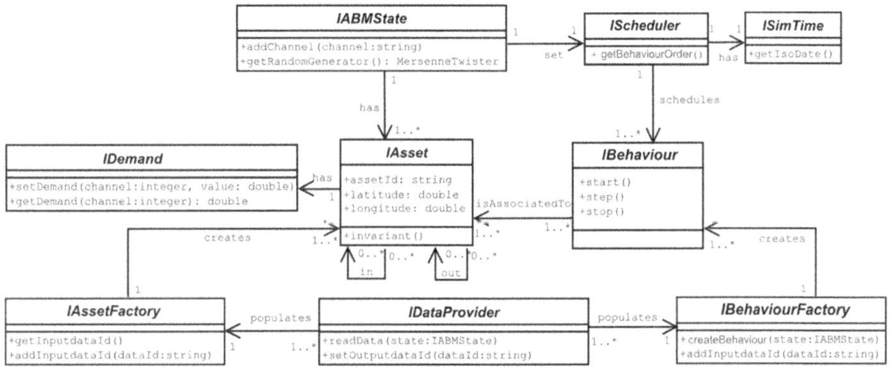

Figure 5 UML diagram of the MODAM framework.

Implementation of an agent

Assets and behaviours are implemented in different ways, according to their requirements.

The assets data model - use of EMF

EMF (Eclipse Modeling Framework (Steinberg et al. 2008)), one of the many available plugins of the Eclipse platform, is a modelling framework that facilitates building code based on structured data models. Particularly well suited to the requirements of our application domain, where the physical infrastructure is to be represented, EMF was chosen to develop a data model describing the assets and their relationship to one another. EMF has the advantage that the implementation of the objects specified in an *ecore* model, described in XMI (XML Metadata Interchange), can have their classes automatically generated, facilitating the implementation of the model within an application.

In addition to this, EMF can handle extension of models, a feature that was of particular interest to us. Any object declared in a data model can be extended or referenced by any other that has been defined in a new child model. Any EMF model can thus be created in a separate plugin and extend one or many models allowing the overall model to keep on expanding. Child models can be in distinct plugins, which makes it possible to choose one model over another at any time, if it describes the problem better, allowing for flexibility in the M&S application.

In our implementation, a few models were defined, which all extend a base model where two entities are defined: *IAsset* and *IDemand,* along with their properties. These entities are implemented as interfaces, as shown in Figure 5. Any other EMF model extending this one can then implement these interfaces, and define others as required. One of the main features of the *IAsset* interface is that it contains two properties (in and out) that define an in-out relationship that specify a directed graph over the assets, to represent the networked structure of our domain problem. In addition, each *IAsset* has a string as a unique ID, as well as a longitude and latitude to give its geographical location. Additional attributes can easily be added to assets, or to particular subclasses of assets, simply by extending the EMF model. This generates new subclasses of *IAsset* that have additional private fields plus getter and setter methods. This is highly customisable, but because it involves code generation and compilation, it is best used for relatively static models and is not sufficiently dynamic to support the kind of runtime composition of behaviours that we want.

Where dynamic composition of behaviours is required, a model designer can use the MODAM *data channels* feature, which is provided by an *IDemand* object associated with each asset. This provides an expandable set of named real-valued attributes for each asset. During the model initialisation phase, the behaviour factories register the channel names that they wish to use, and the MODAM framework maps these to integer indices, so the set of variable channels depends upon which behaviours are included in the model. As the model runs, behaviours can read and write the channel values of any asset. This allows behaviours written by different authors to communicate via data channels simply by using a common name for a channel. It also allows MODAM to provide a generic graphing and logging facility that can graph and save any channels from any set of assets. This can be used for visualisation in Google Earth to provide a platform for stakeholders' engagement.

Two EMF models have been implemented so far in our M&S application that both extend this base model, and have been used together to define the overall asset model. The first model contains the different assets that define a base network, as captured in Figure 4, in the Network Assets module, as well as solar PV and batteries. The second one contains one asset that describes the way a premise would consume electricity depending on the tariff it is subject to. This second model was implemented to test our hypothesis that it is possible to extend the data model using EMF and that this can be done within other plugins. This implementation was successful and the tariff asset was easily added to the model and integrated within the agent-based model.

The behaviours – use of the strategy pattern

Assets and behaviours are implemented independently; however, each behaviour has a reference to its asset in order to retrieve the necessary state variables and make its decisions during a simulation. The implementation of the behaviours' information and rules is contained in the *start()*, *step()* and *stop()* methods of its class that extend an *IBehaviour* interface. The *start()* method belonging to the corresponding asset allows initialising the behaviour, while the *stop()* ends it; the *step()* method updates the behaviour at every timestep.

To obtain flexibility in the behaviour implementation, we used the strategy pattern (Gamma 2009), using interfaces and defining classes that implement the *start()*, *step()*, and *stop()* methods. Any number of classes can implement the *IBehaviour* interface, and be called at runtime to specify which behaviour is to be used. One advantage of this approach is that a new plugin can easily define new behaviours, as long as access to the interface is provided; there is no need to access a behaviour class previously defined, only the interface.

Building an agent-based model

Building an agent-based model requires bringing together the assets and the behaviours that form agents and relate them to one another to form the complex system over which they will evolve. This is done within factories, using data from corporate datasets, as explained below.

Factories

Assets and behaviours are created separately, which is done automatically through the use of the *factory* pattern. A plugin can contain one or more asset factories that can control which assets need to be created; many factories can be defined for a given type of asset creation for example, with slight variations depending on the aim of the factory. The same is true for the behaviours.

These factories implement the *IAssetFactory* and *IBehaviourFactory*, depending on whether assets or behaviours need to be implemented. Using the *factory* pattern ensures that the action and interactions of the agents are taken care of in a consistent manner.

To answer our goal of flexibility, each factory typically creates assets or behaviours for one specific type of asset or type of behaviour. This means that each asset defined in an EMF model can implement its own *IAssetFactory*; similarly each behaviour type can implement its own *IBehaviourFactory*. It is however also possible to have a factory that will define many different entities and put them in relation to one another if judged appropriate; this is most likely to happen for the assets only, and is not recommended for the behaviours. An example of this is for the network asset module which contains one factory implementing *IAssetFactory* where lines, buses, switches, transformers and premises are created and related to one another, as assets form the underlying directed graph over which behaviours act.

Factories and data to populate the model

In each of these factories, data can be used to populate the assets and the behaviours. Different types of data can be used, that can define the number of assets to be created, how they are in relation to one another, or what their properties are. One of the requirements of our ABM was that the distribution network be built from corporate data. This means that different types of datasets needed to be handled, and that allowance be made for new types of dataset formats to be input to the simulation. For this, an interface called *IDataProvider* is accessible by both the *IBehaviourFactory* and the *IAssetFactory* during the model instantiation and is implemented by different readers that access various data formats.

Data offers flexibility in setting up simulations as it can be used to represent different areas of study, to compare different trajectories of possible futures by setting different methods of technology uptake, or different methods to describe behaviours. For example a demand behaviour which is associated to a premise and represents its electricity consumption for every half-hour of a day can be set using three types of data in our implementation: half-hourly profile data from a sample of premises, half-hourly profile data from some feeders, and profiles derived from a weather-driven model of consumption. Any of these methods can be chosen to populate the behaviour of a premise consumption individually, and can also be combined using weights to obtain a desired proportion of profile methods.

Depending on the provenance of the data, the format will change, which is handled by different implementations of the same interfaces, providing flexibility in composing the agent-based model. As an example, two types of networks can be used in our current implementation of the agent-based model: a three phase urban network and a SWER (Single Wired Earth Return) network, with data coming from different corporate databases with different formats.

How the ABM comes together

Large-scale agent-based models can be built using dynamic agent composition, either:

- Via explicit Java code, or
- Via command line configuration or GUI, and the use of an automated model builder.

When using explicit Java code, the programmer needs to instantiate the factories, link them to the desired data providers, and execute the factories to build the model. However, this process was automated to answer our goal of code-free construction of agent-based models. This automation is the subject of another paper, but we are giving here an overview of the automation here. A *Module Manager* automatically discovers the plugins and weaves them together. This means that it finds all the available plugins in the registry and enables those chosen by the user. If there are missing plugins, these are found and added automatically. The assets and behaviours are then created with the required data by the Module Manager who instantiates and parameterises the asset and behaviour factories. Each factory handles its own options that have been given in the command line or GUI panel (parameters, and data). Then methods are called on these factories to populate the assets and behaviours with the required data, using reflection, by just knowing the type of interface they implement.

Discussion

Dynamic agent composition: challenges and responses

While the concept of dynamic agent composition is quite simple, its application to the implementation of large-scale agent-based models with an underlying networked structure has its challenges. These are described below with a discussion about the way we responded to them.

Ordering of the assets' creation

Extending an existing structure, which can be represented by a directed graph, such as the electricity grid where the nodes are assets and the edges their connections, requires a notion of reference. This can be challenging especially as the assets may be defined in separate factories, in separate plugins, and come together to form the complex system only at runtime. For example, if adding battery assets (B), they need to be created and attached to the relevant node (N) (e.g. a premise) in the network. This means that N need to have been created first, and put in relation to the other assets in the initial network. Only then will the battery asset factories be called to create B and attach them to the right nodes N.

To satisfy this requirement of reference over the assets, precedence of the creation of some assets over others was determined. This requires the asset factories to be ordered. If the assets are all created within one factory, their ordering can be handled by the developer within that factory. However, if the assets are defined in independently developed factories, there is a need to mention the order in which the assets need to be created, and consequently the order in which the factories are being called, which can be automatically ordered by the Module Manager. This type of dependency is one of the consequences of aiming at creating an extensible framework, and is the equivalent of inheritance in object programing.

In order to solve this problem, partial ordering of asset factories is used where an attribute (*Predecessors*) allows defining which other factories need to be called before this one. Predecessors that are not included in the current model are ignored, so that maximum flexibility is allowed. For example, if factory F has predecessors A and B, it is possible to run models with any combination of F, A and B. If only A and F are part of the model, then F will automatically be run after A, while B is ignored.

Because of the separation of the *Assets* and the *Behaviours*, this ordering is only necessary on the assets which describe the underlying network structure of the model. Behaviour factories are called after all the asset factories, in any order, since behaviour creations are independent of each other - they communicate only via the assets.

Ordering of the agents' execution

In a "classic" agent-based implementation, agents are often instantiated and scheduled in a central location, e.g. in Repast all the agents are defined in a class implementing *ContextBuilder* (Collier 2014), and in MASON (Luke et al. 2005) in a class extending *SimState*. This means that it is possible to not only create the agents in a given order but also order the agents execution within a time step in relation to other agents. The same capabilities needed to be available with our dynamic composition. As mentioned previously, ordering the behaviours' creation is not important in MODAM, only the ordering of the assets' creation is. However, ordering the execution of the behaviours (i.e. within one step of a simulation) is extremely important, and is necessary to enable deterministic simulation results, which is one of our M&S application requirements for verification purposes.

When talking about scheduling of agents, we do not mean that there exists a central planner that will decide on the actions of the agents but rather on the timing of these actions, which corresponds to the execution of the behaviours. The actions themselves are still undertaken in an autonomous manner by the various behaviours. At the start of the simulation, a global scheduler analyses the dependencies between the behaviours and sorts them into a safe execution order.

Each *BehaviourFactory* creates a set of behaviours, and groups them into named *behaviour groups,* which are used to help order the execution of the behaviours within one timestep. One or many behaviour groups can be defined within a single factory. For example, a BatteryBehaviourFactory could create two types of battery behaviours with different battery control algorithms but that can be executed within the same time step with no specific order. In this case, both kinds of battery behaviours will be assigned the same group. However, if one type of behaviours needs to be executed before the other within the same timestep, an order needs to be specified. For example, if the premise batteries need to be executed first, followed by the grid battery to support the network voltage, then two groups, a PremiseBattery group and a GridBattery group, would be created. This would then allow the ordering to be fully managed by the user who can specify when these groups of behaviours need to be executed. The ordering of the behaviour groups is set through an argument in the command line, followed by the name of the behaviour factory.

An example of behaviour group ordering is given in Figure 6. It shows three behaviour groups, two of which can be run in parallel (*BehaviourGroup A* and *BehaviourGroup B*), that is with no particular order, with the third one requiring its behaviour to be called after both of them. In each of them we have three behaviours: *BehaviourGroup A* and C have their behaviours ordered sequentially, and *BehaviourGroup B* has two behaviours that are ordered sequentially (b_1 and b_2) and one that can be called anytime (b_3). In the example given in Figure 6, the ordering argument in the command line looks like:

$$-order = (down\ (BehaviourGroup\ A)|up(BehaviourGroup\ B))\ ;\ up(BehaviourGroup\ C)$$

where *up* stands for bottom-up, and *down* for top-down ordering; | shows that *BehaviourGroup A* and *BehaviourGroup B* can be ordered in parallel, and the semi-colon (;) is to show sequential ordering.

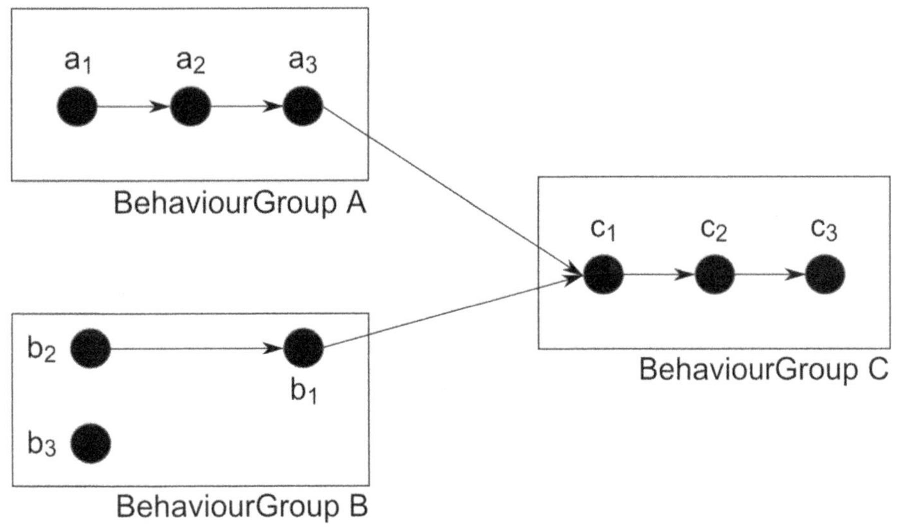

Figure 6 Example of ordering of behaviours within and amongst behaviour groups.

State of the agents at a global scale

Because agents can be developed independently and only come together at runtime to create the agent-based model, keeping track of the state variables can be challenging. Indeed, with a fragmented approach, it is expected that different types of state variables are defined by the developers; however, they still need to be accessed within the *ABMState* to allow tracking the state of the system, and also for other agents to be able to access their value to make their own decisions. To remedy this challenge, we used channels, as explained above, which are effectively globally-named and are typically used as state variables of the simulation. This has the additional advantage that being global variables, they do not necessarily need to be state variables but can also be used for other purposes, such as environmental observation (e.g. local temperature, humidity) or globally observable behaviour attributes (e.g. battery charging strategy).

Explosion of the number of assets and behaviours options

The flexibility in separating Assets and Behaviours can have its downfall. It can quickly become very difficult to know what types of assets and behaviours are currently implemented and which ones are able to come together to form meaningful agents. In addition, knowing what types of parameters or data providers are settable from the command line can also quickly become overwhelming.

One of the principles to follow in this case is that, just because it is possible to break down the system into many simpler components does not mean that it should be done. In some cases, it might be advantageous to keep some groups of assets together, or have many rules within one behaviour especially if it will not be reused elsewhere in the future. This concerns mainly the implementation of the factories, and still implies the separation of assets and behaviours, however many assets or behaviours can be created and put in relation within one factory. An example of this, previously mentioned in this paper, is the asset network factory, which creates many different types of assets (lines, buses, switches, transformers, and premises) and puts them in relation to one another. This implementation was chosen because information about these assets and their connections were available at once, and contained in files. Also, within the context of our domain application, this represents the current configuration of the distribution network over which transformations will happen. While a factory was defined in this way, it would still have been possible to create each set of assets independently and create the network using partial order.

Despite using this principle, it is still expected that a large number of factories and datasets will be defined and it can be difficult to keep track of which ones are available. This can be supported by good documentation of the software. To facilitate this, we automated the process of documentation, so that as the model is growing, so is the documentation. Documentation is not only useful for the programmer who would like to add a new plugin for example, but also for the user who will not have access to the code, and who will not want to have to go through it. We have used annotations on asset and behaviour factories and on the methods within the classes that are used for input parameters. These annotations are discovered automatically by the documentation generator, and used to generate consistent documentation for the parameters and required data providers of each factory.

Finally, as the model grows, so will the command line. To prevent having too many parameters to define, and also build on previous simulation runs, it is possible to use a configuration file previously saved, to which additional factories, data providers and parameters are specified. This further extends our goal of flexibility in setting up ABM simulations.

Benefits in using a dynamic agent composition

Benefits in having a clear structure of large-scale agent-based models, through the application of the dynamic agent composition, can be identified from the point of view of the software developers for which it was initially designed. Additional benefits can be identified from the perspective of the clients, or users, as well as for agent-based modelling as a scientific approach. These are discussed below.

Benefits from the point of view of the software developers

The dynamic agent composition was adopted to answer shortfalls initially identified when using existing model building approaches and software systems such as Repast and MASON, see Table 1. This approach enabled flexibility and extensibility of both the model definition and its implementation.

Thanks to the distinction between Assets and Behaviours it is easy to change the behaviour of the entities represented in the model. This sometimes needs to happen not only during the model creation but also during the simulation run where they can be added to an existing asset. In ((North and Macal 2007), chapter 7), the authors explain four model growth paths when building agent-based models: the addition of compatible behaviours, contentious behaviours, compatible agents and contentious agents. These four model growth paths are fairly common, and are supported by the dynamic agent composition approach.

Further, the separation into Assets and Behaviours allows gathering certain entities into groups, when relevant, where all assets that are in relation to one another can be defined at once, while different behaviours can be tried separately over each of them without additional development time. This mix-and-match of entities to create agents can be done at runtime without the need to modify any code; only the command line argument will be altered.

Additionally, having the data separated from the assets and behaviours also makes it easier and faster to extend the model. For example, if a different network is to be modelled, only the asset classes will need to be informed by different data. A new data reader might be implemented if the data format is different; the rest of the code describing the asset and behaviour properties will remain the same. This new data provider will then be called with its associated file at simulation setup, the rest of the command argument remaining the same. This simplifies handling complex options as these are only set in the command line and do not need to be defined in a central class within the code.

Finally, this separation of the agent's aspects into assets, behaviours and data, allows starting an implementation of a model without needing access to all the required information, whether it is data, assets or even a rule that defines the agents' behaviour. This enables an agile implementation (Thomas and Hansmann 2010; Dingsøyr et al. 2010) of the agent-based model. Also, because the dynamic agent composition is implemented in plugins, it is possible for independent authors to contribute to the model in parallel, which can greatly increase implementation time.

Benefits from the point of view of the users

MODAM was also developed to answer requirements from the users' perspective, which the dynamic agent composition enabled.

Various scenarios can be created with ease by simply changing parameters values, data defining the underlying structure of the network but also behaviours. Having the behaviours independent of the assets makes it easy to add and remove behaviours and assess their impact on the system by setting multiple simulation runs. Further, independent teams can have different implementations of a behaviour, providing customisation of their model to better answer the needs of their analyses. In both cases, there is no need to get into the code; the user can simply bring the building blocks together, by specifying them in the command line.

It is also possible to have a mix of behaviour methods to inform a specific type of assets. For example, if it is expected that a given proportion of the population will behave in a certain way, and the rest in another when using a given asset, these two behaviour types can be applied to the model with varying levels by simply calling on these two behaviours and setting a proportion parameter in the command line. This allocation can be done evenly over the population, or be influenced by known factors such as demographic or geographic characteristics. This has the advantage to represent more accurately how the system might evolve if these proportions are known. When unknown, sensitivity analyses over these allocation levels can be performed to find out what mix would be best for the system. This might be useful for educators, for example, who are trying to bring behavioural change and need to find out the population size to target. Simply varying behaviour calls and parameters values without coding will enable them to quickly set up scenarios.

Finally, the use of data to populate the agents offers the advantage of accurately representing a system, which is not widely done especially for large-scale ABMs, and can be of great benefit for the user. Indeed, many large-scale ABMs are currently developed using taxonomies of agents and probability distributions to represent the system. While such an approach can still be taken using MODAM, we are able to use specific data of the domain of study, which offers the additional benefit of greater fidelity to the system represented. For our electricity network, our asset factories constructed assets based on a data file extracted directly from the Ergon Energy database, so had access to the real attributes of each asset, and their connection to one another. This was a requirement of our client, who is interested in knowing as accurately as possible what is likely to happen on their network at specific locations.

Benefits from the point of view of ABM as a scientific approach

Taking the dynamic agent composition approach when developing MODAM also highlighted advantages in terms of agent-based modelling as a scientific approach.

The separation of an agent into asset and behaviours creates a natural simulation space over which ABMs of networked structures are represented. Indeed, the assets and their connections form a complex network representing the overall structure, which becomes the space over which the behaviours interact with one another. In many simulations such as the Schellings' model (Schelling 1971), a grid is defined as a 2D matrix, over which the behaviours will evolve and get information to make their decision. Here, the idea is similar where the environment is represented by a scale-free network made of the assets which are publicly available within the model and allow every entity to know their relationship to one another. The behaviours are not bound by structure directly but access the underlying network through their asset. The behaviours only contain private data on which they make their decisions. The assets hold the state variables of the ABM simulation which are modified by the behaviours as they make their decisions. While the assets' connections form a network of their own, references to their geo-location (longitude and latitude) are maintained, allowing displaying the network using spatial information software.

MODAM allows replicability of simulations results, thanks to its deterministic capability through each random seed. While independence of the execution of the agents is still ensured through randomisation of their decisions output, having a deterministic order of their execution allows replicability of the experiment and reproducibility of the results.

Finally, MODAM facilitates model comparisons and validation of behavioural sub models. Indeed, data can be used to set the parameters of sub models, that some behaviours use to inform their decisions. As an example, our implementation of solar PV behaviours can be informed by historical solar PV output data or weather data, see Figure 4. The weather data is used to populate a weather-driven model of expected electrical output of solar PV subject to weather with the passage of clouds, while the historical solar PV data simply gives the electrical output of specific solar PVs recorded over a period of time. Because these two approaches require different data input, two data readers were written to be used by each method. When the user chooses their preferred behaviour method, the required data provider will then be called upon. This has the advantage of extending the model, but also offers model comparison and validation of the behavioural sub models. Indeed, the weather-driven solar PV output model could be validated by comparing its output with the actual observations of solar PV output. The dynamic agent composition therefore has the additional benefit that validation of models can be done easily by setting two simulations and simply changing the datasets and data providers, and comparing their outputs.

Finally, MODAM was developed in Java using open-source frameworks (Eclipse, EMF) and standards (OSGi), ensuring that it is soundly constructed, but also enabling it to be reproduced or extended by interested users.

Related work

Agent-based modelling, a bottom-up modelling technique, describes autonomous agents and their relationships at a fine level of detail with the view of capturing the dynamics of a complex system (Macal and North 2010; Bonabeau 2002). Many toolkits are available to support the development of agent-based models (Nikolai and Madey 2009; Berryman 2008; Railsback et al. 2006; Najlis et al. 2001; North 2013; Luke et al. 2005) with most containing the following features: agents, a scheduler, an interaction space, random number streams, logging and a user interface (North 2013). In these toolkits, agents are generally made up of a unique identifier, behaviours that can be activated and attributes that can be modified. Both the static information and the behaviours are then held in one place, often defined within a class, because object-oriented programing is well suited for agent-based modelling implementation. While held in one class, however, behaviour implementations might still be the result of the composition of behaviours extending others, as is the case for example for the JADE architecture (Bellifemine et al. 2007), where these are extending behaviour classes (Bellifemine et al. 2010). However, these behaviours still are to be added within an agent's implementation, whose architecture is partly hidden, and which requires coding to define the agent. The dynamic agent composition presented in this paper distinguishes itself from these ones as behaviours can be combined without the need to access the agent's code. The behaviours are combined with an asset to form an agent, which can be done by a non-programmer, through a command line, and this composition of the agent happens at runtime.

Our dynamic approach of composing the agents rather follows some of the principles described in MALEVA (Briot et al. 2006) where composability of behaviours is described. While similarities exist between our conceptual frameworks, with behaviours being composed to form a more complex one, they differ in some aspects. MALEVA uses connectors and has output interfaces so that the output of one behaviour is the input of another one to form a chain of complex behaviours. This is not our chosen approach as we are rather more interested in alternative behaviours. While we can use many behaviours to compose one, we do not have this element of precedence of the way the behaviour is executed. Further, our composition concentrates on bringing asset characteristics and behaviours together to form an agent rather than bringing behaviours together, whose need arose from the requirements of growing models, especially when dealing with large-scale ABMs. Because our behaviours communicate via the assets, we also have more flexibility to mix-and-match of the different combinations than MALEVA. Our approach is rather closer to the one specified in (Bae et al. 2012) where a hierarchy of models composed of an action model, an agent model and a multi-agents model has been defined. This is done so that agent-based models can be built incrementally and in a flexible manner, which goal is the same as ours. The authors have presented a formal specification using DEVS (Discrete Event System Specification) formalism (Zeigler 1976), which they have applied to show that two models could be formalised independently and brought together. It is unclear however that more than two models could be brought together easily, and how large the agent-based models can be. Also, to our knowledge, there is currently no implementation platform to support this specification.

Growing large-scale ABMs is not well documented in the literature, which mainly concentrates on the speed of execution of simulations rather than the modelling needs. However, Parry, in (Parry 2012) differentiates two problems when increasing the scale of a multi-agent system, which includes the computational resources but also the increase in difficulty with a growing agent-based model. Despite this distinction, most of the paper however focusses on how to deal with large-scale simulations which suggests optimising existing code, considering simple solutions such as upgrade of the hardware or evaluating the suitability of the chosen scaling solution on a simplified version of the model. Other approaches to dealing with large-scale simulation requirements is the use of alternative computational techniques, such as considering distributed or parallel implementations of the agent-based model. To this end, simulation toolkits now offer parallel and distributed implementations of their initial implementations as is the case for Repast and MASON for example (Cordasco et al. 2013; Collier 2013), amongst many others. These allow running larger simulations while still getting reasonable execution time, as described in (Parker 2007), where an epidemic simulation runs up to 6 billion agents using a distributed simulation. While the focus of this paper is not on large-scale simulations but rather large-scale model, such challenge is also at the heart of our problem. For this, we have implemented a parallel implementation of our ABM scheduler, as described in (Boulaire et al. 2013b), however, this is not the subject of this paper.

Conclusion

This paper has defined *dynamic agent composition*, a novel approach to build large-scale ABMs. This approach had for goal to extend, with ease, an agent-based model with an underlying networked structure. Also, it aimed at having it flexible so that many scenarios could be created using large corporate databases, without the need for a programmer to build the simulations. By breaking down the model into components containing the data, the assets and the behaviour descriptions, and providing a mechanism to bring them together at runtime to compose the agent-based model, this was achieved.

This approach has many advantages over the way agent-based models are traditionally built for the user as well as the developer. Developers can extend the model without the need to access or modify previously written code; they can develop groups of assets and behaviours independently and add them to those already defined to extend the model. The model can then evolve as new agents need to be modelled, which facilitates the models to be used and extended after previously-defined goals are modified. Users can mix-and-match already implemented asset and behaviour components to form large-scales ABMs. This allows them to quickly setup simulations and easily compare various scenarios without the need to program.

Further, using data extracted from corporate databases to populate the ABM enables to represent accurately the physical infrastructure over which the agents evolve. This aspect of our approach contributes to the application of ABMs to the electricity domain as most ABMs use taxonomies or probability distribution to represent the network under study.

Future work includes continued expansion of the model to include more asset types as well as behaviours of assets' usage, such as small-scale generators other than solar PVs, and electric vehicles with different charging methods. While currently applied to the electricity distribution grid only, it is expected this approach can be used more broadly and be of benefit to other applications, especially those that have a networked structure, such as water or gas networks.

Competing interests

The authors declare that they have no competing interests.

Authors' contributions

FB: has been lead author in drafting the manuscript, has designed and implemented the dynamic agent composition method, has conducted experiments, and has revised the paper. MU: has made key contributions to the paper conception, has aided in the design of the dynamic agent composition method and co-implemented it. RD: has reviewed and revised the paper for important intellectual content. All authors read and approved the final manuscript.

Acknowledgement

The authors gratefully acknowledge the funding through the NIRAP grant, which is making this research possible, the contributions of diverse partners on this project, and especially Ergon Energy for supplying us with data of their network and guidance in terms of tools requirements.

Author details

[1]Queensland University of Technology, Gardens Point – P Block level 8, Brisbane Qld 4000, Australia. [2]University of the Sunshine Coast, Sippy Downs QLD 4556, Australia. [3]Work undertaken at QUT, now at University of the Sunshine Coast, Sippy Downs QLD 4556, Australia.

References

Argonne National Laboratory. The Repast Suite. 2014. http://repast.sourceforge.net/. Accessed 11/11/2014 2014.

Bae JW, Lee G, Moon I-C. Formal specification supporting incremental and flexible agent-based modeling. 2012 Winter Simulation Conference; 2012. Berlin, Germany: IEEE; 2012.

Batten DF, Grozev G, NEMSIM. Finding ways to reduce greenhouse Gas emissions using multi-agent electricity modelling. Complex science for a complex world: exploring human ecosystems with agents. vol Book. Canberra: ANU E Press; 2006. p. 227–52.

Bellifemine FL, Caire G, Greenwood D. Developing multi-agent systems with JADE. Wiley series in agent technology, vol book, whole. Hoboken, New Jersey, USA: Wiley-Blackwell; 2007.

Bellifemine F, Caire G, Trucco T, Rimassa G. JADE Programmer's Guide 2010. Free Software Foundation. Boston, MA, USA

Berryman M. Review of Software Platforms for Agent Based Models. In: Department of Defense, editor. Edinburgh, South Australia, Australia: Defence Science Technology Organisation; 2008.

Bonabeau E. Agent-based modeling: methods and techniques for simulating human systems. Proc Natl Acad Sci U S A. 2002;99(Suppl 3(3)):7280–7. doi:10.1073/pnas.082080899.

Boulaire F, Utting M, Drogemuller R, Abeygunawardana A, Ledwich G, Bell J. Planning for the impact of distributed solar energy on the grid. Solar 2012 Conference; 6–7 December 2012. Melbourne: Swinburne University of Technology; 2012a.

Boulaire F, Utting M, Drogemuller R, Ledwich G, Ziari I. A hybrid simulation framework to assess the impact of renewable generators on a distribution network. 2012 Winter Simulation Conference; 9–12 December 2012. Berlin, Germany: IEEE; 2012b.

Boulaire F, Utting M, Drogemuller R, editors. MODAM: A MODular Agent-based Modelling Framework. 2nd International Workshop on Software Engineering Challenges for the Smart Grid as part of 35th International Conference on Software Engineering (ICSE 2013); 2013a 18–26 May 2013; San Fransisco, CA, USA: IEEE Press

Boulaire F, Utting M, Drogemuller R. Parallel ABM for electricity distribution grids: a case study. 1st Workshop on Parallel and Distributed Agent-Based Simulations, Euro-Par 2013; 26/08/2013. Aachen, Germany: Lecture Notes in Computer Science; 2013b.

Briot J-P, Meurisse T. A Component-based Model of Agent Behaviors for Multi-Agent-Based Simulations. Proceedings of the 7th International Workshop on Multi-Agent-Based Simulation (MABS'06), 5th International Joint Conference on Autonomous Agents and Multiagent Systems (AAMAS'2006). New York, NY, USA: Association for Computing Machinery (ACM); 2006. p. 183–90.

Cai C, Jahangiri P, Thomas AG, Zhao H, Aliprantis DC, Tesfatsion L. Agent-based simulation of distribution systems with high penetration of photovoltaic generation 2011 24-29/07/2011. San Diego, CA: Power and Energy Society General Meeting, IEEE; 2011.

Castiglione F. Agent based modeling. Scholarpedia. 2006;1(10):1562. doi:10.4249/scholarpedia.1562.

Collier, N. Interface ContextBuilder<T>. 2014 http://repast.sourceforge.net/docs/api/repast_simphony/index.html.

Collier, N. Repast HPC Manual. 2013. http://repast.sourceforge.net/docs/RepastHPCManual.pdf.

Cordasco G, Chiara RD, Raia F, Scarano V, Spagnuolo C, Vicidomini L. Designing computational steering facilities for distributed agent based simulations. Proceedings of the 2013 ACM SIGSIM conference on Principles of advanced discrete simulation. Montreal, Quebec, Canada. 2486147: ACM; 2013. p. 385–90.

del Valle Y, Venayagamoorthy GK, Mohagheghi S, Hernandez JC, Harley RG. Particle swarm optimization: basic concepts, variants and applications in power systems. IEEE Trans Evol Comput. 2008;12(2):171–95. doi:10.1109/tevc.2007.896686.

Dingsøyr T, Dybå T, Moe N. Agile Software Development: An Introduction and Overview. In: Dingsøyr T, Dybå T, Moe NB, editors. Agile Software Development. Berlin, Germany: Springer Berlin Heidelberg; 2010. p. 1–13.

Ergon Energy. Corporate profile. 2013. https://www.ergon.com.au/about-us/who-we-are/our-company/corporate-profile. Accessed 02/06/2013 2013.

Gamma E. Design patterns: elements of reusable object-oriented software. vol Book, Whole. Boston: Addison-Wesley; 2009.

Hamill L. Agent-based modelling: The next 15 years. JASSS. 2010;13(4):7.

Institute for Energy and Transport. Agent Based Modelling for Smart Grids. European Comission. 2014. http://ses.jrc.ec.europa.eu/agent-based-modelling-smart-grids. Accessed 20/07/2014 2014.

Klügl F, Bazzan ALC. Agent-based modeling and simulation. AI Mag. 2012;33(3):29–40.

Luke S, Cioffi-Revilla C, Panait L, Sullivan K, Balan G. MASON: a multi-agent simulation environment. Simulation. 2005;82(7):517–27.

Macal CM, North MJ. Tutorial on agent-based modelling and simulation. J Simul. 2010;4:151–62.

Macal CM, North MJ, editors. Tutorial On Agent-Based Modeling And Simulation Part 2: How To Model With Agents. Winter Simulation Conference; 2006 December 3–6, 2006; Monterey, California, USA; Sl.l: Omni press

Morton A. A fast 'do-it-yourself' load flow algorithm for power systems with sparse topology. AUPEC 2003 Australasian Universities Power Engineering Conference. Christchurch, New Zealand: University of Canterbury, NZ; 2003.

Najlis R, Janssen MA, Parker DC, editors. Software Tools and Communication Issues. Proceedings of a Special Workshop on Land-Use/Land-Cover Change; 2001 04-07/10/2001; Center for Spatially Integrated Social Science, University of California at Santa Barbara: Irvine, California 2002.

Nikolai C, Madey G. Tools of the trade: a survey of various agent based modeling platforms. J Artif Soc Soc Simul. 2009;12(2):2.

North MJ. A theoretical formalism for analyzing agent-based models. Complex Adaptive Systems Modeling. 2013;2(1):3. doi:10.1186/2194-3206-2-3.

North MJ, Macal CM. Managing Business Complexity. vol Book, Whole. New York, NY, USA: Oxford University Press; 2007.

North M, Conzelmann G, Koritarov V, Macal C, Thimmapuram P, Veselka T. E-laboratories : agent-based modeling of electricity markets. American Power Conference; 15-17/04/2002. Chicago, IL (US): Argonne National Lab., IL (US); 2002.

Parker J. A flexible, large-scale, distributed agent based epidemic model. 2007 Winter Simulation Conference; 2007. Washington, DC, USA: IEEE Press; 2007.

Parry HR. Agent based modeling, large scale simulations. New York, NY: Springer New York; 2012. p. 76–87.

Railsback SF, Lytinen SL, Jackson SK. Agent-based simulation platforms: review and development recommendations. Simulation. 2006;82(9):609–23.

Schelling TC. Dynamic models of segregation. J Math Sociol. 1971;1:143–86.

Steinberg D, Budinsky F, Paternostro M, Merks E. EMF: Eclipse Modeling Framework Eclipse Series. Boston, MA, USA: Addison-Wesley Professional; 2008.

The Eclipse Foundation. About the Eclipse Foundation. The Eclipse Foundation. 2012. http://www.eclipse.org/org/. Accessed 27/02/2012 2012.

Thomas S, Hansmann U. Overview of Agile Software Development. Agile Software Development: Best Practices for Large Software Development Projects. vol Book, Whole. Berlin, Heidelberg: Springer Berlin Heidelberg; 2010. p. 35–9.

Weidlich A. Engineering interrelated electricity markets: an agent-based computational approach. vol Book, Whole. Heidelberg: Springer; 2008.

Zeigler BP. Theory of modelling and simulation. vol Book, Whole. New York, N.Y: Wiley; 1976.

9

Human development dynamics: an agent based simulation of macro social systems and individual heterogeneous evolutionary games

Mark Abdollahian[1], Zining Yang[1*], Travis Coan[2] and Birol Yesilada[3]

* Correspondence:
zining.yang@cgu.edu
[1]School of Politics and Economics, Claremont Graduate University, 170 E. Tenth St. Claremont, CA 91711, USA
Full list of author information is available at the end of the article

Abstract

Purpose: In the context of modernization and development, a complex adaptive systems framework can help address the coupling of macro social constraint and opportunity with individual agency. Combining system dynamics and agent based modeling, we formalize a simulation approach of the Human Development (HD) perspective to explore the interactive effects of economics, culture, society and politics across multiple human scales.

Methods: Based on a system of asymmetric, coupled nonlinear equations, we first capture the core qualitative logic of HD theory, empirically validated from World Values Survey (WVS) data. Using a simple evolutionary game approach, second we fuse endogenously derived individual socio-economic attribute changes with Prisoner's Dilemma in an agent based model of the interactive political-cultural effects of heterogeneous, spatial intra-societal economic transactions. We then explore a new human development dynamics (HDD) model behavior via quasi-global simulation methods to identify paths and pitfalls towards economic development, cultural plasticity, social and political change behavior.

Results: Our preliminary results suggest strong nonlinear path dependence and complexity in three areas: adaptive development processes, co-evolutionary societal transactions and near equilibrium development trajectories, with significant implications for anticipating and managing positive development outcomes. Strong local epistatic interactions characterized by adaptive co evolution, shape higher order global conditions and ultimately societal outcomes.

Conclusions: Techno-social simulations such as this can provide scholars and policymakers alike insights into the nonlinear, complex adaptive effects of societal co-evolution. We believe complex adaptive or evolutionary systems approaches are necessary to understand both near and potentially catastrophic, far-from-equilibrium behavior and societal outcomes across all human scales of modernization.

Keywords: Economic development; Modernization; Cultural shift; Democratization; Game theory; Agent based model; Techno-social simulation; Co-evolution; Complex adaptive systems

Background

Since the turn of the 20th century, social scientists have identified dynamic linkages between economic development, cultural change, and political development with popularized arguments for and against modernization (Huntington 1996; Chua 2003; Sachs 2005). However, recent international events amplify the dynamism of social forces. Revolutions, international terrorism with economic development (Barnett 2004), instability via globalization (Rodrik 1998), modernization's "clash of civilizations" (Huntington 1996), the interlocks between politics and economics (Acemoglu and Robinson 2012) and the interdisciplinary Human Development (HD) perspective (Inglehart and Welzel 2005) are but some examples. Across all, positive feedback mechanisms shift seemingly stable system phenomena towards complexity and catastrophic, far-from-equilibria conditions (Thom 1977; Cambel 1993; Arthur 1994a, b;). Such events force our current scientific understanding to change perhaps radically, considering new possibilities and outcomes that integrate explanations across previously disparate fields.

Rooted in comparative political economy, the HD perspective is a qualitative, trans-disciplinary approach to understanding modernization and development through the lens of interdependent economic, cultural, social and political forces at multiple scales, from individuals to institutions and society as a whole. Here we extend previous work by Abdollahian et al.'s (2012) novel, quantitative systems dynamic representation of HD theory at the societal level towards integrated macro–micro scales in an agent based framework. As macroscopic structures emerging from microscopic events lead to entrainment and modification of both, co-evolutionary processes are created over time (Prigogine and Stengers 1984). Quek et al. (2009) also design an interactive macro–micro agent based framework, which they call a spatial Evolutionary Multi-Agent Social Network (EMAS), on the dynamics of civil violence. We posit a new, Human Development Dynamics (HDD) approach where agency matters: individual game interactions, strategy decisions and outcome histories determine an individual's experience. These decisions are constrained or incentivized by the changing macro economic, cultural, social and political environment via HD theory, conditioned on individual attributes at any particular time. Emergent behavior results from individuals' current feasible choice set, conditioned upon past behavior and macro outcomes. Conversely, progress on economic development, the formation of cultural mores, societal norms and democratic preferences emerge from individuals' behavior interactions.

In order to create a robust techno-social simulation (Vespignani 2009) platform, first we instantiate a system of asymmetric, coupled nonlinear difference equations that capture the core logic of HD macro-social theory. These are then empirically validated with five waves of data from the World Values Survey (2009). Similar to Quek et al. (2009), second we then fuse HD endogenous systems to agent attribute changes with a generalizable, non-cooperative Prisoner's Dilemma game following Axelrod (1987, 1997a, b), Nowak and Sigmund (1993, 1998, and Osterkamp (1997) to simulate intra-societal, spatial economic transactions. Understanding the interactive political-cultural effects of macro-socio dynamics and individual agency in intra-societal transactions are key elements of a complex adaptive systems approach. Finally, we explore the model's behavioral dynamics via simulation methods to identify paths and pitfalls towards economic, social, cultural and political development as well as societal cooperation across different stages of development. We find strong epistatic interactions, where strategies are interdependent, and local social co-evolution (Kauffman 1993), help determine global-macro development outcomes in a particular society.

HD dynamics background

HD postulates a modernization process that is neither inevitable nor unidirectional. Following Maslow's (1954) hierarchy of needs, individual value orientations drive an individual's level of existential security and change in predictable ways given shifts in existential security. These value orientations are shaped by economic development and societal wealth. HD theory provides a framework in which economic development interacts with basic human needs and facilitates generalizable shifts in cultural predispositions and potentially revolutionary political behavior (Inglehart 1997; Inglehart and Baker 2000; Welzel et al. 2003; Inglehart and Welzel 2005). Robust empirics support that individual value orientation are represented by two primary dimensions— rational-secular and self-expression value orientations (Inglehart 1997; Inglehart and Baker 2000; Welzel et al. 2003; Norris and Inglehart 2004; Inglehart and Welzel 2005). Both are strongly influenced by economic progress and are more prevalent at different stages of development. HD theory naturally lends itself to a complex adaptive systems approach given the theoretical entrainment of macroscopic structures and microscopic individual agency, with individual value orientations shaped by and impacting macro societal outcomes.

While not a strict, formal economic growth model, HD theory leverages off growth theory. Neoclassical growth models (Solow 1956; Swan 1956; Barro 1991) specify three key economic drivers as the engine of growth: increasing capital relative to labor, higher rate of returns for underdeveloped societies and finally diminishing returns to capital. Together these factors propel middle income societies on high growth paths towards convergence and steady state equilibrium – measured as *sigma* convergence which is the reduction in income dispersion levels across economies and *beta* convergence when developing societies growth rates slow down as they become developed (Mankiw et al. 1992; Sala-i-Martin 1996). Such approaches specify detailed and interactive vectors of economic determinants, with country and time specific effects separate and in a more auxiliary manner (Caselli et al. 1996). Dissatisfied with a theoretically exogenous and *ad hoc* role of technology decoupled from economic factors, endogenous growth scholars (Romer 1986; Lucas 1988) explicitly link technology and human capital's effects to productivity. One major finding is that increasing rates of return to human capital do not produce steady state outcomes nor convergence. This allows highly developed societies to continue growing.

Similar to the endogenous growth literature, HD theory expands the breadth of development processes to include cultural, social and political effects to account for modernization across all scales. First, rational-secular values correspond to individuals' growing emphasis on technical, mechanical, rational, and bureaucratic views of the world. During economic industrialization phases, cultural dispositions tend to progress from an emphasis on traditional pre-industrial values—often measured in terms of religious ceremony attendance and the importance of religion in an individual's life—to secular world views, transferring authority from traditional religious figures to technological progress and bureaucratic political life.

The second dimension, self-expression, corresponds to the post-industrial phase of economic development where economic progress and an advanced welfare system provide many individuals in with an overwhelming sense of existential security (Bell 1973). As economic activity changes from primary, to secondary and tertiary sectors, individuals no longer work on farms, but in factories, while some pursue careers in the service industry. The reduction in human constraint—via increased productivity and wealth, advances in education, and service related economic activities—brings about a new sense of autonomy. Because the primary focus is no longer on survival, individuals are free to emphasize a general need for self-expression, question authority, and demand political participation. Rising self-expression values lead to the emergence of effective political institutions (Welzel et al. 2003). Mass tendencies toward self-expression facilitate a political climate conducive to potentially elite-challenging activity and a civic culture consistent with genuine democratic governance.

Self-expression values promote liberal political institutions through two mechanisms. First, to the extent that there is incongruence between cultural demand for, and political supply of liberal institutions, individuals are more or less prone to elite-challenging activity (Gurr 1970; Eckstein and Gurr 1975). Second, self-expression values support the social acceptance of basic democratic norms such as trust and political participation. The end result is a gradual transition toward democratization in autocratic nations and more effective political representation in democratic nations (Inglehart and Welzel 2005). Lastly, based on work on liberal institutions and economic development (Diamond 1992; Boix 2003; Feng 2003; Acemoglu and Robinson 2012) HD theory expects positive feedbacks between democratic institutions and economic progress. Declining economic conditions reintroduce the primacy of basic economic needs, fueling the structural conditions for more traditional value orientations, reducing society's ability to seek self-actualization, and increasing the likelihood of political change. Disequilibrium between culturally defined political expectations and political rights or civil liberty realities promotes elite challenging activity and provides motivation for revolutionary change. Equations 1 through 4 below transform HD's qualitative theoretical specification into a nonlinear, first order interdependent system (Abdollahian et al. 2012).

$$\frac{dY}{dt} = \lambda_1(1-Y) + \lambda_2 D \tag{1}$$

$$\frac{dRS}{dt} = [\alpha_1 Y - \alpha_2(RS-Y)]RS(1-RS) \tag{2}$$

$$\frac{dSE}{dt} = \beta_1 YSE(1-SE) \tag{3}$$

$$\frac{dD}{dt} = [\gamma_1 SE(SE-D) + \gamma_2 D](1-D) \tag{4}$$

Y economic progress

RS rational-secular values
SE self-expression values
D effective democracy
α_1 maps the interactive effects of Y and RS onto growth in level of RS
α_2 regulates the dynamic effects of economic incongruence
β_1 maps the interactive effects of Y and SE onto growth in the level of SE
γ_1 regulates the dynamic effects of political incongruence
γ_2 maps past governance structure onto the growth in effective democracy
λ_1 parameter for economic progress
λ_2 parameter of the democratic impact on economic development

Figure 1 above identifies four macro HD system phase portraits (Abdollahian et al. 2012) under equilibrium conditions. The HD perspective suggests a staged process in which rising level of existential security via economic development leads to an increased emphasis on rational-secular and self-expression values. Specifically, individuals tend to emphasize rational-secular values during the industrialization phase of development and self-expression values during the post-industrial phase. Second, the rise in self-expression values strengthens democratic norms and promotes effective democracy, implying a positive relationship between self-expression and liberal political institutions. However, these effects are neither linear nor monotonic as we see strong reversion towards autocratic institutional preferences in low self-expressive, survival

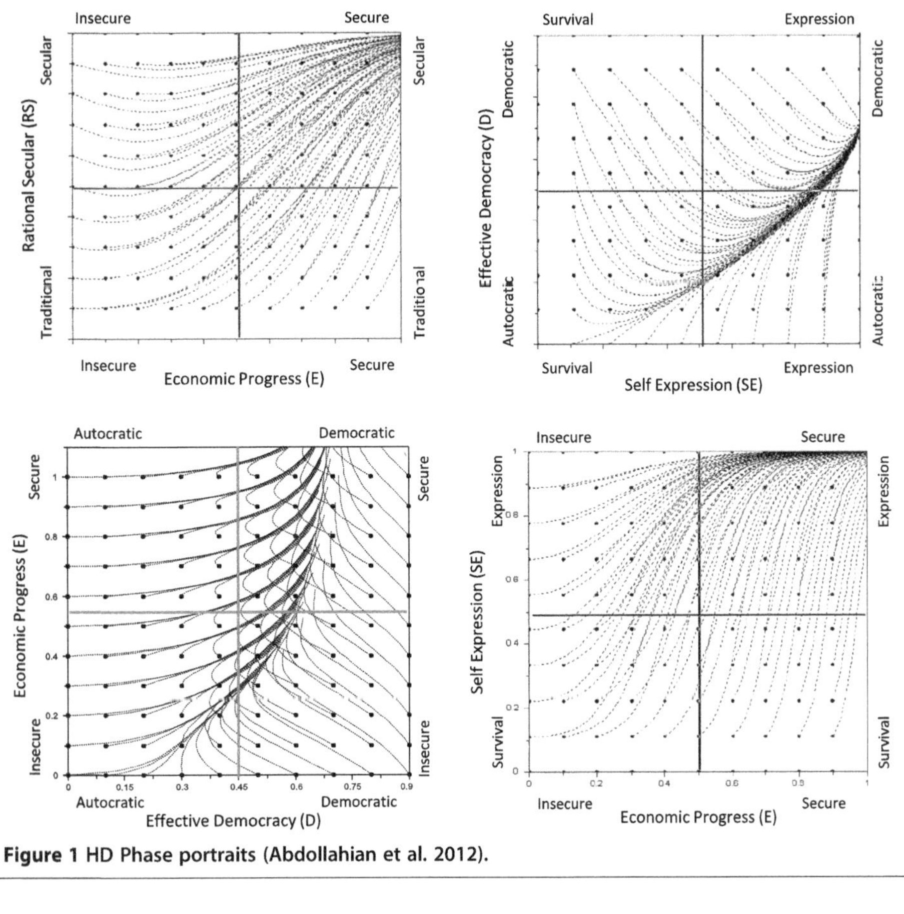

Figure 1 HD Phase portraits (Abdollahian et al. 2012).

dominated societies. Key to driving effective democracy are sustainable levels of expressive values. Democratic norms and institutions that outpace economic progress are inherently unstable with a persistent, turbulent reversion processes. This can occur even at high levels of democratic norms and existential security. This suggests societies experiencing democratization can frequently expect punctuated reversals and revolutions towards more autocratic institutions until more sustainable economic growth and democratic institutions re-emerge.

Methods

While innovative and the first to formalize a systems approach for HD theory, a limitation of Abdollahian et al's (2012) work lacks, coupling and interdependence across human scales, from individuals to institutions and finally the societal outcomes they generate. Our HDD model uniquely combines the interactive effects and feedbacks between individual human agency as well as the macro environmental constraints and opportunities that change over time for any given society. Decisions by individuals are affected by other individuals, social context, and system states. These decisions have variegated first and second order effects, given any particular system state or individual attributes. Such an approach attempts to increase both theoretical and empirical verisimilitude for some key elements of complexity processes – emergence, connectivity, interdependence and feedback (Miller and Page 2007) – found throughout several disciplines across all scales of modernization and human development.

Implemented in NetLogo (Wilensky 1999), Figure 2 depicts the high level process and multi-module architecture. We maintain individual agent attribute relationships and postulated changes of *RS*, *SE*, *D* and *Y* following HD theory. These endogenously derived, individual agent attributes (RS^i, SE^i, D^i and Y^i) impact how economic

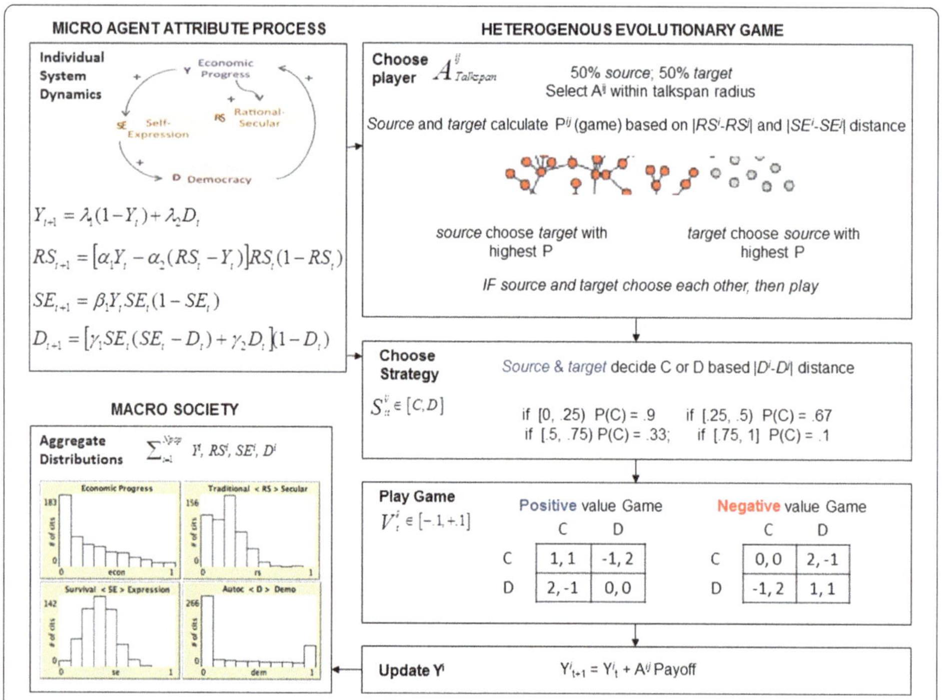

Figure 2 HDD architecture.

transaction games occur, either increasing or decreasing individual wealth and at increasing scales, ultimately societal productivity (Binmore 1998). Thus we create the tessellated pattern of heterogeneously mixed populations that cooperate and compete in a changing political economy backdrop to model social co-evolutionary processes

Capturing individual agent endogenous processes, we first transform Abdollahian et al.'s system of equations from differential to discrete equations for NetLogo tractability and use their empirically validated parameter values as a good first approximation. Given individual citizen attributes and HD processes at each iteration, we sum up each agent attributes across *Y, RS, SE* and *D* to find resulting societal distributions for each variable, yet are mindful of ecological correlation. This allows us to explore the interactive effects of income inequality, cultural schisms, social complexity or highly polarized political dialogues in any given society as the emergence of individual efforts and patterns of interactions.

Evolutionary game theory provides insights to understanding individual, repeated societal transactions in heterogeneous populations (Fudenberg and Maskin 1986; Sigmund 1993; Maynard-Smith and Szathmary 1995). Social co-evolutionary systems allow each individual to either influence or be influenced by all other individuals as well as macro society (Snijders et al. 2007; Zheleva et al. 2009), perhaps eventually becoming coupled and quasi-path interdependent. In our case, we do not have well mixed populations, but explicit spatial contact networks given population density, technology diffusion and agent attributes. Thus we explicitly recognize that the differential impact of heterogeneous, spatial structures matters (Santos et al. 2006). This captures various individual preferences and their socioeconomic attributes. Accordingly, we instantiate a non-cooperative, socio-economic Prisoner's Dilemma (PD) transaction game given agent *i*'s attribute vector (A^i) of social, cultural, political and economic preference (RS^i, SE^i, D^i and Y^i) similarity to agent *j* (A^j) for any A^{ij} pairs. The motivation behind this is that individuals are more likely to interact, engage and conduct transactions with other agents of similar religious, secular and expressive norms (Schelling 1971) and produce different co-evolutionary behavior via frequency and rate dynamics (McKelvey 1999).

At each iteration *t*, we first randomly choose 50% of spatially proximal agents as sources who can select a partner. The remaining targets are chosen by other agents based on symmetric preference rankings and asymmetric neighborhood proximity distributions. Here we look at communications reach, social connectivity and technology diffusion as constraining the potential set of A^{ij} game pairs, called talk span. Low values constrain games to local neighborhoods among spatially proximate agents, while higher values expand potential A^{ij} pairs globally, modeling socially compressed space.

Social Judgment Theory (Darity 2008; Griffin 2009) describes how the positions of two agents can be conceived along a Downsian continuum while distance between these positions affects the likelihood of one accepting the other's position. Agent *i* evaluates the likelihood of conducting a simple socio-economic transaction with agent *j* based on similarity of socio-cultural preferences $|RS^i\text{-}RS^j|$ and $|SE^i\text{-}SE^j|$ within a given local neighborhood. This captures communications and technology diffusion for frequency and social tie formation (McPherson et al. 2001). This also reflects recent work on the importance of both dynamic strategies and updating rules based on agent attributes affecting co-evolution (Kauffman 1993; Moyano and Sanchez 2008). Source agents evaluate *RS* and *SE* distances separately between themselves and all target agents within a given neighborhood radius. Shorter preference distances increase the probability that A^{ij} will enter into a socio-economic transaction and play the PD game. After each source agent calculates its probability of playing a game with all possible target agents, it chooses the target with the highest probability as its partner. Target agents also repeat the same process symmetrically. We then choose the A^{ij} pairing with the highest probability derived from its preference-proximity function at a particular iteration.

After agents decide to play at each *t*, each select strategies probabilistically based on similarity of political preferences as expressed by $|D^i\text{-}D^j|$. Siero and Doosje (1993) among others show that messages close to a receiver's position has little effect, while those far from a receiver's position is likely to be rejected, capturing the notions of bi-partisanship or conversely hyper-partisanship. So when *D* distance is short, there is a higher probability of cooperating while larger distances results in a higher probability of defecting. Relative payoffs for each agent are based on simple PD, non-cooperative game theory (Nowak and Sigmund 1993; Fehr and Fischbacher 2004; Dixit et al. 2009) where T > R > P > S, with T= 2, R=1, P = 0 and S= −1. When both agents cooperate, each agent receives a payoff of T. When one plays cooperate but the other plays defect, the cooperating agent loses with payoff S while the defecting agent gains with payoff T. When both play defect, neither gain from the transaction, and each agent receives a payoff of P.

Following Nowak and Sigmund (1993), we randomly assign game transaction values. However we do not asymmetrically constrain such values, instead allowing any particular game transaction value between pairs, V^{ij}, to lie in between [−.1, .1], randomly distributed. This instantiation allows for different potential deal sizes, costs, benefits, or synergies for any agent to gain or lose. We specifically model socio-economic transaction games as producing either positive or negative values as we want to capture behavioral outcomes from games with both upside gains or downside losses. Subsequently, A^{ij} games' V^{ij} outcomes condition agent E^i_{t+1} values, modeling realized costs or benefits from any particular interaction. The updated $E^i_{t+1} = E^i_t + A^{ij}$ game payoff for each agent then gets added to the individual's attributes for the next iteration. We then repeat individual endogenous processing, aggregated up to society as a whole and repeat the game processes for *t*+*n* iterations, where *n* is the last iteration.

In our HDD framework, A_i strategies are adaptive, which affect A_{ij} pairs locally within a proximate radius as first order effects. Other agents, within the system but outside the talkspan radius, are impacted through cascading higher orders. In HDD, we explicitly model epistatic interactions (Kauffman 1993) to capture co-evolutionary behavior in a simple, yet elegant manner. Although easily done, we specifically do not model mathematically complex, individual agent memory or learning from V^{ij} outcomes as many others do (Axelrod 1987; Sigmund and Nowak 1999). However, memory and history still matter. The sum of all prior individual behavioral histories, evolutionary through iterations, does contribute to each individual and societal current states. As an initial effort at a scale integrated framework, this frees us to focus on the coupling of structure and agency first, before enriching subcomponent process detail. Thus agents simultaneously co-evolve as strategy pair outcomes CC, DC/CD or DD at *t* to increase Y^i at *t*+1, thus driving both positive and negative *RS, SE* and *D* feedback process through $t+n$ iterations. These shape A^i attributes which spur adaptation to a changing environment, summing Y^i, RS^i, SE^i and D^i vector values. Feedback into subsequent A^{ij} game selection networks and strategy choice yields a complex adaptive system representation across multiple scales. The accompanying pseudo code is shown below.

```
for each agent
 {
  update four endogenous attributes
  decide whether to play game with another agent within
talkspan based on distance between RS and SE
  if probability < 0.05
    no interaction between the two agents
  else if probability >= 0.05 then
    for each game
     {
     randomly set game value
     }
    choose strategy based on D distance
   add payoff from the game to Y
  end if
 }
aggregate agent attributes to macro environment
```

Results

We use Abdollahian et al's parameter estimates from their genetic algorithm non-linear least squares (GANLS) estimation procedure on WVS data to populate system coefficients to increase model validity. The data spans from 1981 to 2007 from five waves for over 45 countries. The GANLS three step approach (Coan 2011) first employs a genetic algorithm to search global parameter space, a simplex algorithm to search local parameters space, and finally a bootstrapping procedure to derive the uncertainty associated with the parameter estimates. Given empirical β coefficient estimates for the HD system, we then conduct a quasi-global sensitivity analysis on the entire HDD agent based model input and initial condition parameters for 700 time steps using pooled OLS regression. Here we approximate one time step as one month given past data calibration (Abdollahian et al. 2012) for a simulated time span of almost 60 years. We performed over 40,000 runs given various input parameter combinations for a low resolution, initial sensitivity approximation.

Figure 3 depicts our HDD interface and a single sample run. The interface shows physical output space—heterogeneously mixed agents distributed spatially, where agent size indicates individual economic wealth Y, agent color represents democratic preference D, ranging from fully autocratic in red to fully democratic in blue. A^{ij} game transactions V^{ij} are shown as links at each t. For all agents' Y, RS, SE, D attributes, we setup the following initial conditions: societal mean $(\bar{Y}, \bar{RS}, \bar{SE}, \bar{D})$ and standard deviation, population density and techno-social connectivity via talkspan. Although not detailed

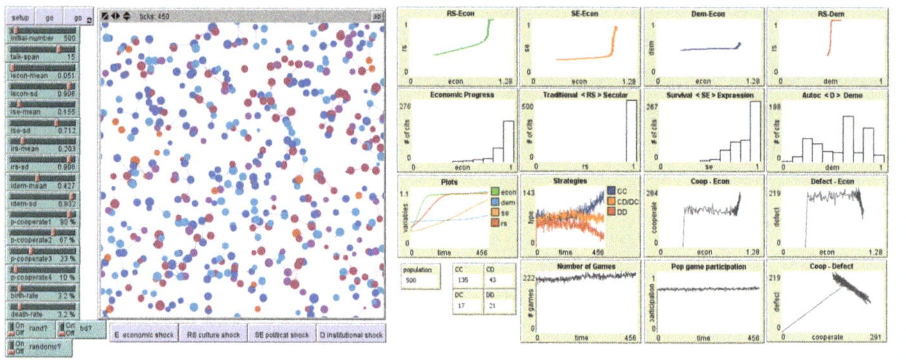

Figure 3 Sample HDD run.

herein, we also parameterize different agent birth and death rates and probability of game interactions in addition to strategies that can be randomized individually or in combinations to isolate dynamics in any specific subcomponent module.

Monitors and plots include time series and dynamic phase portraits for the societal average of *Y*, *RS*, *SE* and *D* trajectories and dynamic histograms across all attributes. This allows us to trace the development of wealth, income inequality, cultural dispositions, and political polarization. We also track the number of transactions games, the percentage of population interacting, and the number of different cooperative, mixed, or non-cooperative strategy pairs under different macro environmental conditions.

Before turning to our sensitivity results, we detail our notional run. Here a lesser developed society, with a mean low income level but high degree of inequality, escapes the poverty trap through high growth and increasingly moderating democratic institutions. Individual productivity and wealth, driven by successful CC and DC/CD strategy outcomes of individual transactions, help accelerate the emergence of modern secular *RS* and expressive *SE* norms and values, while democratic *D* preferences solidify, but are not engrained throughout all of society. Figure 4 depicts spatial details from the above simulation at t =0, 150, 300 and 450 snapshots. Initially, we find our low developed society with high income inequality and polarization along autocratic and democratic preferences. By t=150, agents quickly increase in individual wealth as several economic transaction games are played between individuals of similar *R* and *S* preferences while the strategy history shows not one dominant strategy emerges as society goes through a high growth phase of development. By t=300, we see continued heterogeneously mixed populations in terms of income and democratic preferences, while by t=400 we see increased convergence toward democratic preferences. Although just one particular simulation, what is critical is that co-evolutionary behavior results in path dependence of economic and cultural change as well as being a key determinant for development outcomes. Moreover, changes towards democratic values leads to increasing cooperative strategy pairs over time.

Sensitivity analysis

In order to make more generalizable model inferences, Table 1 details the interactive parameter effects on economic prosperity *Y*, as well as strategy choice pairs CC, CD/DC and

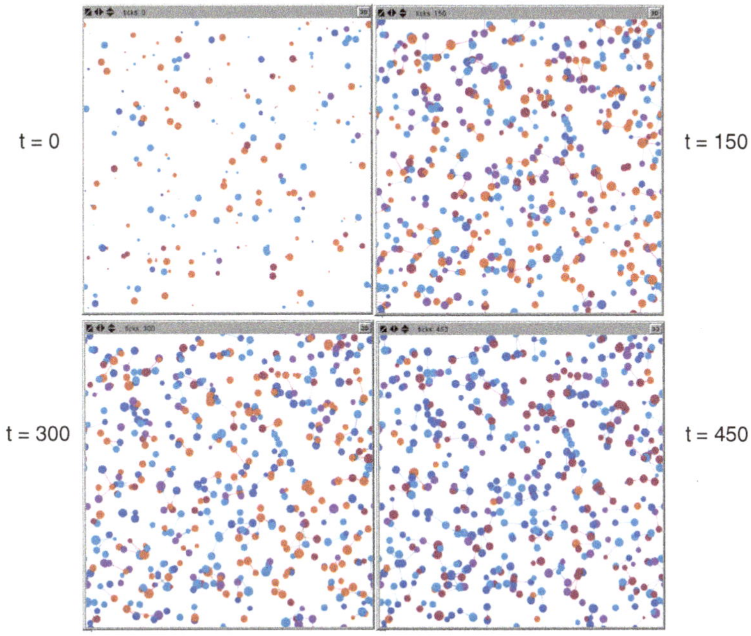

Figure 4 Spatial outcomes at t = 0, 150, 300 and 450.

Table 1 Impact on economic development and strategy pairs

Model	Economic	CC	CD	DD
Economic		0.3846*	0.2467*	0.1032*
		(0.0019)	(0.0015)	(0.0015)
Rational secular	0.3298*	−0.2756*	0.0598*	0.0643*
	(0.0022)	(0.0022)	(0.0020)	(0.0021)
Self expressive	0.1488*	0.2139*	−0.0120*	−0.1571*
	(0.0015)	(0.0016)	(0.0014)	(0.0014)
Democratic	0.1161*	0.0623*	−0.0790*	−0.0337*
	(0.0016)	(0.0018)	(0.0015)	(0.0015)
Cooperate	0.1606*			
	(0.0006)			
Defect	0.1088*			
	(0.0011)			
Talk-span	0.1615*	0.5484*	0.3024*	0.1306*
	(0.0005)	(0.0010)	(0.0007)	(0.0007)
Time	0.0558*	0.1742*	−0.2414*	−0.1958*
	(0.0009)	(0.0013)	(0.0009)	(0.0007)
Constant	0.1972*	−0.2481*	0.0028	0.1353*
	(0.0019)	(0.0017)	(0.0015)	(0.0016)
# of obs	634459	634459	634459	634459
Prob > F	0.0000	0.0000	0.0000	0.0000
R-squared	0.6441	0.6468	0.5171	0.3442
Root MSE	0.1285	0.2081	0.1500	0.1291

Numbers in parentheses are corresponding robust standard errors.
* Significance at 1% level.

DD. Our quasi-global sensitivity analysis generated over 27 million observations that we randomly down sample to approximately 634,000 for pooled OLS tractability. As all variables are relatively scaled, we can interpret magnitude and substantive effects across OLS β coefficients.

Our first model on economic development *Y* confirms HD theory that positive values of mean societal *RS*, *SE* and *D* values do significantly speed the pace of economic development, with *RS* providing the most substantive effect (β = .3298). Looking at the impact of evolutionary games, we see the number agents both choosing cooperation has a stronger positive impact (β = .1606) than defection (β = .1088) in increasing societal economic value. This suggests that cooperation does pay higher social dividends on average. Talkspan spatial proximity is positive and significant (β = .1615), confirming priors that increasing technology and compressing potential social space also speed development processes. Time is also slightly positive (β = .0558), indicating that economic prosperity is quasi self-reinforcing and path dependent. Model fit (R^2=.6441) is acceptable given the highly complex and non-linear dynamics and pooled nature of sensitivity analysis data.

Next we explore the potential for co-evolutionary behavior across micro and macroscopic scales. Turning to HDD parameter effects on heterogeneously mixed evolutionary games, we first focus on the conditions associated with CC strategy pairs - number of agent interactions where both *i* and *j* chose to cooperate. Talkspan is the most substantial (β = .5484), indicating increasing individual agents' ability to reach other like-minded agents spurs cooperation dramatically based on first order local interactions. Not surprisingly, economic progress *Y* (β = .3846) and self-expression *SE* (β = .2139) are also influential on increasing cooperative societal transactions showing coupled and cascading emergence at increased scale. Unexpectedly, secular values *RS* (β = –.2756) significantly decrease societal cooperation but still exhibit strong co-evolution, perhaps supplanting contract enforcement in traditional societies. Time is significantly positive (β = .1742), showing how cooperation tends to evolve over time with path dependence. This is especially interesting in light of *RS*'s positive impact on growth. This suggests that non-Pareto, asymmetric payoffs might be a necessary condition for development, regardless of path dependency. Model fit (R^2=.6468) is acceptable.

Looking at the number of agent interactions where either *i* or *j* cooperates when the other defects (CD or DC), we find strong co-evolution but with the different effects. While talkspan still dominates locally (β = .3024) and the economic environment (β = .2467) influences the nature of strategic interactions, there is less relative impact in driving asymmetric agent strategies. Consistently, increasing secular values *SE* (β = .0598) slightly contributes to asymmetric strategic behavior while increasing democratic norms *D* (β = –.0790) and expressive *RS* behavior (β = –.0120) curtails such unilateral advantages. More importantly, time (β = –.2414) is highly negatively related to asymmetric payoffs. Under asymmetry, agents cannot sustain such transaction outcomes as economic gains motivate preferences to become more closely aligned. Model fit (R^2=.5171) is also acceptable.

Finally, focusing on the number of agent interactions where both *i* and *j* chose defection (DD), we find self expressive *SE* behavior (β = –.1571) and time (β = –.1958) have the largest impact in dampening asymmetric payoffs in societal transactions. Democratic institutions *D* (β = –.0790) also still deter such behavior. Talk span

(β = .1306) and economic progress Y (β = .1032) as expected increase co-evolving DD strategy pairs but surprisingly at much lower levels than either in CD/DC or CC transactions. Expanding the feasible spatial set and individual's wealth does increase the temptation to defect for both agents. It also raises the immediacy of first order effects. Model fit (R^2=.3442) is relatively poor on DD strategy pairs compared to our other results, suggesting more complex processes need further exploration to increase result confidence.

Discussion

Figure 5 depicts a panel of average run trajectories across sampled parameter space. Here we focus on the interactive effects of low, medium and high mean self expressive ($\bar{SE} = .05, .5, .95$) behavior plots across societies with differing low, medium and high mean democratic norms ($\bar{D} = .05, .5, .95$) given various levels of economic development and mutually cooperative (CC) behavior. Each line represents the average iteration value between t= 0 to 700 of our 40,000 runs. Low, middle and high mean developed ($\bar{Y} = .05, .5, .95$) societies trajectories are in blue, orange and green respectively. Regardless of initial macro social conditions and consistent with neoclassical economic growth theories, we see both *sigma* income levels and *beta* growth rates conditional convergence with similar trajectory slopes and phase shifts. We also find

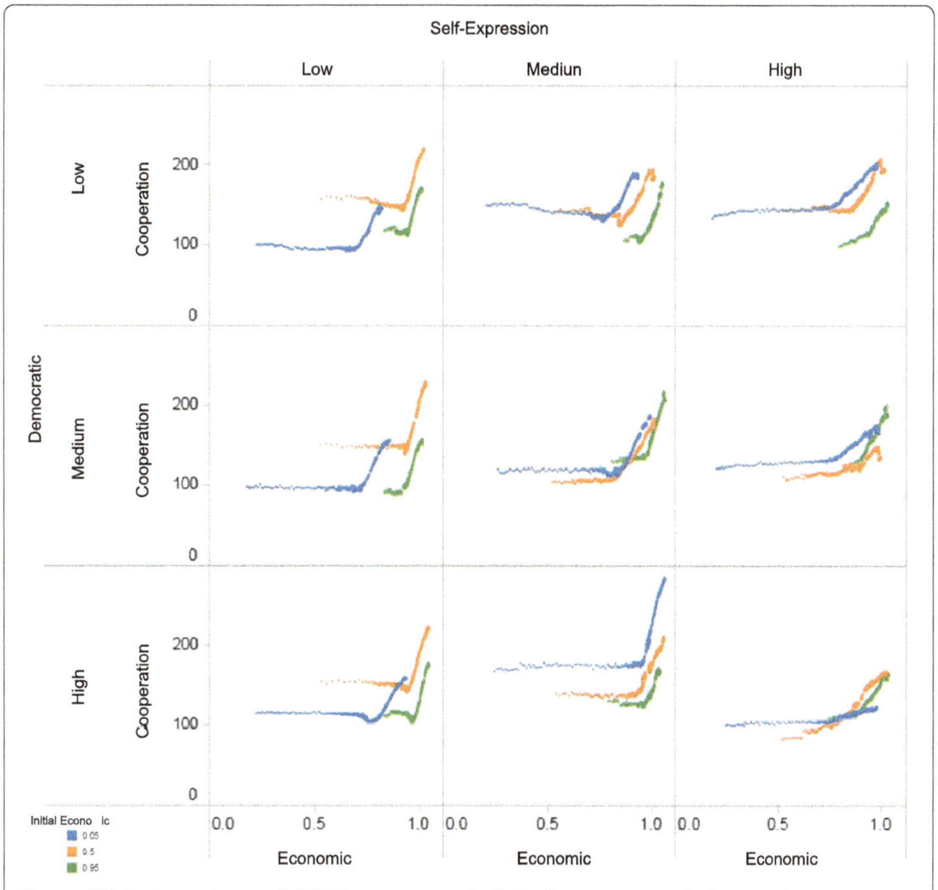

Figure 5 **Path dependency of 40,000 average run trajectories across sampled parameter space.**

cooperation does take time to emerge across all plots, indicating increasing entrainment of social co-evolution. However, focusing on highly developed societies (in green) across all plots, we see on average, intra-societal cooperative transactions emerge faster than for low income societies (in blue). This does not occur necessarily at higher development levels as our previous results suggest income growth, not levels, is a key driver of CC strategy pairs. Looking across initial democratic societal conditions, development outcomes and cooperative norms are most varied in low *SE* societies and tend to homogenize as individual and societal expressive culture increases. More interestingly, democratic institutions actually provide the largest impetus, fostering more cooperation and increasing growth in low and middle-income societies. These effects are both time and path dependent as they diminish in highly expressive, democratic societies where economic factors once again become the key determinant.

Conclusions

Consistent with qualitative HD theory and empirical reality, our HDD model finds complexity and nonlinear path dependence in three areas: adaptive development processes, social co-evolutionary transactions and near equilibrium development trajectories. From a complex adaptive system perspective on HD theoretical processes, economic progress is a necessary condition for successful secularization and expressive political behavior, which are antecedents for lasting democratic institutions. While modernization is not inevitable, our results support empirical observations for a staged process where increasing existential security via economic development leads to increased emphasis on rational-secular and self-expressive values that results in societal development. Here we find that rational-secular norms strongly impact economic growth and speed up the pace of development more than self-expressive societal values alone. Beyond supporting HD theory, agents do adapt interactively with their environments as mutual cooperation does result in higher societal wealth than defection alone and is self reinforcing over time.

Exploring the impact of societal conditions on transactions, in CC strategy pairs, compressing techno-social space by increased spatial proximity promotes co-evolutionary entrainment and mutual cooperation. Surprisingly, high secular values actually dampen CC outcomes suggesting the necessary role of asymmetric, Nash dominant strategies as one potential engine of high growth during industrialization phases. For CD/DC strategy pairs, we find consistently similar factors but with converse impacts. More interestingly, increasing macroscopic secular values over time has a dampening effect on individual non-Pareto outcomes. Higher macroscopic societal wealth and expanded social connectivity also dampen DD outcomes initially.

Turning to the path dependence of development trajectories, our results are consistent with neo-classical growth income convergence and the evolution of cooperation (Axelrod 1987, 1997a). We find convergence despite various initial income levels, inequality, social and cultural conditions on both *sigma* levels and *beta* growth rates while cooperation emerges over time. Social co-evolution of Pareto outcomes as seen by CC strategy pairs, tends to materialize faster in low income countries then dissipate in high growth, middle income societies only to return towards more predominately Pareto transaction outcomes in high income societies. One possible explanation is that asymmetric payoffs and non-cooperative behavior might be a necessary condition for high growth, emerging market economies. Epistatic Pareto outcomes and resulting wealth creation are most varied in low self-expressive societies, consistent with characterizations of a poverty trap (Sachs 2005). Democratic norms and institutions are a key driver, providing the largest impact in fostering cooperation and speeding economic growth in low and middle income societies. However, we find diminishing returns to democratization's benefits in advanced, highly expressive and already democratic societies, when economics once again takes primacy.

As economic activity pushes into the quaternary phase of development, we can only begin to ponder potential implications. Transitioning knowledge based services, and the multiplier impact that technology has on culture, politics and human development, only accelerates discontinuous change. Such forces are likely to produce new positive feedbacks across previously disparate and isolated domains, amplifying the coupling of technology, culture and society, to induce far-from-equilibrium conditions that will surprise us the most. Punctuated equilibria changes are the norm, not the exception, to modern human history – spanning the rise of the nation state, industrialization, access to mass transportation, the internet and most recently the social media revolution. Given the strategic, adaptive nature of human beings, we can only speculate as to new possible streams of norms, institutions and organizational complexity with subsequent, catastrophic system order shifts and dynamics that might emerge.

Our next research step is to thoroughly empirically verify the entire HDD model with upcoming available WVS data. Once validated, a natural extension is to specifically incorporate agent memory and learning following Quek et al. (2009) and multi-player PD from evolutionary game theory to map the complex transaction networks that emerge over time. Another is to tackle far-from-equilibrium challenges by exploring the full dynamic characteristics of both societal systems and individuals. Key will be identifying behavioral symmetry breaks and formally testing for catastrophic or chaotic responses (Wolf et al. 1985) to economic, cultural, social and political crisis (Abdollahian and Kang 2008), coupling or decoupling. One approach is to endogenize perturbations and shocks across Y, RS, SE and D values and agent space, to map equilibrium stability and system phase transitions. As domain linkages couple and morph, differential patterns of individual behavioral and system response can be identified through multiple local and global solution trajectories. This can allow us to parameterize the entire HDD macro–micro model to explicitly test for fold, cusp, swallowtail and higher order catastrophic (Çambel 1993) or chaotic, observed system behavior via Lyapunov exponents (Pecora and Carroll 1990), driven by micro-level interactions.

While only an initial, rough approximation at the truly complex, interdependent and highly nonlinear nature of modernization, our HDD approach provides insights into the interactivity of individual agency and societal outcomes seen through the lens of evolutionary games. While high risk for implementing properly, the knowledge rewards from fusing more detailed and leading theories from economics, cultural studies, sociology and political science via complex adaptive systems methods can be substantial. We hope such work motivates others to extend potential inquiries and insights with even greater theoretical fidelity and empirical resolution. Perhaps techno-social simulations like HDD can assist policy makers and scholar alike, to better understand, anticipate and shape positive social outcomes for all.

Competing interests
The authors would like to declare that there are no competing interests.

Authors' contributions
The authors of this paper contributed to the work in the following way: MA, ZY, TC and BY were involved in the conceptualization of HDD. MA and ZY architected the agent based model. ZY instantiated and coded the software implementation in NetLogo. All authors contributed to the manuscript.

Acknowledgments
The authors would like to acknowledge Hana Oh for her contributions to previous HD system dynamics work, support from the Trans Research Consortium and the anonymous referees for their excellent suggestions.

Author details
[1]School of Politics and Economics, Claremont Graduate University, 170 E. Tenth St. Claremont, CA 91711, USA. [2]School of Law, Harvard University, 1563 Massachusetts Ave. Cambridge, MA 02138, USA. [3]Hatfield School of Government, Portland State University, 1825 SW Broadway Portland, OR 97201, USA.

References

Abdollahian M, Kang K: **In search of structure: the nonlinear dynamics of power transitions.** *International Interactions* 2008, **34**(4):333–357.

Abdollahian M, Coan T, Oh H, Yesilda B: **Dynamics of cultural change: the human development perspective.** *Int Stud Q* 2012, **56**(4):1–17.

Acemoglu D, Robinson J: *Why nations fail: the origins of power, prosperity, and poverty.* New York, NY: Crown Business; 2012.

Arthur WB: **Positive feedbacks in the economy.** *Scientific American* 1990, **262**(2):92-99.

Arthur WB (Ed): *Increasing returns and path dependence in the economy.* Ann Arbor, MI: University of Michigan Press; 1994.

Axelrod R: **The evolution of strategies in the iterated prisoner's Dilemma.** In *Genetic algorithms and simulated annealing.* Edited by Davis L. Los Altos: CA: Morgan Kaufman; 1987:32–41.

Axelrod R: *The complexity of cooperation: agent-based models of competition and collaboration.* Princeton, NJ: Princeton University Press; 1997a.

Axelrod R: **The dissemination of culture: a model with local convergence and global polarization.** *J Confl Resolut* 1997b, **41**:203–26.

Barnett T: *The pentagon's new map: war and peace in the twenty-first century.* New York, NY: J.P. Putnam's Sons; 2004.

Barro R: **Economic growth in a cross section of countries.** *Q J Econ* 1991, **106**(2):407–444.

Bell D: *The coming of postindustrial society.* New York, NY: Basic Books; 1973.

Binmore KG: *Game theory and the social contract.* Cambridge, MA: MIT Press; 1998.

Boix C, Stokes SC: **Endogenous democratization.** *World Politics* 2003, **55**(4):517-549.

Çambel AB: *Applied chaos theory: a paradigm for complexity.* San Diego, CA: Academic Press Inc; 1993.

Caselli F, Esquivel G, Lefort F: **Reopening the convergence debate: a new look at cross-country growth empirics.** *J Econ Growth* 1996, **1**(3):363–389.

Chua A: *World on fire.* New York, NY: Doubleday; 2003.

Coan T: **Participation at the edge of chaos: a study of the nonlinear dynamics of protest politics.** PhD thesis. Claremont Graduate University, School of Politics and Economics; 2011.

Darity W: *Social judgment theory.* Detroit, MI: Macmillan; 2008.

Diamond L: **Economic development and democracy reconsidered.** In *Reexamining Democracy.* Edited by Diamond L, Marks G. London, UK: Sage; 1992.

Dixit A, Reiley D, Skeath S: *Games of strategies.* New York, NY: Norton & Company; 2009.

Eckstein H, Gurr TR: *Patterns of authority: a structural basis for political inquiry.* New York, NY: John Wiley & Sons; 1975.

Fehr E, Fischbacher U: **Social norms and human cooperation.** *Trends Cogn Sci* 2004, **8**:185–190.

Feng Y: *Democracy, governance, and economic performance: theory and evidence.* Cambridge, MA: The MIT Press; 2003.

Fudenberg D, Maskin E: **The folk theorem in repeated games with discounting or with incomplete information.** *Econometrica* 1986, **50**:533–554.

Griffin E: *A first look at communication theory.* Boston, MA: McGraw-Hill Higher Education; 2009.

Gurr TR: *Why men rebel.* Princeton, NJ: Princeton University Press; 1970.

Huntington SP: *The clash of civilizations and the remaking of world order.* New York, NY: Simon & Schuster; 1996.

Inglehart R: *Modernization and postmodernization: cultural, economic and political change in 43 societies.* Princeton, NJ: Princeton University Press; 1997.

Inglehart R, Baker WE: **Modernization, cultural change, and the persistence of traditional values.** *Am Sociol Rev* 2000, **65**:19–51.

Inglehart R, Welzel C: *Modernization, cultural change, and democracy: the human development sequence.* New York, NY: Cambridge University Press; 2005.

Kauffman SA: *The origins of order: self-organization and selection in evolution.* Oxford, UK: Oxford University Press; 1993.

Lucas RE Jr: **On the mechanics of economic development.** *Journal of monetary economics* 1988, **22**(1):3–42.

Mankiw NG, Romer D, Weil DN: **A contribution to the empirics of economic growth.** *Q J Econ* 1992, **107**(2):407–437.

Maslow A: *Motivation and personality.* 3rd edition. New York, NY: Harper and Row; 1954.

Maynard-Smith J, Szathmary E: *The major transitions in evolution.* Oxford, UK: Freeman; 1995.

McKelvey B: **Avoiding complexity catastrophe in coevolutionary pockets: strategies for rugged landscapes.** *Organizational Science* 1999, **10**:294–321.
McPherson M, Smith-Lovin L, Cook J: **Birds of a feather: homophily in social networks.** *Annu Rev Sociol* 2001, **27**:415–44.
Miller JH, Page SE: *Complex adaptive systems: an introduction to computational models of social life.* Princeton, NJ: Princeton University Press; 2007.
Moyano LG, Sanchez A: **Spatial prisoner's dilemma with heterogeneous agents: cooperation, learning and co-evolution.** *Arxiv preprint arXiv* 2008:0805-2071.
Norris P, Inglehart R: *Sacred and secular: religion and politics worldwide.* Cambridge, UK: Cambridge University Press; 2004.
Nowak MA, Sigmund K: **Strategy of win-stay, lose-shift that outperforms tit-for-tat in the prisoner's dilemma game.** *Nature* 1993, **364**:56–58.
Nowak MA, Sigmund K: **Evolution of indirect reciprocity by image scoring.** *Nature* 1998, **1998**(393):573–577.
Osterkamp KM: **The stability of cooperation in evolutionary games played by finite automata**. PhD thesis. Claremont Graduate University, School of Politics and Economics; 1997.
Pecora LM, Carroll TL: **Synchronization in chaotic systems.** *Phys Rev Lett* 1990, **64**(8):821–824.
Prigogine I, Stengers I: *Order out of chaos: man's new dialogue with nature.* New York, NY: Bantam Books; 1984.
Quek HY, Tan KC, Abbass HA: **Evolutionary game theoretic approach for modeling civil violence.** *IEEE Trans Evol Comput* 2009, **13**(4):1–21.
Rodrik D: **Globalization, social conflict, and economic growth.** *The World Economy* 1998, **21**:143–158.
Romer PM: **Increasing returns and long-run growth.** *The Journal of Political Economy* 1986, **94**:1002–1037.
Sachs JD: *The end of poverty: economic possibilities for our time.* New York, NY: Penguin Books; 2005.
Sala-i-Martin XX: **Regional cohesion: evidence and theories of regional growth and convergence.** *Eur Econ Rev* 1996, **40**(6):1325–1352.
Santos FC, Pacheco JM, Lenaerts T: **Evolutionary dynamics of social dilemmas in structured heterogeneous populations.** *Proc Natl Acad Sci USA* 2006, **103**(9):3490–3494.
Schelling T: **Dynamic models of segregation.** *The Journal of Mathematical Sociology* 1971, **1**:143–186.
Siero FW, Doosje BJ: **Attitude change following persuasive communication: integrating social judgment theory and the elaboration likelihood model.** *European journal of social psychology* 1993, **23**(5):541–554.
Sigmund K: *Games of life: Oxford.* UK: Oxford University Press; 1993.
Sigmund K, Nowak MA: **Evolutionary game theory.** *Current Biology* 1999, **9**(14):503-505.
Snijders TA, Steglich CE, Schweinberger M: **Modeling the co-evolution of networks and behavior.** *Longitudinal Models in the Behavioral and Related Sciences;* 2007:41–71.
Solow R: **A contribution to the theory of economic growth.** *The Quarterly Journal of Economics* 1956, **70**(1):65–94.
Swan T: **Economic growth and capital accumulation.** *Economic Record* 1956, **32**(2):334–361.
Thom R: **Structural stability and morphogenesis.** *Bull Math Biol* 1977, **39**(5):629–632.
Vespignani A: **Predicting the behavior of techno-social systems.** *Science* 2009, **325**:425–428.
Welzel C, Inglehart R, Klingemann HD: **The theory of human development: a cross-cultural development.** *European Journal of Political Science* 2003, **42**:341–380.
Wilensky U: *NetLogo.* Northwestern University, Center for Connected Learning and Computer-Based Modeling; 1999. http://ccl.northwestern.edu/netlogo/.
Wolf A, Swift JB, Swinney HL, Vastano JA: **Determining lyapunov exponents from a time series.** *Physica D: Nonlinear Phenomena* 1985, **16**(3):285–317.
Zheleva E, Sharara H, Getoor L: *Co-evolution of social and affiliation networks*, Proceedings of the 15th ACM SIGKDD international conference on knowledge discovery and data mining. New York: NY: ACM; 2009.

10

Agent-based modeling and simulation of emergent behavior in air transportation

Soufiane Bouarfa[1*], Henk AP Blom[1,2], Richard Curran[1] and Mariken HC Everdij[2]

* Correspondence:
s.bouarfa@tudelft.nl
[1]Delft University of Technology, Faculty of Aerospace Engineering, Kluyverweg 1, 2629, HS Delft, The Netherlands
Full list of author information is available at the end of the article

Abstract

Purpose: Commercial aviation is feasible thanks to the complex socio-technical air transportation system, which involves interactions between human operators, technical systems, and procedures. In view of the expected growth in commercial aviation, significant changes in this socio-technical system are in development both in the USA and Europe. Such a complex socio-technical system may generate various types of emergent behavior, which may range from simple emergence, through weak emergence, up to strong emergence. The purpose of this paper is to demonstrate that agent-based modeling and simulation allows identifying changed and novel rare emergent behavior in this complex socio-technical system.

Methods: An agent based model of a specific operation at an airport has been developed. The specific operation considered is the controlled crossing by a taxiing aircraft of a runway that is in use for controlled departures. The agent-based model includes all relevant human and technical agents, such as the aircraft, the pilots, the controllers and the decision support systems involved. This agent-based model is used to conduct rare event Monte Carlo (MC) simulations.

Results: The MC simulation results obtained confirm that agent based modeling and simulation of a socio-technical air transportation system allows to identify rare emergent behavior that was not identified through earlier, non-agent-based simulations, including human-in-the-loop simulations of the same operation. A typical example of such emergent behavior is the finding that alerting systems do not really reduce the safety risk.

Conclusions: Agent based MC simulations of commercial aviation operations has been demonstrated as a viable way to be evaluated regarding rare emergent behaviour. This rare emergent behaviour could not have been found through the more traditional simulation approaches.

Keywords: Agent-based modeling and simulation; Complex socio-technical systems; Air transportation; Airport operations; Safety risk analysis

Background

The Air Traffic Management (ATM) community is continuously seeking to improve ATM system performance and identify best practices (PRC & ATOS PBU 2009). Within the Single European Sky ATM Research program (SESAR), a performance framework for the future European ATM system has been proposed (SESAR Consortium 2006). This framework aims at improving ATM system performance, in particular with regard to the Key Performance Areas (KPAs) safety, capacity, cost efficiency, and environment. These

KPAs are important to accommodate the expected air traffic growth and respond to user's concerns (Bouarfa et al. 2012; Fron 2001). As part of this framework, airport performance is considered to have an impact on the entire ATM system performance, as airports make up the fixed nodes on which the ATM system is built (Ball et al. 2007; PRC 2009).

Civil air transportation is an example of a complex socio-technical system. Each airport comprises of interactions between a variety of facilities, users, technical systems, human resources, rules, and procedures, and is embedded in a large network of other airports, multiple airlines, and ATM centers. This type of complex socio-technical systems is characterized by a large number of interconnected parts, the difficulty to predict the behavior, and the existence of many different stakeholders (Forrester 1971; Sussman 2007a-b). Moreover, in civil air transportation, different parties are involved in the operation of an airport with varying boundaries of responsibility per region or country. At Amsterdam Airport Schiphol for instance, the parties involved in air transport related operations include the airport operator, the airport authority, the ministry of infrastructure and environment, the slot coordinators, the air navigation service provider, the airlines, as well as the general public (Deregee 2006). All these stakeholders have a certain viewpoint on performance which adds to the airport complexity (Bouarfa et al. 2012). In addition, each airport is characterized by a number of runways, taxiways, navigational aids, stop-bars, markings, and so on. Technology plays a central role in the airport system as does the social context within which the system is operating. This makes major airports by their very own nature complex socio-technical systems.

Past experience has shown that improving airport and ATM performance is very difficult to achieve. For example, Bar-Yam (Bar-Yam 2005) discusses the failure of design and implementation of the Advanced Automation Systems (AAS) in the previous century. A centerpiece of AAS was the replacement of the air traffic control system near airports. This process faced so many problems in terms of cost overruns, program delays, and safety issues, that it could only be partially completed after an FAA emergence decree. Bar-Yam argues that due to the level of complexity, different parts of the AAS design are so interdependent that changes in one part may have unforeseen effects on other parts.

Holland (Holland 2006) argues that there are few points through which the behavior of complex socio-technical systems can be changed to a desired state. These points named "lever points" are not where designers typically expect them to be. They are points where an action has an amplifier effect on the entire performance. If these points can be discovered, designers can make best use of them. There is however no theory that tells where or how to look for these points (Holland 2006).

Airport operators are faced with different trade-offs every day. These trade-offs are created by the complexities inherent to the processes managed and the finite resources of operational systems (Hollnagel 2009). Potentially, there are conflicting goals leading to dilemmas and bottlenecks that must be dealt with. An example would be how to make sure that a new environmentally-oriented procedure is safe without putting aircraft and people at risk? The widely established system engineering approach has not been developed to capture the socio part of a socio-technical system well. Then it should not come as a surprise when this creates unforeseen behavior that goes unnoticed during the development and implementation of a complex socio-technical system. Instead, such socio-technical behavior should be identified early on in the development process. This requires an approach to identify emergent behavior early on in the development of changes in civil air transportation operations.

Emergent behavior by an air transportation system results from interactions between the various human operators, technical systems, and procedures. The emergence concept is central to complex socio-technical systems and refers to how collective properties arise from the properties of the parts. Examples in air transportation include the impact on air traffic safety by new air traffic control techniques such as time-based spacing, the impact on airport capacity or airline business by policies restricting airport noise, the propagation of delays through the air transportation network, or the consequences that multiple actors can have when coordinating together. The Tenerife airport disaster for instance resulted from an interruption of routines among aircraft crew and air traffic control, loss of communication accuracy, and low visibility conditions (Weick 1990). At the same time, one should be aware of the fact that current aviation also works thanks to the explicit use of emergent behavior for the better. Examples of this are the various control loops that are working in current ATM, within each aircraft itself and also those formed by the interplay between the aircraft crew and each ATM center on its path. And there is no doubt that the role of control loops will only increase for advanced ATM. Logically, one should expect that emergent behavior that is not well understood is often characterized by poor performance, missed opportunities, and the inability to quickly adapt to disturbances. Only once emergent behavior is well understood, it may be exploited for the better.

In (Shah et al. 2005), it is well explained that the key difficulty of evaluating advanced ATM operations is to address emergent behavior, i.e. behavior which emerges from the combined dynamic actions and reactions by individual systems and humans within the overall ATM system. This emergent behavior cannot be foreseen and evaluated by examining the individual behaviors alone. To understand the behavior of a complex socio-technical system we must understand how the parts act together to form the behavior of the whole (Bar-Yam 2003). It is the goal of this research to study the emergence concept in air transportation, and develop a deeper understanding of system-wide performance issues and/or benefits arising from a change in design. In this paper we aim to identify key lever points through which the safety of runway crossing operations at airports can be significantly affected. Our approach is to embrace Agent-Based Modeling and Simulation (ABMS) because it has been extensively used to: a) analyze complex socio-technical systems and their emergent behavior (Shah et al. 2005; Chen & Cheng 2010; Stroeve et al. 2003; Wolfe et al. 2009); and b) address cases where agents need to collaborate and solve problems in a distributed fashion (Klein et al. 2004). ABMS provides a platform to integrate multiple heterogeneous components at different levels. Models of actors, technological systems, and the operating environment as well as the interactions between them can be naturally covered. Therefore, it is expected that an ABMS approach will help: a) predicting airport system-wide behavior emerging from the interactions between individual components; b) identifying and understanding key lever points; and c) managing dependencies between the activities of multiple actors both at the organizational and operational level.

This paper is organized as follows. The next section shortly discusses different perspectives on emergent behaviour, and identifies examples in the air transportation domain using a comprehensive taxonomy. This is followed by the methodology section where the agent-based modeling approach used is described. Then, the subsequent section positions the runway crossing application in the broader context of agent-based modeling. Here, the main agents, their entities, as well as their interactions are described. The Monte Carlo Simulation results and conclusion are provided in the subsequent sections. A short version of this paper has been published in (Bouarfa et al. 2013).

Emergent behavior in air transportation

There is a wide consensus that it is essential for future ATM developments to study and understand emergent behavior (Shah et al. 2005; SESAR Consortium 2007; European commission and Eurocontrol 2010; Everdij et al. 2011). In (European commission and Eurocontrol 2010), it is explained that with the introduction of advanced ATM concepts as considered in SESAR, yet unknown emergent risk may appear. Hazards that were not anticipated before could rise as a result of new concepts, tools, or procedures. Next to this negative aspect of emergence, positive emergent behavior would also be possible. (Woods et al. 2010) explain that as much as new concepts could give rise to new vulnerabilities, they could also remove existing ones. This positive aspect of emergence was also emphasized by Beart (Beart 2012) who claims that there are things that emergent systems can do that other systems cannot:

– They are robust and resilient: There is no single-point of failure, so if a single unit fails, becomes lost or is stolen, the system still works.

– They are well-suited to the messy real world: Human-engineered systems may be 'optimal' but often require a lot of effort to design and are fragile in the face of changing conditions. Importantly, they don't need to have complete knowledge/understanding to achieve a goal.

– They find a reasonable solution quickly and then optimize: In the real world, time matters because decisions need to be taken while they are still relevant. Traditional computer algorithms tend to not produce a useful result until they are complete (which may be too late in case of avoiding an obstacle for instance).

Different perspectives on emergence

Philosophers have long been interested in the concept of emergence, and especially in trying to establish a common definition for this vague yet very useful concept. The term has been in use since at least the time of Aristotle (Wikipedia 2012b) who referred to emergence as "the whole is something over and above its parts, and not just the sum of them all". Two thousand years later, (Mill 1872) used the example of water to illustrate the same idea. The term emergent was said to be coined by Lewes in his multi-volume problems of life and mind (Lewes 1874–1879). Lewes argued that certain phenomena produce "qualitative novelty", or material changes that cannot be expressed in simple quantitative terms. Quoting Lewes: "The emergent is unlike its components insofar as these are incommensurable, and it cannot be reduced to their sum or their difference." (Casti 1997), like Lewes associates emergence with dynamic systems whose behavior arises from the interaction among its parts and cannot be predicted from knowledge about the parts in isolation. Crick explains in (Crick 1994) that: "The scientific meaning of emergent, or at least the one I use, assumes that, while the whole may not be the simple sum of its separate parts, its behavior can, at least in principle, be understood from the nature and behavior of its parts plus the knowledge of how all these parts interact".

(Bedau 1997) distinguishes between two types of emergence namely, strong and weak emergence. Strong emergence was defined as nominal emergence in which the emergent properties are supervenient properties with irreducible downward causal powers. In the second type "weak emergence," the system's global behavior derives from the operation of micro-level processes, but the micro-level interactions are interwoven in such a complicated network that the global behavior has no simple explanation. Bedau argues that 'strong' emergence has had a prominent place in the philosophical discussions but that its scientific credentials are very poor, whereas 'weak emergence' is consistent with materialism and scientifically useful. Bedau proceeds to defend one version of weak emergence (noting that there are other versions), which is: "A nominally emergent property of a locally reducible system is called weakly emergent if it is derivable from all of the micro facts of this system, but only by simulation."

(Chalmers 2002) includes a notion of "unexpectedness" or "surprise" to the definition of emergence, when providing alternative definitions for strong and weak emergence. Other authors also refer to the notion of surprise, like (Sanz 2004) who defines emergence as "just systemic behavior — nothing more, nothing less— that is difficult to predict in advance." (Bedau 2002) explains that he left the notion of surprise absent on purpose, due to it being rather subjective. Instead, Bedau claims that with his definition of weak emergence in terms of simulation he is presenting objectivist approaches to emergence, though he notes that his classification is not exhaustive.

The diverse writings on emergence show that the term emergence captures a broad spectrum of system behavior. (Goldstein 1999) notes that "emergence functions not much as an explanation but rather as a descriptive term pointing to the patterns, structures, or properties that are exhibited on the macro-scale". Editor Lissack acknowledged in his inaugural article on emergence and complex systems theory that "it is less than an organized, rigorous theory than a collection of ideas that have in common the notion that within dynamic patterns there may be underlying simplicity that can, in part, be discovered through large quantities of computer power and through analytical, logical, and conceptual developments."

Identifying emergence in air transportation

Air transport operations are feasible thanks to a complex socio-technical system involving interactions between human operators, technical systems, and procedures. These interactions generate various types of emergent behavior, ranging from simple emergence up to strong emergence. Understanding these types of emergence is critical for effective decision-making. The more we learn about these types of emergence, the more opportunities we identify to improve the performance of the air transportation system, and prevent system failure. In this paper, we use the taxonomy proposed by (Fromm 2005; CAS wiki 2013) to illustrate various types of emergence in the air transportation system. This taxonomy builds upon a simplified formulation for cellular automata based on (Wolfram 1984). (Fromm 2005; CAS wiki 2013) distinguish between four primary classes (Types I-IV) based on the type of feedback observed in the phenomena. Table 1 shows that the stronger the emergence, the less predictable are the emergent

Table 1 Types of emergence identified by Fromm (Fromm 2005; complex adaptive systems wiki 2013)

Type of emergence	Name	Type of feedback	Predictability
Type I	Nominal or Simple Emergence	No Feedback	Predictable
Type II	Weak Emergence	Top-down feedback	Predictable in principle
Type III	Multiple Emergence	Multiple feedback loops	Not Predictable (Chaotic)
Type IV	Strong Emergence	Multiple feedback loops	Not Predictable, even in principle

properties, patterns, and structures. Type I corresponds to the simplest form of emergence whereas type IV corresponds to strongest form of emergence.

Type I – Nominal or simple emergence

This type of emergence is totally predictable due to the controlled and planned interaction of the individual components. A machine for instance has a function which is different from the function of its parts, but the overall function is well-known. In air transportation, one can think of the multitude of technical systems either on-board the aircraft or on the ground. There are no unpredicted or unexpected behavior patterns in these systems.

Type II – Weak emergence

This type of emergence describes emergence with top-down feedback, which is predictable in principle, but not in every detail. The roles of the elements and agents are flexible. Coherent global structures appear and become visible on a higher level of organization through the local interaction of several autonomous agents. The top-down feedback from the group imposes constraints on the local interactions. Examples from air transportation include a sequence of flying aircraft, which limits the possible speed adjustments of individual aircraft. The agents (e.g. flight crew) adjust their behavior and their role in the group according to the actual context and situation (e.g. following an ATC instruction, or TCAS warning). Feedback from the environment or group to the agent is possible through this form of context dependency.

Type III – Multiple emergence

This type of emergence is characterized by multiple positive and negative feedback loops appearing in complex systems with many agents. The behavior is not predictable and can be chaotic. Completely new roles can appear while old ones disappear. Although air transportation operations are feasible thanks to such feedback loops, chaotic behavior may arise following some exceptional events. There are two categories of such exceptional events: 1) catastrophic accidents involving one or two aircraft; and 2) events that push the dynamics of the air transportation system far away from its point of operation and therefore dramatically affect the performance of the system. Examples of the latter are a terror action causing closing down of air travel in large areas (e.g. 9/11 in 2001), a disease causing passengers to change their travel behaviour (e.g. SARS in 2003) or volcanic ashes blocking air travel in a large area (e.g. Iceland volcano in 2010). Examples of the former are fatal runway incursions (e.g. Linate runway collision in 2001), fatal mid-air collisions (e.g. Ueberlingen mid-air in 2002), loss of control of an aircraft flying through a hazardous weather system (e.g. Air France crash in Atlantic Ocean in 2009).

Type IV – Strong emergence

This type of emergence is not predictable, even in principle, because it describes the appearance of a completely new system in a multi-level or multi-scale system with many levels. Combinatorial explosion renders any attempt of explaining emergent macroscopic phenomena in terms of microscopic low-level phenomena useless and futile. An intermediate or mesoscopic level often protects the macroscopic level from the microscopic level, i.e. the microscopic level is irrelevant to the behavior of the macroscopic level. Life is a strongly emergent property of genes, genetic code and nucleic/amino acids, as is culture a strongly emergent property of language and writing systems. In the air transportation domain, one can think of the safety culture which is described as the product of routine aspects of everyday practice as well as organizational structure and rules (Leveson et al. 2005; Ek et al. 2007). However the causal relations are not yet understood (Sharpanskykh & Stroeve 2011).

Agent-Based Modeling

In order to understand complex socio-technical systems and their emergent behavior, rigorous models are needed that allow us exploring their properties (Holland 2006). One of these properties, is what Holland (Holland 2006) calls 'lever points.' These are points where a simple intervention causes an important effect on system performance. According to Holland, all complex systems that have been studied carefully exhibit such points. In the same article, Holland suggests using ABMS to study and analyze complex systems and capture their emergent behavior.

ABMS and complex socio-technical systems

The agent-based modeling paradigm is increasingly recognized as a powerful approach to model and simulate complex socio-technical systems exhibiting emergent behavior (Holland 1997). This is because it can represent important phenomena resulting from the characteristics and behaviors of individual agents and their interactions (Railsback & Grimm 2012). (Burmeister et al. 1997) discuss the benefits of using an agent-based approach in domains that are functionally or geographically distributed into autonomous subsystems, where the subsystems exist in a dynamic environment, and the subsystems have to interact more flexibly. According to Burmeister, agent-based modeling can be used to structure and appropriately combine the information into a comprehensible form. For a large complex system such as a traffic system, they provide the tools for analyzing, modeling, and designing the whole system in terms of its subsystems, each with its own set of local tasks and capability. The integration can then be achieved by modeling the interactions among the subsystems. So agent-based modeling provide abstraction levels that make it simpler and more natural to deal with the scale and complexity of problems in these systems. Agent components can be described at a high level of abstraction, yet the resulting systems are very efficient (Burmeister et al. 1997). (Burmeister et al. 1997) conclude that agent-based modeling reduce the complexity in systems design by making available abstraction levels that lend themselves to a more natural way of modeling the problem domain. They enhance the robustness and adaptivity of systems by virtue of increasing the autonomy of subsystems and their self-organization. In the same vein, Jennings (Jennings 2000) outlines that ABMS and complex system development requirements are highly compatible. (Jennings 2000) shows that agent-based modeling techniques are particularly well suited to complex systems because: a) they provide an effective way of partitioning the problem space of a complex system; b) they provide a natural means of modeling complex systems through abstraction; and c) they capture the interactions and dependencies.

Agents in air transportation

In the ABMS domain, although there is no widely agreed definition for an agent, there is general consensus that autonomy is central to the notion of agency. (Wooldridge 2009) explains that part of the difficulty is that beyond this autonomy point various attributes associated with agency are of different importance for different domains.

Among the various definitions in the literature for an agent are:

- An agent is anything that can be viewed as perceiving its environment through sensors and acting upon that environment through effectors (Russel & Norvig 2006).
- An autonomous agent is a system situated within a part of an environment, which senses that environment and acts upon on it, over time, in pursuit of its own agenda and so as to effect what it senses in the future (Franklin & Graesser 1997).
- An agent is a computer system that is situated in some environment, and that is capable of autonomous action in this environment in order to meet its delegated objectives (Wooldridge 2009).
- An agent is a system with the following properties (Tessier et al. 2002):
    - It lives in an artificial world *W*
    - It has facilities to sense *W* and to manipulate *W*
    - It has a (at least partial representation of *W*
    - It is goal-directed, and as a consequence it has the ability to plan its activities
    - It can communicate with other agents

In the context of air transportation, in particular where different actors, hardware, and software are interacting elements of a complex socio-technical system, we consider agents as autonomous entities that are able to perceive and act upon their environment. These agents may be humans, systems, organizations, and any other entity that pursues a certain goal. For instance, an air traffic controller can be viewed as an agent observing his/her environment (displays, alerting systems, runway availability, etc.) and acting upon this environment (e.g. through communicating with other agents like pilots/other controllers, or turning off runway stop-bars remotely). The agent environment is understood as all surrounding human and non-human agents.

Agent-based safety risk analysis

The aim of this paper is to position an active runway crossing application in the broader context of agent-based modeling. This was motivated by the capability of this rather new approach to obtain known and unknown emergent behaviors (Chan et al. 2010). Traditional safety approaches assume well defined cause-effect links that propagate the effects of events contributing to the safety risk (e.g. sequential or epidemiological safety models). However, recent views indicate that such models may not be adequate to represent the complexity of modern socio-technical systems (Hollnagel et al. 2006). Instead, agent-based modeling forms a logical choice for the safety-risk analysis from a socio-technical perspective. By having distinguished a number of agents and their interactions, the overall process can be analyzed as emerging from the individual agent processes. This not only provides a transparent way of structuring the model, which supports the analysis both conceptually and computationally, but also makes the model easier to maintain, resulting in local model refinements instead of global changes. For the runway crossing operation, a systematic comparison of the agent-based approach against a sequence based approach has been made in (Stroeve et al. 2011). The study revealed many advantages of the former approach, including considerable differences in the risk results obtained. The only disadvantage however, is that the agent-based approach requires computational modeling experience that differs from the expertise of most current safety analysts. Finally, compared to the major simulation paradigms, agent-based modeling can be used across all abstraction levels (see Figure 1) which is necessary to cover all relevant agents who directly control the hazardous process in the context of the runway crossing application.

TOPAZ safety risk assessment methodology

Motivated by the need to model the dynamics, the stochastic, and the interactions of safety critical multi-agent systems, NLR has developed the TOPAZ safety risk assessment methodology, e.g. (Blom et al. 2001a; Blom et al. 2006). The quantitative part of TOPAZ develops and evaluates an agent-based model through running Monte Carlo simulations (Blom et al. 2009) in combination with bias and uncertainty analysis (Everdij et al. 2006). Next to these techniques, TOPAZ also integrates human performance modeling (Blom et al. 2001b), and powerful petri net modeling syntax (Everdij & Blom 2010). Applications of these modeling techniques requires dedicated expertise from safety analysts. However, when the TOPAZ toolset is available, normal safety expertise is sufficient.

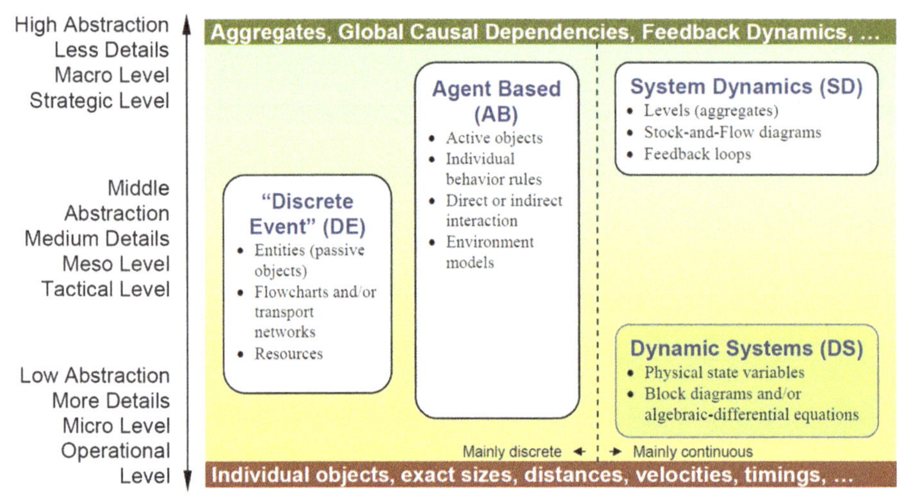

Figure 1 Approaches (Paradigms) in simulation modelling on abstraction level scale (Borshchev & Filippov 2004).

Case study: ABMS of an active runway crossing operation

In this section we explore ABMS to study emergent behavior in an active runway crossing operation. Our study is performed in three main steps, namely:

1. Model the agents and their interactions with the environment: For example human performance models of pilots and controllers interacting with technical systems.
2. Run Monte Carlo Simulations of both nominal and off-nominal scenarios: In order to assess the impact of agents' individual performance on system behavior.
3. Analyze system behavior: Explore an array of behaviors or parameters of system performance to guide alternative selection and future development.

Active runway crossings and incursions

In many airports around the world, runway crossings are used by taxiing aircraft from the apron area to the runway and vice versa. These crossings are attractive because they reduce the taxiing time and save fuel. However, they also have safety implications, namely the risk of having a runway incursion. A runway incursion is defined by the International Civil Aviation Organization (ICAO) (ICAO 2007) as "Any occurrence at an aerodrome involving the incorrect presence of an aircraft, vehicle or person on the protected area of a surface designated for the landing and take-off of aircraft". While a runway incursion does not imply a collision, the probability of an accident is not nil. One of the most famous aviation accidents is the Tenerife airport disaster that occurred on March 1977. Two Boeing 747 aircraft, operated by KLM and Pan American World Airways, respectively collided. While the Pan American aircraft was taxiing, the KLM aircraft took off, resulting in a collision causing 583 fatalities. This accident is the deadliest aviation accident in history (Wikipedia 2012a). 35 years later, runway incursions are still frequently reported in many countries. In the united states alone, preliminary data (FAA 2012) through the end of August 2012 shows a total number of 1010 runway incursions in the year 2012, a 17 percent increase over the same span in 2011.

Researchers and planners operating from different perspectives have proposed many options to address this problem, such as new technology (e.g. in aircraft, ATC tower, or Airport) and new procedures such as ICAO compliant procedures. These proposals aim to reduce the probability of runway incursions, and reduce the accident risk in case runway incursions occur. However, evaluating the impact of these proposals is a demanding task, given the large number of human operators and technical systems that closely interact at the airport. This paper evaluates the safety risk of an active runway crossing operation from an agent-based modeling perspective.

Agent-based model of the active runway crossing operation

An agent-based model of the active runway crossing operation has been developed in a series of studies (Stroeve et al. 2011; Stroeve et al. 2013). The agents in this operation and their interactions are shown in Figure 2. The developed agent-based model

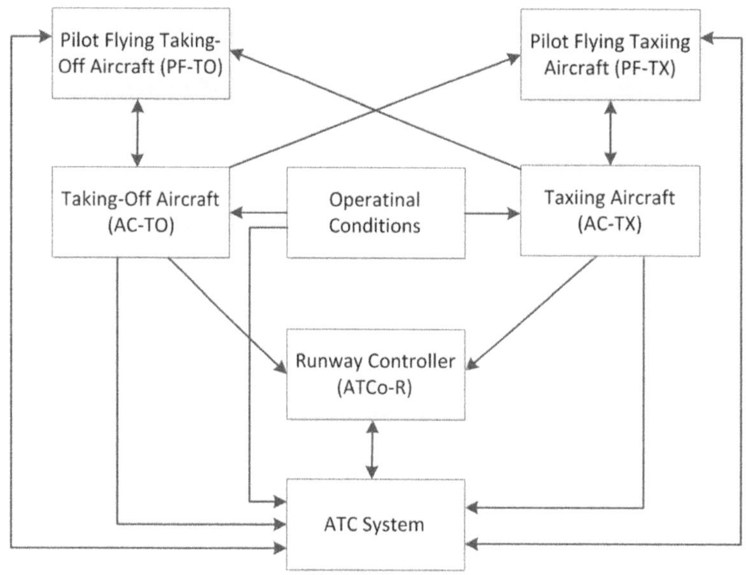

Figure 2 **Agent-based model of the active runway crossing operation showing the interactions between the agents (Stroeve et al. 2011).**

considers accidents as emerging phenomena from the interactions between multiple agents involved. In these previous studies, it has been shown that the level of safety in sociotechnical systems depends on the interactions between organizational entities in their contextual conditions. In this paper we present the model in the wider context of socio-technical systems and run new Monte Carlo Simulations of non-nominal scenarios using this model.

In order to define the stochastic dynamics of the agents and their interactions unambiguously, the agent-based model has been specified in terms of Stochastically and Dynamically Colored Petri Nets (SDCPNs). SDCPNs are a powerful extension over normal Petri nets in that they are able to represent general stochastic hybrid processes in a form that supports powerful stochastic analysis (Everdij & Blom 2010). In view of the scope of the current paper, it suffices to describe this agent-based model in normal language rather than in SDCPN language.

At the highest hierarchical level, the relevant agents in the active runway crossing operation are identified. These agents are concurrently interacting with each other and include human and non-human agents. The human agents are the flight crew operating the aircraft, and the runway controller handling the traffic on the runway and its crossings. The non-human agents represent the aircraft and ATC system. In the context of the active runway crossing operation, the ATC system is comprised of three components where each component is performing a number of functions.

These components include: 1) the Radio/Telecommunication (R/T) system used for communication between the controller and flight crew; 2) the Surveillance system used for providing radar track data; and 3) the alerting system used to generate ATC alerts in safety critical situations.

In Figure 2 we use the following abbreviations for the different agents:
Human Agents

- PF-TX: represents the Pilot Flying the Taxiing Aircraft.
- PF-TO: represents the Pilot Flying the Taking-Off Aircraft.
- ATCo-R: represents the Runway Controller.

Non-Human Agents

- AC-TX: represents the Aircraft Taxiing.
- AC-TO: represents the Aircraft Taking-Off.
- ATC System: represents the ATC system.

The main entities within these agents as well as their interactions are described below.

Human agents
The human operators' function in the distributed air transportation system includes visual monitoring, perception, spatial reasoning, planning, decision-making, communication, procedure selection, and execution (Corker et al. 2008). In order to account for these cognitive and perceptual functions of the human operators, human performance modeling has been used as a complementary technology (Blom et al. 2001b). Figure 3 gives an overview of the main entities within the human agent's internal model in the context of advanced aviation concepts. This model applies to both the ATCo-R and pilots.

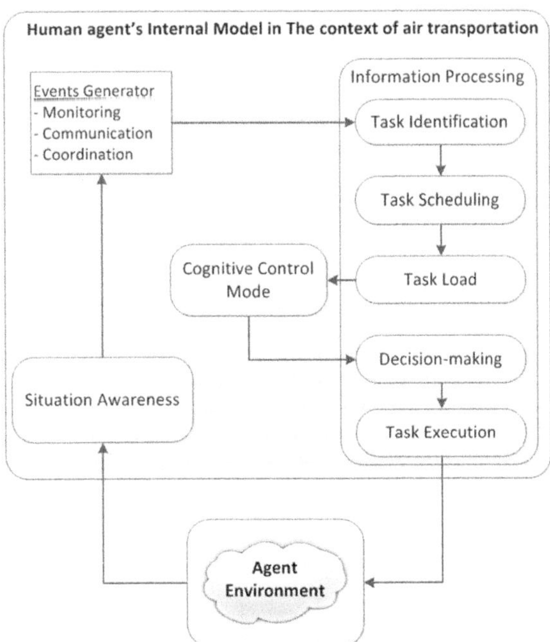

Figure 3 **Overview of the internal model of a human agent and its interactions with its environment.**

The concept of situation awareness (SA) addresses perception of elements in the environment, their interpretation, and the projection of their future status (Endsley 1995). Stroeve et al. (Stroeve et al. 2003) have captured these perception, interpretation, and project notions of SA mathematically in terms of three components namely, State SA, Mode SA, and Intent SA. The state SA represents the awareness by one agent of other agents. In the context of the active runway crossing operation, this might be the awareness of the flight crew about the state (position and speed) of their own aircraft and other aircraft, or the awareness of the controller about traffic state at the airport. The mode SA represents the awareness by one agent of the mode of other agents. This mode could be for instance the flight phase of an aircraft, or status of an ATC alert. The intent SA represents the awareness by one agent of the intent of other agents. The intent includes continuous states of agents as well as the related times at which these states are expected to be achieved. For example, intent SA may represent the expectations by a runway controller of aircraft destination, and the time it will arrive at a certain waypoint. The timing of the situation awareness updates depends on the initiation and duration of related tasks (e.g. monitoring, communication, coordination).

The information processing entity considers processing of information from the environment that leads to actions that may influence the environment. This entity is based on task analysis, which takes into account the multiple resources model (Wickens 1992). The idea reflected by this model is that humans have several different mental capacities with resource properties. In this view, task interference depends on the extent to which tasks use the same resources: two difficult tasks may be time-shared easily if they use different types of resources (Blom et al. 2001b). The principal idea behind the model is that human cognitive effort can be divided over several activities. This may account for failures in time-sharing between competing activities, since the human cognitive effort is limited. So, the underlying assumption is that the human is an information processing system with limited processing capacity. The human information processing entity focuses on how this limited processing capacity can be used to time-share several processing tasks. Its sub-entities include:

- Task Identification: Considers the ways the human operator identifies the tasks that need to be performed at a particular time instance.
- Task Scheduling: Determines which tasks may be performed concurrently, as well as a priority among the tasks that cannot be performed concurrently.
- Task Load: Describes the number of tasks to be performed and/or the resources required by tasks at the level of visual, auditory, cognitive and motor performance.
- Decision Making: Decision Making processes are based on decision rules dedicated to the scenario considered. In an active runway crossing operation, this could be how to react to a conflict situation.
- Task execution: Considers the dynamic and stochastic performance characteristics of tasks.

This Cognitive Control entity considers that humans can function in a number of cognitive control modes such as: Strategic, Tactical, Opportunistic, and Scrambled (Hollnagel 1993). The cognitive control mode may depend on the taskload sub-entity which describes the range of tasks to be done or the situation awareness of the human.

It influences human aspects such as the planning horizon and the accuracy of task performance.

Non-human agents

The model of the aircraft agent represents aircraft dynamics in different flight phases, which is a function of the aircraft type (see Figure 4). For AC-TX the model represents aircraft movement during taxiing, including braking as a means to avoid a collision. For AC-TO, it represents the ground run, airborne transition and airborne climb-out phases during takeoff. It also includes the possibility of a rejected takeoff by the pilot.

The main entities within the ATC system agent include the surveillance system, the alerting system, and R/T communication system (see Figure 5).

- The model of the surveillance system provides position and velocity estimates for both aircraft. The surveillance data is used by the alerting system.
- The alerting system generates two types of alerts in case surveillance data indicate a safety critical situation. The alerts include 1) a stopbar violation alert in case AC-TX has passed the stopbar, and 2) a runway incursion alert in case AC-TX is within a critical distance of the runway center-line and AC-TO has exceeded a velocity threshold in front of the runway crossing. There is a chance that the surveillance system is not available, resulting in track loss and no alerts being generated.
- The model for the R/T communication system between ATCo-R and pilots accounts for the communication system of the aircraft, the communication system of the ATCo-R, the tower communication system, and the frequency selection of the aircraft communication system. The nominal status of these communication systems accounts for direct non-delaying communication. The model accounts for the chance of delay or failure of the communication systems.

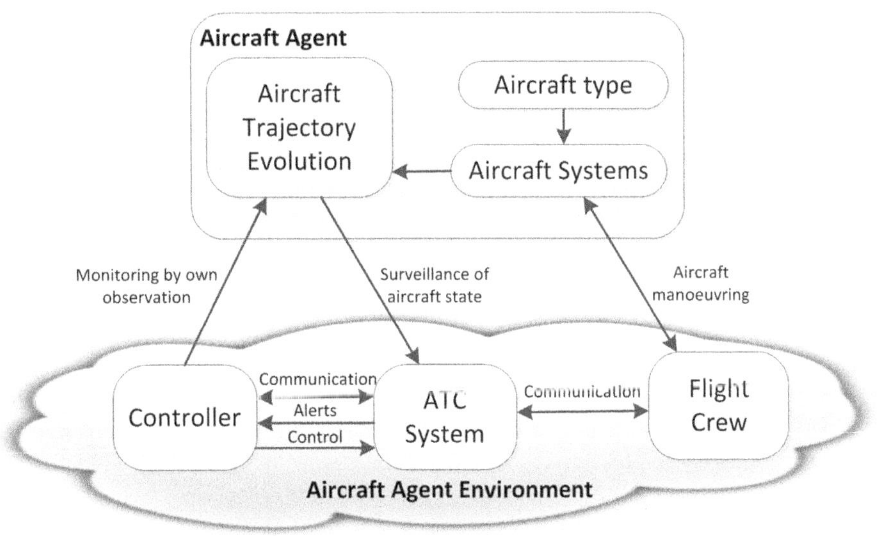

Figure 4 **Overview of the internal model of the aircraft agent and its interactions with its environment.**

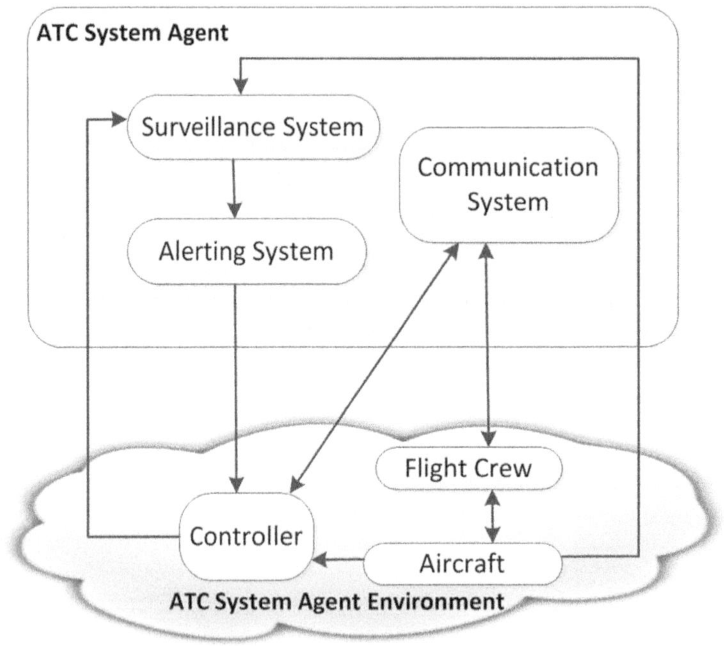

Figure 5 Overview of the internal model of the ATC system agent and its interactions with its environment.

Operational conditions

The operational conditions in the active runway crossing operation are characterized by two elements namely the runway configuration and visibility conditions. For the first element, runway characteristics such as runway length, width, location of stopbars, and number of branches are specified. Each branch is of the class holding, crossing, or exit. For the second element, four different visibility conditions are represented.

Environment of each agent

Each agent is situated in its environment, from which this agent perceives information and acts upon its environment (See Figures 4, 5 and 6).

Active runway crossing operation

In the active runway crossing operation, we consider a departure runway that is used by AC-TO and has a crossing taxiway at a distance of 1000 m from the runway threshold (see Figure 6). The runway crossing is used by the taxiing aircraft (AC-TX) to cross the active runway. In this experiment, we focus on the scenario that the taxiing aircraft is crossing the runway while it should not, due to a wrong intent situation awareness of its flight crew. Such condition has been shown in previous work to have a strong effect on accident risk (Stroeve et al. 2003; Stroeve et al. 2009). Evaluating the conditional collision risks associated with this condition provides more insight on the collision risk contributions.

In the active runway crossing operation, AC-TX enters the taxiway leading to the runway crossing at a position close to the stopbar, whereas AC-TO initiates take-off

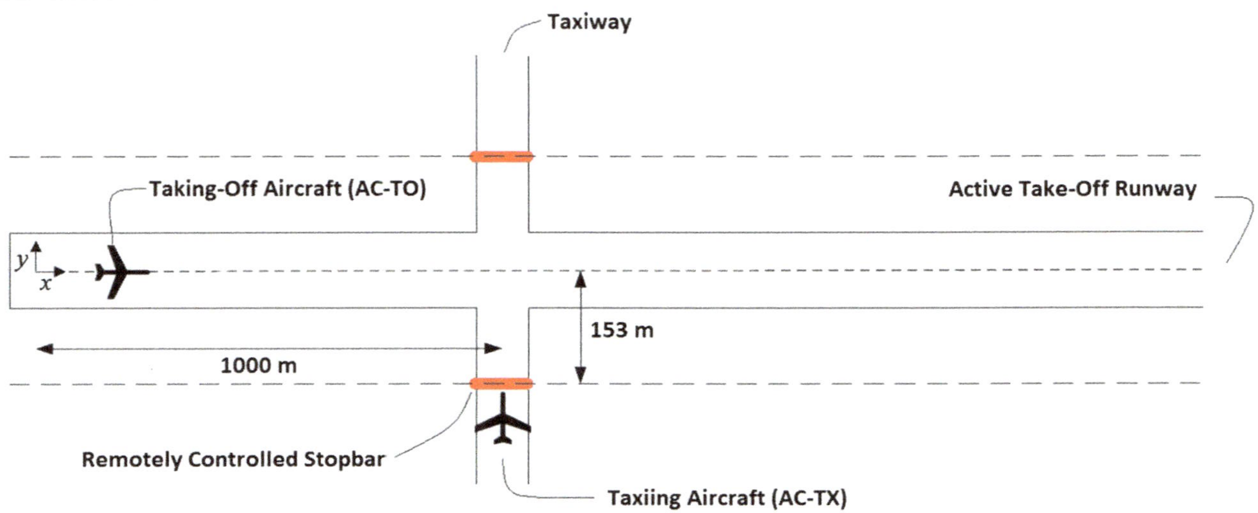

Figure 6 Schematic overview of the traffic situation considered. The taking-off aircraft accelerates along the runway while the crew of the taxiing aircraft intends to proceed along the taxiway towards the active runway (Stroeve et al. 2009).

from a position near the runway threshold (Figure 6). The entrance time of AC-TX is uniformly distributed around the take-off time of AC-TO. In addition, both AC-TX and AC-TO may be medium-weight or heavy-weight. The operation is assumed to be without restricted visibility range in line with visibility condition 1 of ICAO (ICAO 2004). This implies that the involved human operators can visually observe the traffic situation.

The pilots PF-TO and PF-TX are both responsible for the safe conduct of the flight operations and should actively monitor for potential conflicting traffic situations. ATCo-R is responsible for the safe and efficient traffic handling on the runway and the runway crossings.

During taxiing and take-off, both PF-TX and PF-TO visually monitor the traffic situation at stochastically distributed times. Both pilots may detect a safety critical situation by their own observation. PF-TO observes a conflict when AC-TX is observed to be within a critical distance to the runway centerline. PF-TX observes a conflict when AC-TO is observed to be approaching AC-TX with an increasing speed, and AC-TX is within a critical distance from the runway centerline.

Next to their own monitoring actions, pilots of both AC-TX and AC-TO may also detect a conflict following a call from ATCo-R. In this case, the R/T communication system is used for communication between ATCo-R and the flight crew. ATCo-R may detect a conflict during monitoring if AC-TX is observed to have passed the stopbar, or following an ATC alert. The ATC alerting system may generate two types of alerts to warn ATCo-R: 1) A runway incursion alert for the situation that AC-TX is crossing the runway in front of AC-TO that has initiated take-off; and 2) A stopbar violation alert for the situation that AC-TX crosses an active stopbar.

These alerts consist of audible warnings and an indication on the ground surveillance display. The alerts are based on radar tracking data (Aircraft position and velocity estimates) provided by the surveillance system.

Following conflict detection, PF-TX may decide to start a full braking action or continue crossing in case AC-TX is within a critical distance of the runway center line. Likewise, PF-TO may decide to start a full braking action or continue taking off if it is too late to brake.

Results and discussion

Monte Carlo simulation results

To better understand the potential of agents to restrict the risk in cases where the performance of other agents is affected, one or more agents can be placed out of the monitoring role in the Monte Carlo simulations. This is done for all agents that are capable of detecting a conflict in the active runway crossing operation, namely PF-TO, PF-TX, ATCo-R, and the ATC alerting system. The conditions for placing these agents out of the monitoring role are:

- PF-TX does not actively monitor the traffic situation visually, such that he may only detect a conflict via a call of ATCo-R.
- PF-TO does not actively monitor the traffic situation visually, such that he may only detect a conflict via a call of ATCo-R.

- ATCo-R does not actively monitor the traffic situation visually, such that he may only detect a conflict through ATC alerts.
- The ATC alerting system does not generate any alerts.

The conditions above refer to the situation at the start and during the active runway crossing operation, and they are not assumed to hold prior to the occurrence of the crossing operation. Combining these conditions results in sixteen cases where agents are either monitoring or not monitoring the traffic situation. A total of 1 million Monte Carlo simulation runs were performed for each of the sixteen combinations. The Monte Carlo method was used to capture the stochastic nature of the various agents in the active runway crossing operation. The conditional collision risk was obtained by dividing the number of collisions by the number of performed simulations for each case.

Figure 7 shows the conditional collision risk results corresponding to the sixteen cases. The conditional collision risk lies between 1,7E-4 and 9,5E-2. The lowest value corresponds to the nominal case C1 where the performance of agents is not affected. Here all agents operate according to the active runway crossing operation described in the previous section. The highest collision risk corresponds to the hypothetical case C16 where none of the agents is actively involved in recognizing the conflict and avoiding a collision. In this case, the risk increases by a factor of 556 with respect to case C1.

The conditional collision risk results in Figure 7 can be divided into two main categories. The first category is characterized by low to medium risk increase factors and includes the first seven cases C1 to C7 on the left-hand side. The second category is characterized by high risk increase factors and includes cases C8 to C16 on the right-hand side.

In the first category, the risk increase factors are low for all cases except for case C6 which has a medium risk increase factor. The low factors are between 1.1 and 2. This means that agents can restrict the conditional collision risk in some cases where the performance of one or more agents is affected. These cases are:

- Case C2 - ATC alerts are not generated.
- Case C3 - ATCo-R is not actively monitoring the traffic situation visually.
- Case C4 - ATC alerts are not generated and ATCo-R is not actively monitoring the traffic situation visually.
- Case C5 - PF-TO is not monitoring the traffic situation visually.
- Case C7 - Both PF-TO and ATCo-R are not actively monitoring the traffic situation visually.

In the second category, the risk increase factor with respect to C1 is between 60 and 556. The risk levels are high and are in the range between 1.0E-2 (case C9) and 9.5E-2 (case C16). In almost all cases of the second category, PF-TX was not actively monitoring the traffic situation (cases C9-C16). The only exception is Case C8 where the conditional collision risk is around 1.8E-2. In this case however, the performance of all other agents was affected to attain similar risk levels corresponding to cases without active monitoring of PF-TX.

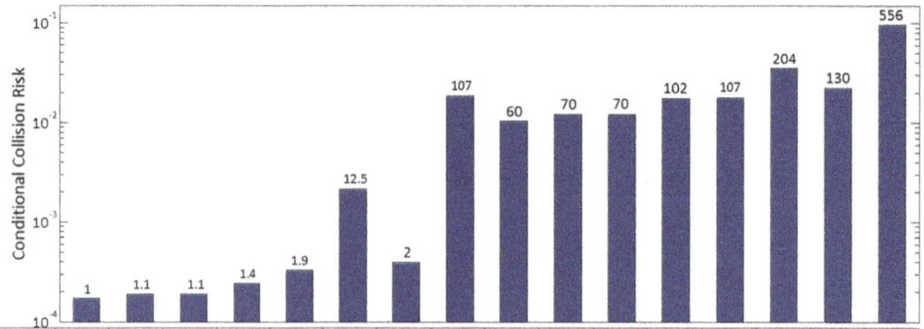

Case / Agents	C1	C2	C3	C4	C5	C6	C7	C8	C9	C10	C11	C12	C13	C14	C15	C16
PF-TX	yes	Yes	Yes	Yes	Yes	Yes	Yes	Yes	No	No	No	No	No	No	No	No
PF-TO	Yes	Yes	Yes	Yes	No	No	No	No	Yes	Yes	Yes	Yes	No	No	No	No
ATCo-R	Yes	Yes	No	No	Yes	Yes	No	No	Yes	Yes	No	No	Yes	Yes	No	No
Alerts	Yes	No	Yes	No	Yes	No	Yes	No	Yes	No	Yes	No	Yes	No	Yes	No

Figure 7 Conditional collision risk results corresponding to different cases where agents are monitoring (white cells with yes) or not monitoring (grey cells with no). The value on top of each bar represents the risk increase factor with respect to case C1[a]. PF-TX refers to Pilot-Flying Taxiing Aircraft. PF-TO refers to Pilot-Flying Taking-Off Aircraft. ATCo-R stands for Runway Controller.

Figure 7 gives some insights on the role of each agent in restricting the conditional collision risk in case the performance of other agents is affected. The role of each agent is explained below:

PF-TX

The results show that an actively monitoring PF-TX restricts the risk in cases where the performance of one or two agents is affected (cases C1 to C7). The risk increase factor with respect to case C1 is between 1.1 and 12.5. In addition, all cases without active monitoring of PF-TX, are characterized by a high conditional collision risk. The risk increase factor with respect to C1 varies between 60 and 556.

PF-TO

The conditional collision risk is increased by a factor of 1.9 in the hypothetical case C5 where PF-TO is the only agent not monitoring the traffic situation. In case C8, where no ATC alerts are generated and both PF-TO and ATCo-R are not monitoring, the risk is higher by a factor of 76 with respect to case C4. This indicates that an active monitoring of PF-TO considerably restricts the risk in cases that involve both the lack of monitoring of ATCo-R, and malfunctioning of ATC alerts (Case C4 versus C8). The risk is also reduced by the active monitoring of PF-TO (factor of 11) in case no ATC alerts are generated (Case C2 versus C6). The role of PF-TO in restricting the risk is however limited when PF-TX is not actively monitoring the traffic situation. This can be noticed when comparing cases C9 with C13, C10 with C14, C11 with C15, and C12 with C16.

ATCo-R

The comparison of cases with and without active traffic monitoring by ATCo-R indicate that in most cases, the controller's observation does not significantly restrict the risk. This can be noticed when comparing case C1 with case C3. The active monitoring by ATCo-R does reduce the risk with only a factor of 1.1. Close factors can be found when comparing cases C2 with C4, C5 with C7, C9 with C11, C10 with C12, C13 with

C15, C14 with C16. However, an interesting remark can be made when comparing case C6 with C8. Here the controller's own visual observation does play a role in restricting the conditional collision risk when the PF-TO is not monitoring the traffic situation and the ATC alerts are not functioning.

ATC alerts

The comparison of cases with and without alerts indicates that in most cases ATC alerts barely reduce the collision risk. This can be noticed when comparing cases C1 and C2. In case C1, ATC alerts are generated in case of safety-critical situations, and can be detected by ATCo-R. In case C2, No ATC alerts are generated as described in the experiment set-up. Looking at both cases, the risk increases by a factor of 1.1 meaning that ATC alerts do not significantly reduce the conditional collision risk. A similar remark can be made when comparing C3 with C4, C9 with C10, and C11 with C12. In all these cases, the presence of ATC alerts does reduce the risk by a factor no larger than 1.5. This indicates that alerts to ATCo-R are often too late resulting in late instructions to the PF-TX as manifest from C9 and C11. Only in cases without active monitoring of PF-TO and at the same time with active monitoring of PF-TX, do alerts restrict the risk by higher factors. This can be noticed when comparing cases C5 with C6, and C7 with C8. Here, the conditional collision risk is reduced by a factor of 6.5 and 46 respectively due to ATC alerts. However, if the PF-TX is out of the monitoring role, the contribution of ATC alerts in restricting the risk is reduced to a factor of 1.9 and 4.2 for the cases C13 versus C14, and C15 versus C16 respectively. These results indicate that ATC alerts are of little help to ATCo-R in cases where PF-TX is not actively monitoring the traffic situation. They are often too late to prevent collisions.

Monte Carlo simulation events

In the Monte Carlo simulation, a number of event occurrences was defined and recorded in order to have more insight regarding the interactions between agents and their relation to accident risk. These event occurrences included conflict detection by the agents, their actions, as well as the consequences of these actions. In a Monte Carlo simulation run, the time τ_q of the first occurrence of event E_q together with the position of the aircraft at time τ_q were recorded. This is helpful to understand how a collision emerges from the interactions between agents, through tracing back possible sequences of events preceding a collision. The authors came up with a novel method to visualize possible sequences of events resulting from the interactions between the agents (Figure 8). For instance, Figure 8 indicates that an ATC alert (event E4) may result in warnings specified by ATCo-R towards the flight crew of both taxiing and taking-off aircraft (event E8). The possible sequences of events with the corresponding simulated scenarios are summarized in Table 2.

Figures 9 and 10 show probability distribution functions of relevant event times for case C1 described in Figure 7. Figure 9 corresponds to simulation runs that have resulted in a collision, whereas Figure 10 corresponds to simulation runs that have not resulted in a collision. Comparing the two figures show that, for the nominal case C1, early detection of a potential conflict by the human agents play a key role in restricting the accident risk.

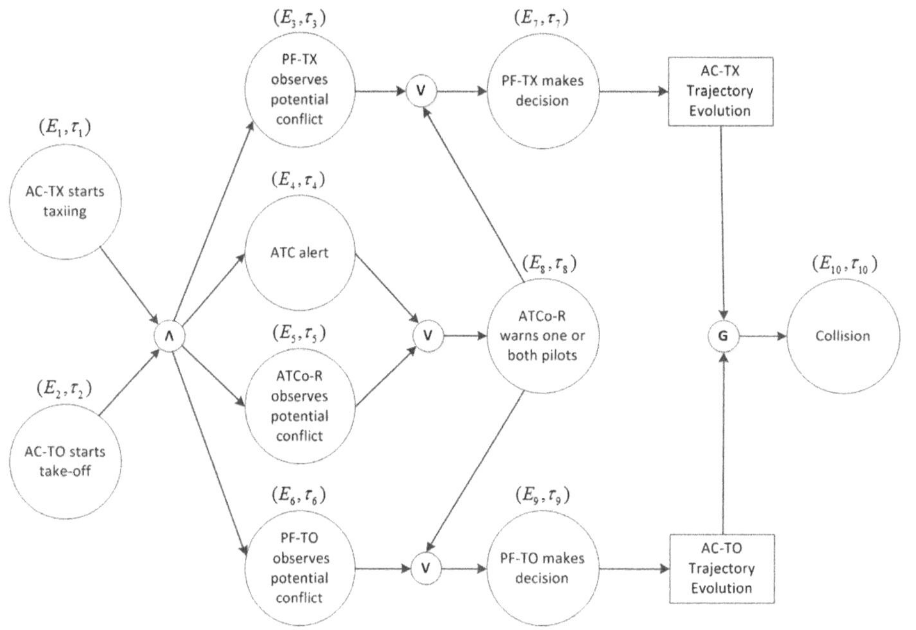

Figure 8 Recorded events in the Monte Carlo simulation and their relation with collision in the active runway crossing operation. E_q (q = 1,2,...,10) represents the event number, τ_q represents the time of the first occurrence of event E_q, '∧ ' means that both events E_i and E_j should have occurred in order for the next event to happen such that $\tau_i \wedge \tau_j = \max\{\tau_i, \tau_j\}$, '∨ ' means that as soon as one of the (competing) events occur, the next event will happen such that $\tau_i \vee \tau_j = \min\{\tau_i, \tau_j\}$, G is a guard function evaluating the first time τ_{10} that trajectories of AC-TO and of AC-TX are such close that the aircraft shapes hit each other.

Results discussion

The results suggest that if PF-TX is not actively monitoring the traffic situation, then the conditional collision risk is high, and then the role of PF-TO, ATCo-R, and alerts in restricting the risk is rather limited. This can be explained by the condition considered in this experiment regarding the intent situation awareness of PF-TX. Here, PF-TX crosses the runway without contacting ATCo-R because he is not aware about the runway crossing. PF-TX can only detect a conflict in two cases: 1) when PF-TX observes that AC-TO is approaching AC-TX (Event E3 in Figure 10) or 2) when PF-TX timely receive an R/T call from ATCo-R (Event 8 in Figure 10). This means that if PF-TX is not actively monitoring the traffic situation, the only way to detect a conflict is through an R/T

Table 2 Possible sequences of events before a collision

Seq. nr.	Possible sequences of events	High risk restriction cases	Low risk restriction cases
1	(E1 ∧ E2) → E3 → E7 → E10	C1-C7	C8
2	(E1 ∧ E2) → E6 → E9 → E10	C1-C4	C9-C12
3	(E1 ∧ E2) → E4 → E8 → E7 → E10	C1, C3, C5, C7	C9, C11, C13, C15
4	(E1 ∧ E2) → E5 → E8 → E7 → E10	C1, C2, C5, C6	C9, C10, C13, C14
5	(E1 ∧ E2) → E4 → E8 → E9 → E10	C1, C3, C5, C7	C9, C11, C13, C15
6	(E1 ∧ E2) → E5 → E8 → E9 → E10	C1, C2, C5, C6	C9, C10, C13, C14
7	(E1 ∧ E2) → E4 → E8 → (E7 ∧ E9) → E10	C1, C3, C5, C7	C9, C11, C13, C15
8	(E1 ∧ E2) → E5 → E8 → (E7 ∧ E9) → E10	C1, C2, C5, C6	C9, C10, C13, C14
9	(E1 ∧ E2) → E10	NA	C16

The table maps different possible sequences of events with corresponding cases as defined in Figure 8.

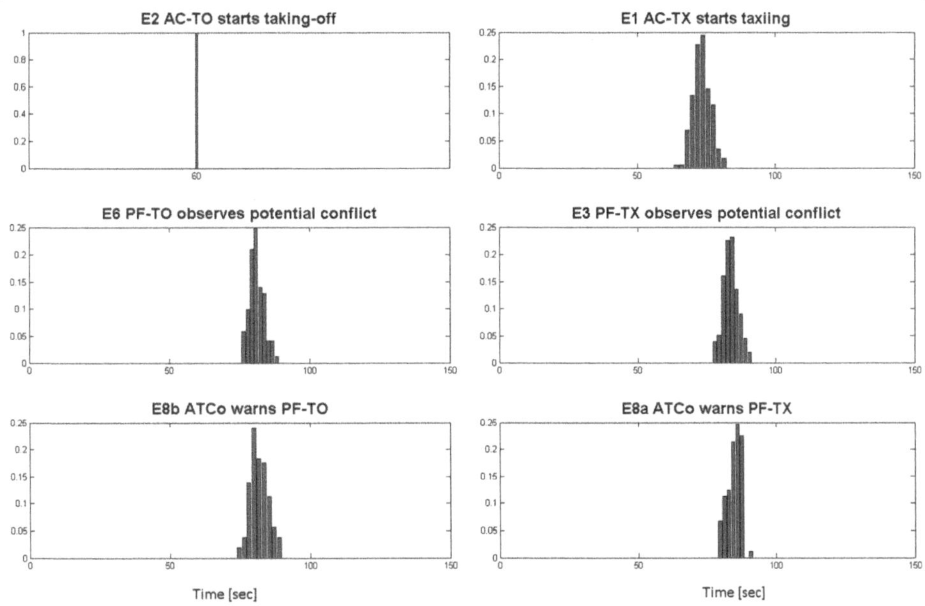

Figure 9 Conditional probability distribution functions of relevant event times for runs corresponding to case C1 that have resulted in a collision.

call from ATCo-R (Event E8). For PF-TX, this call often comes too late because AC-TX has already started crossing. A similar evaluation of the results for the PF-TO shows another picture. For PF-TO, the call from ATCo-R does restrict the collision risk considerably. This can be noticed by comparing cases C5 and C8. This is mainly because PF-TO has more time compared to PF-TX to start a collision avoidance braking action, as is manifest from the recorded aircraft positions. The comparison of both of these cases also shows that the call from ATCo-R was mainly triggered by the ATC alerting system.

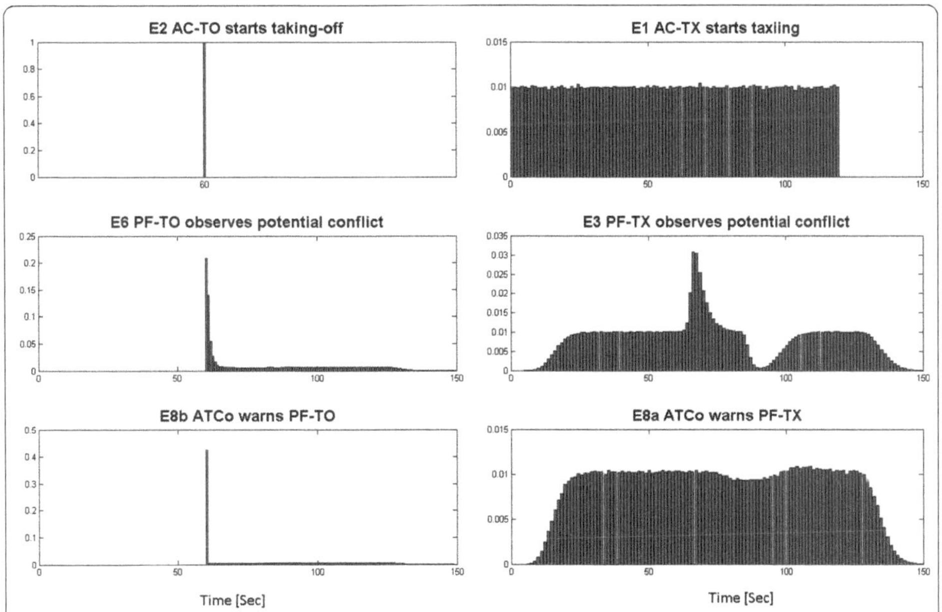

Figure 10 Conditional probability distribution functions of relevant Event times for runs corresponding to case C1 that have not resulted in a collision.

Identifying weak emergence of type II (see Table 1) was revealed due to the development and simulation of the agent-based model in the wider context of socio-technical systems. The ABMS approach covered both the socio and technical parts through: a) developing the internal model of both human operators and technical systems; and b) capturing the totality of their interactions through Monte Carlo simulations. This is an example where emergent behavior of type II has been predicted thanks to agent-based modeling and simulation (Figure 7). In addition, the events recorded in the Monte Carlo simulation revealed unpredicted behavior regarding the sequence of conflict detection by the agents. Normally, in the context of the runway crossing operation, one would expect according to the ConOps, that conflicts would be first detected by ATCo-R who will then warn the pilots. However, this might not necessarily always be the case. It was shown that PF-TX might detect a conflict even before ATCo-R which could lead to an accident if detection by the pilot is too late (e.g. Figure 9). Such unpredictable behavior can be chaotic and falls under the category multiple emergence type III. In such type of emergence, complete new roles can appear while old ones disappear.

Conclusion

Emergent behavior in the open socio-technical air transportation system results from the interactions between the constituent elements and the operating environment. For an airport, these elements include human operators and their organizations, working procedures, technical systems, and weather. Changing the characteristics of these elements or the way they interact may have an impact on the overall system behavior. Both the operating environment and the socio-technical system elements influence the evolution over time. On the one hand, weather conditions for instance may be a factor in deciding which flight plan to file or which runway configuration to use. On the other hand, flight progress may change as a result of pilots and controller's decisions. In order to identify which changes at the "elements" level do have a significant impact on the emergent behavior of the socio-technical system, models of constituent elements and their interactions are required.

In this paper, Monte Carlo Simulations of an agent-based model were performed to assess the safety risk of an active runway crossing operation. The safety performance of such an operation depends on the coordination between the runway controller, pilots, and technology all functioning together. The agent-based model has been developed in a hierarchical way. At the highest level, the relevant agents including human operators and technical systems were identified. Then the interactions between these agents were captured and included deterministic and stochastic relationships, as is appropriate for the human performance or the technical systems considered. The impact on the emergent safety behavior was studied through running simulations of both nominal and off-nominal scenarios.

It has been explained that ATM is a complex socio-technical system that exhibits various types of emergent behavior ranging from simple emergence, through weak emergence, up to strong emergence. It has been demonstrated that Agent-Based Modeling and Monte Carlo Analysis provide a platform to integrate and simulate multiple heterogeneous components at different levels, and identify different types of emergence through modeling the interactions between technical systems, human operators, working procedures, and environment.

One of the main findings is that the role of alerting systems in restricting the conditional collision risk is limited in cases without active monitoring of the PF-TX. Another interesting result is that although the role of the controller's own observation does not restrict the risk in some cases, it plays a significant role in cases without active monitoring of PF-TO and no ATC alerts being generated. These findings are revealed due to the development and simulation of the agent-based model that covers the totality of interactions of components and their variability in performance over time. The Monte Carlo simulations make it possible to understand the potential of agents in restricting the risk in off-nominal scenarios, through capturing their stochastic nature.

Endnote

[a]In this paper, the aggregated simulation included extra combinations compared to (Stroeve et al. 2013). This way the role of ATC alerts in restricting the risk becomes clearer in cases where the controller is not actively monitoring the traffic situation visually (e.g. cases C3, C7, C11, and C15).

Competing interests

The authors declare that they have no competing interests.

Authors' contributions

SB: has been lead author in drafting the manuscript, has conducted additional research in positioning the runway crossing application in the broader context of ABMS and emergent behaviour, has been running the Monte Carlo simulations and has done the interpretation of the results, and has revised the paper for important intellectual content. HB: has made key contributions to the paper conception as well as to the paper writing and the interpretation of the results, provided continuous support to the first author during all stages of the paper drafting, and gave final approval of the version to be published. RC: has significantly supported the first author during regular meetings, critically revised the paper for important intellectual content, provided new ideas, and gave final approval of the version to be published. ME: has significantly contributed to the research of philosophical discussion in literature on emergent behaviour, critically revised the paper, and gave approval of the version to be published. All authors read and approved the final manuscript.

Acknowledgments

The first author would like to acknowledge SESAR Joint Undertaking for partly funding and supporting his research through a ComplexWorld PhD project; National Aerospace Laboratory NLR for making the TOPAZ TAXIR toolset for runway crossing operations available to conduct this study; Ir. Bert Bakker (NLR) for learning the first author how to use the TOPAZ TAXIR toolset in running the MC simulations and in analyzing the results; Dr. Sybert Stroeve (NLR)for valuable discussions on the agent based models used within the TOPAZ TAXIR tool set; and Prof. Jacco Hoekstra (Delft University of Technology) for stimulating discussions on emergent behavior.

Author details

[1]Delft University of Technology, Faculty of Aerospace Engineering, Kluyverweg 1, 2629, HS Delft, The Netherlands.
[2]National Aerospace Laboratory NLR, Anthony Fokkerweg 2, 1059, CM Amsterdam, The Netherlands.

References

Ball M, Barnhart C, Nemhauser G, Odoni A: **Air transportation: irregular operations and control.** In *Handbook in Operations Research and Management Science, Vol. 14.* Edited by Barnhart C, Laporte G. Amsterdam: The Netherlands: Elsevier; 2007:1–73.

Bar-Yam Y: *Dynamics of Complex Systems.* Boulder, Colorado: Westview Press; 2003.

Bar-Yam Y: **About Engineering Complex Systems: Multiscale Analysis and Evolutionary Engineering.** In *Engineering Self-Organizing Systems: Methodologies and Applications.* Edited by Brueckner SA. Springer; 2005:16–31.

Beart P: *Emergent Behaviour;* 2012. http://beart.org.uk/?page_id=63

Bedau MA: **Weak Emergence.** *Philosophical Perspectives* 1997, **11**:375–399.

Bedau M: **Downward causation and the autonomy of weak emergence.** *draft for prinipia* 2002.

Blom HAP, Bakker GJ, Blanker PJG, Daams J, Everdij MHC, Klompstra MB: **Accident Risk Assessment for Advanced Air Traffic Management.** In *Air Transportation Systems Engineering.* Edited by Donohue GL, Zellweger AG. Progress in Astronautics and Aeronautics: AIAA; 2001a:463–480.

Blom HAP, Daams J, Nijhuis HB: **Human Cognition Modeling in Air Traffic Management Safety Assessment.** In *Air Transportation Systems Engineering.* Edited by Donohue GL, Zellweger AG. Progress in Astronautics and Aeronautics: AIAA; 2001b:481–511.

Blom HAP, Stroeve SH, de Jong HH: **Safety Risk Assessment by Monte Carlo Simulation of Complex Safety-critical Systems.** In *Developments in Risk-Based Approaches to Safety: Proceedings of the Fourteenth Safety-Critical Systems Symposium.* Edited by Redmill F, Anderson T. Bristol, UK: Springer; 2006.

Blom HAP, Bakker GJ, Krystul J: **Rare Event Estimation for a Large-Scale Stochastic Hybrid System with Air Traffic Application.** In *Rare event simulation using Monte Carlo methods.* Edited by Rubino G, Tuffin B. J.Wiley; 2009:193–214.

Borshchev A, Filippov A: **From System Dynamics and Discrete Event to Practical Agent-Based Modeling: Reasons, Techniques, Tools.** In *Proceedings of the 22nd International Conference of the System Dynamics Society.* Oxford, England; 2004.

Bouarfa S, Blom HAP, Curran R: **Airport Performance Modeling using an Agent-Based Approach.** In *Proceedings of the Air Transport and Operations Symposium.* The Netherlands: Delft; 2012.

Bouarfa S, Blom HAP, Curran H, Everdij MHC: **Agent-Based Modeling and Simulation of Emergent Behaviour in Air Transportation.** In *proceedings of the Aviation Technology, Integration, and Operations (ATIO).* LA, California: USA; 2013. forthcoming.

Burmeister B, Haddadi A, Matylis G: **Applications of multi-agent systems in traffic and transportation.** *IEE Transactions on Software Engineering* 1997, **144**(1):51–60.

Casti JL: *Would-be Worlds: How simulation is changing the frontiers of science.* New York: John Wiley; 1997.

CAS Wiki: *Emergence.*; 2013. Available at [http://wiki.cas-group.net/index.php?title=Emergence].

Chalmers DJ: *Varieties of emergence.* Granada: Templeton foundation workshop on emergence; 2002.

Chan WKV, Son YJ, Macal CM: *Agent-Based Simulation Tutorial – Simulation of Emergent Behavior and Differences between Agent-Based Simulation and Discrete-Event Simulation.* Baltimore, MD, USA: Proc. Of the Winter Simulation Conference; 2010.

Chen B, Cheng H: **A review of the applications of agent-technology in traffic and transportation systems.** *IEEE Trans Intell Transp Syst* 2010, **11**(2):485–497.

Corker KM, Muraoka K, Verma S, Jadhav A, Gore BF: **Air MIDAS A Closed-Loop Model Framework.** In *Foyle and Hooey, Human performance modeling in aviation.* Boca Raton (FL) USA: CRC Press; 2008.

Crick F: *The astonishing hypothesis: The scientific search for the soul.* New York: Charles Scribner's Sons; 1994.

Deregee E: *Non-Stop Schiphol.* Schiphol, The Netherlands: Contrail Publishers; 2006.

Ek A, Akselsson R, Arvidsson M, Johansson CR: **Safety Culture in Swedish Air Traffic Control.** *Safety Science* 2007, **45**(7):791–811.

Endsley MR: **Toward a theory of situation awareness in dynamic systems.** *Hum Factors* 1995, **37**(1):32–64.

European commission and Eurocontrol: *European Operational Concept Validation Methodology (E-OCVM) version 3.0. Volume II.* Brussels, Belgium: Eurocontrol; 2010.

Everdij MHC, Blom HAP: **Hybrid State Petri Nets which have the Analysis Power of Stochastic Hybrid Systems and the Formal Verification Power of Automata.** In *P. Pawlewski, Petri Nets, Chapter 12.* Vienna: I-Tech Education and Publishing; 2010:227–252.

Everdij MHC, Blom HAP, Stroeve SH: **Structured Assessment of Bias and Uncertainty in Monte Carlo Simulated Accident Risk.** In *Proc. 8th Int. Conf. on Probabilistic Safety Assessment and Management (PSAM8).* New York, USA: American Society of Mechanical Engineers; 2006.

Everdij MHC, Scholte JJ, Blom HAP, Stroeve SH: *An investigation of emergent behaviour viewpoints in literature, and their usability in Air Traffic Management.* Seville, Spain: 1st ComplexWorld Annual Conference; 2011.

FAA: *Runway Incursion Totals by quarter FY2012 vs. FY2011*; 2012. Available at [http://www.faa.gov/airports/runway_safety/statistics/year/?fy1=2012&fy2=2011].

Forrester JW: **Counterintuitive Behavior of Social Systems.** *Technol Rev* 1971, **73**:52–68.

Franklin S, Graesser A: **Is it an agent, or just a program?: a taxonomy for autonomous agents.** In *Intelligent Agents 3, Lecture Notes in Artificial Intelligence, Volume 1193.* Edited by Muller JP, Wooldridge MI, Jennings NR. Berlin: Springer; 1997.

Fromm J: *Types and Forms of Emergence*; 2005.

Fron X: **Performance Review in Europe.** In *Air Transportation Systems Engineering.* Edited by Donohue GL, Zellweger AG. AIAA: Progress in Astronautics and Aeronautics; 2001:49–59.

Goldstein J: **Emergence as a construct: History and Issues.** *Emergence: Complexity and Organization* 1999, **1**(1):49–72.

Holland JH: *Emergence: From Chaos to Order.* Reading, MA: Addison-Wesley; 1997.

Holland JH: **Studying Complex Adaptive Systems.** *Journal of Systems Science and Complexity* 2006, **19**:1–8.

Hollnagel E: *Human Reliability Analysis: Context and Control.* London Academic Press; 1993.

Hollnagel E: *The ETTO Principle: Why things that go right sometimes go wrong.* Aldershot, UK: Ashgate Publishing; 2009.

Hollnagel E, Woods DD, Leveson N: *Resilience Engineering: Concepts and Precepts.* Aldershot: Ashgate Publishing Ltd; 2006.

ICAO: *Advanced Surface Movement Guidance and Control Systems (A-SMGCS) Manual. Doc 9830 AN/452.* 1st edition; 2004.

ICAO: *Manual on the Prevention of Runway Incursions. Doc 9870 AN/463.* 1st edition; 2007.

Jennings NR: **On Agent-Based Software Engineering.** *Artificial Intelligence* 2000, **177**(2):277–296.

Klein G, Feltovich P, Bradshaw JM, Woods DD: **Common Ground and Coordination in Joint Activity.** In *Organizational Dynamics in Cognitive Work.* Edited by Rouse W, Boff K. New York: Wiley; 2004.

Leveson NG, Barrett B, Carrol J, Cutcher-Gershenfeld J, Dulac N, Zipkin D: *Modeling, Analyzing, and Engineering NASA's Safety Culture. Phase 1 Final Report*; 2005.

Lewes GH: *Problems of life and mind.* london: Truebner; 1874-1879.

Mill SJ: *A system of logic ratiocinative and inductive.* London: John W. Parker and Son; 1872:1843.

PRC: **An Assessment of Air Traffic Management in Europe during the Calendar Year 2008.** *Eurocontrol* 2009.

PRC, ATOS PBU: *U.S./Europe Comparison of ATM-related Operational Performance.* Brussels, Belgium: FAA and Eurocontrol; 2009.

Railsback SF, Grimm V: *Agent-based and Individual-based Modeling: A practical Introduction.* Princeton, New Jersey: Princeton University Press; 2012.

Russel S, Norvig P: *Artificial Intelligence: a Modern Approach.* 2nd edition. Hong Kong: Pearson Education Asia Limited and Tsinghua Univ. Press; 2006.

Sanz R: *Converging trends in complex software-intensive control.* Faro, Portugal: 6th Portuguese conference on automatic control; 2004.

Consortium SESAR: *Air Transport Framework. The Performance Target D2.* The European Commission and Eurocontrol: Brussels, Belgium; 2006.

SESAR Consortium: *SESAR Definition Phase, SESAR Safety Management Plan (SMP).WP4.2/Task 4.2.1, Part of DLT-0710-421-01-00*; 2007.

Shah AP, Pritchett AR, Feigh KM, Kalaver SA, Jadhav A, Corker KM, Holl DM, Bea RC: *Analyzing air traffic management systems using agent-based modeling and simulation.* Baltimore: Proc. 6th USA/Europe ATM R&D Seminar; 2005.

Sharpanskykh A, Stroeve SH: **An Agent-Based Approach for Structured Modeling Analysis and Improvement of Safety Culture.** *Comput Math Organ Theory* 2011, **17**:77–117. Springer US.

Stroeve SH, Blom HAP, van der Park MNJ: *Multi-Agent Situation Awareness Error Evolution in Accident Risk Modeling.* Budapest, Hungary: Proc. 5th USA/Europe Air Traffic Management R&D Seminar; 2003.

Stroeve SH, Blom HAP, Bakker GJ: **Systemic accident risk assessment in air traffic by Monte Carlo simulation.** *Safety Science* 2009, **47**(2):238–249.

Stroeve SH, Blom HAP, Bakker GJ: *Contrasting Safety Assessments of a Runway Incursion Scenario by Event Sequence Analysis versus Multi-Agent Dynamic Risk Modelling.* Berlin, Germany: Proceedings of 9th USA/Europe Air Traffic Management Research and Development Seminar; 2011.

Stroeve SH, Blom HAP, Bakker GJ: **Contrasting Safety Assessments of a Runway Incursion Scenario: Event Sequence Analysis versus Multi-Agent Dynamic Risk Model.** *In Reliability Engineering and System Safety* 2013, **109**:133–149.

Sussman J: *Collected Views on Complexity in Systems. Course materials for ESD.04J Frameworks and Models in Engineering Systems*; 2007a. MIT OpenCourseWare; (http://ocw.mit.edu), Massachusetts Institute of Technology. Downloaded on [24 12 2009].

Sussman J: *The 'Clios Process' A User's Guide. Course materials for ESD.04J Frameworks and Models in Engineering Systems*; 2007b. MIT OpenCourseWare; (http://ocw.mit.edu), Massachusetts Institute of Technology. Downloaded on [24 12 2009].

Tessier C, Chaudron L, Muller HJ (Eds): *Conflicting, Conflict Management in Multi-Agent Systems.* Kluwer Academic Publishers; 2002.

Weick K: **The vulnerable system: An analysis of the Tenerife air disaster.** *J Manag* 1990, **16**:571–593.

Wickens CD: *Engineering, Psychology, and Human Performance.* Columbus: Merril; 1992.

Wikipedia: *Tenerife Airport Disaster*; 2012a. http://en.wikipedia.org/wiki/Tenerife_airport_disaster, retrieved 27 September 2012.

Wikipedia: *Emergence*; 2012b. http://en.wikipedia.org/wiki/Emergence. Retrieved 27 September 2012.

Wolfe S, Jarvis P, Enomoto F, Sierhuis M, Putten B: **A Multi-Agent Simulation of Collaborative Air Traffic Flow Management, Chapter 18.** In *Multi-Agent Systems for traffic and transportation engineering.* Edited by Bazzan A, Klugl F. x: IGI Global Publishing; 2009:357–381.

Wolfram S: **Universality and Complexity in Cellular Automata.** *Physica D: Nonlinear Phenomena* 1984, **10**(1–2):1–35.

Woods DD, Dekker S, Cook R, Johannesen L, Sarter N: *Behind Human Error.* 2nd edition. England: Ashgate Publishing Limited; 2010.

Wooldridge M: *An Introduction to Multi-Agent Systems.* 2nd edition. Ltd, Publication: John Wiley and Sons; 2009.

11

Penetration capacity of the wood-decay fungus *Physisporinus vitreus*

Matthias Jörg Fuhr[1,2], Mark Schubert[2*], Chris Stührk[1,2], Francis WMR Schwarze[2] and Hans Jürg Herrmann[1]

* Correspondence:
Mark.Schubert@empa.ch
[2]EMPA, Swiss Federal Laboratories for Materials Science and Technology, Wood Laboratory, Group of Wood Protection and Biotechnology, Lerchenfeldstrasse 5, St Gallen CH-9014, Switzerland
Full list of author information is available at the end of the article

Abstract

Purpose: Bioincising is a biotechnological process for improving the permeability of refractory wood such as Norway spruce heartwood using the wood-decay fungus *Physisporinus vitreus*. The degradation of the bordered pit membranes by *P. vitreus* in its first stage of wood colonization enhances the uptake of preservatives and wood modification substances, whereas the strength of the material is not significantly reduced.

Methods: We propose to study bioincising by means of a mathematical model, because many factors affect the growth and effects of *P. vitreus* in Norway spruce in such a complex way that an evaluation of the optimal incubation conditions (i.e. water activity, temperature or pH) is very expensive or even not possible solely using laboratory experiments.

Results: Using a hyphal growth model we demonstrate here for the first time how to optimize bioincising by linking the microscopic growth behavior of *P. vitreus* with macroscopic system properties of the wood. Moreover, we propose universal measures of wood-decay fungi, i.e., penetration velocity, penetration work and penetration capacity, which may figure as measures for the efficiency of wood colonization. For example, our simulation shows that an increase of the hyphal growth rate (i.e. changing the incubation conditions) from 1 to 2 $\mu m{\cdot}d^{-1}$ results in an increase of the mycelium's growth velocity from 0.8 to 1.75 $\mu m{\cdot}d^{-1}$ and an increase of the penetration capacity from 0.5 to 0.6 $10^{-3}{\cdot}mm^2{\cdot}d^{-1}$ using a pit degradation rate of 2 $\mu m{\cdot}d^{-1}$.

Conclusions: Information about the penetration velocity, penetration work and penetration capacity is of significance for both its biotechnological use and the study of the colonization strategy of wood-decay fungi in general.

Keywords: Filamentous fungus, Fungal colony, Mycelia modeling, Bioincising, Picea abies, Bordered pit

Background

The primary goal of a modeling framework for the optimization of bioincising is to analyze the influence of biotic and abiotic factors on the wood's permeability, which is in combination with the wood's strength, the quality criterion of bioincised wood (i.e. a higher permeability results in a higher treatability of the wood). The chemical composition (including environmental factors such as the water activity, temperature and type of medium) and the physical structure (i.e. the distribution of tracheids and pits) of the resource determine the penetration behavior of a wood-decay fungus as shown in Figure 1. Thereby, wood-decay fungi of the class basidiomycetes and ascomycetes

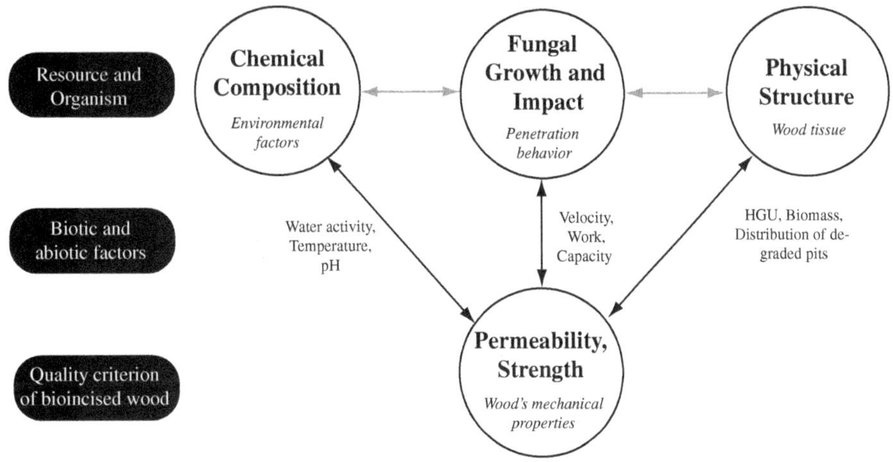

Figure 1 ***Background*: Resource, Organism, biotic and abiotic factors influence the quality of bioincised wood.** We propose to study the optimization of the biotechnological process bioincising by means of mathematical models, because of the complexity of the system (fungus - wood).

colonize wood by forming a root-like network called mycelium. This highly interconnected structure consists of elongated tubular cells called hyphae that explore and colonize a substrate by branching. Characteristics for such filamentous organisms are their highly dynamic and adaptive response to their environment (Rayner & Boddy 1988). Thereby, the processes governing the growing fungus take place on multiple scales from the nanoscopic (e.g., uptake of nutrients) to the macroscopic level (e.g., transport of nutrients) (Fuhr et al. 2011a). It is this complexity that is a challenge for the biotechnological use of fungi such as *P. vitreus*.

Using a discrete modeling approach called the fungal growth model (FGM), in previous studies the metabolism of *P. vitreus* (Fuhr et al. 2011b) and the effect of the wood tissue on its growth (Fuhr et al. 2011a) have been analyzed, whereas the penetration behavior was not studied in detail until now (Figure 1). Thus, the focus of this work is to study the rule of the bordered pits to obtain a complete modeling framework covering all relevant effects for the optimization of the permeability of bioincised wood sample.

The degradation of the bordered pit membranes in its first stage of Norway spruce (*Picea abies* (L.) H. Karst.) wood colonization might be the main reason for the significant increase of the permeability of treated wood (Schwarze et al. 2006). This biotechnological process, which is termed bioincising, can be used to improve the uptake of wood preservatives and wood modification substances (Schwarze & Schubert 2009; Schwarze & Schubert 2011). In this context several studies concerning *P. vitreus* addressed the anatomy of treated Norway spruce wood (Fuhr et al. 2011a; Lehringer et al. 2010; Stührk et al. 2010; Fuhr et al. 2012), the influence of environmental factors to its radial growth rate (Fuhr et al. 2011b; Schubert & Schwarze 2009; Schubert et al. 2010) and the alteration of wood properties (Spycher et al. 2008; Schwarze et al. 2008; Lehringer et al. 2011a; Lehringer et al. 2011b). Despite their potential, the methods mentioned are unable to illustrate the influence of microscopic effects such as the pit degradation rate on the macroscopic behavior of the fungus, because wood is an opaque material and the *in vivo* observation of processes inside a wood sample has not been possible until now. Therefore, we propose the use of mathematical models to study and optimize the penetration behavior of *P. vitreus*, since its growth behavior was successfully simulated by FGM, even in complex physical and chemical structured environments (Fuhr et al. 2011a).

The pits are valve-like structures between the voids within the wood (i.e. pores or lumens) and regulate the transport of nutrients in the tree. For example, in the living wood (i.e. sapwood) of Norway spruce the bordered pits are permeable for liquids. Whereas, in the dead part of the tree (i.e. heartwood) they are closed and lignified (Liese & Bauch 1967). It is this closing, called aspiration, of the bordered pits that makes efficient treatment of this wood species impossible without energy- and cost-intensive technical processing.

The degradation of the bordered pit membranes in their first stage of growth appears to be a common strategy of wood-decay fungi to colonize softwoods (Schwarze et al. 2006; Liese & Schmid 1961; Daniel 1994; Green et al. 1995; Green & Highley 1997). The size, structure and distribution of bordered pits is a significant factor for the accessibility of wood tissue both for fluids and fungal hyphae (Rayner & Boddy 1988). The time required by a hypha to penetrate through a bordered pit either by the support of pressure or enzymes (i.e. pit penetration time) obviously influences the ability of the mycelium to colonize a refractory wood such as Norway spruce heartwood.

The material is presented in three parts. In the first step, we discuss the growth pattern of *P. vitreus* using the FGM and laboratory experiments (Figures 2 and 3). In the second step, we use an analytical model to analyze the penetration velocity of *P. vitreus* to quantify the key factors determining the growth and expansion of the mycelium in wood modeled by the FGM (Figure 4). Moreover, such an analytical model enables us to discuss various modes of pit penetration by *P. vitreus*, i.e. pit penetration by pressure or enzymes (Table 1). However, the analytical model does not provide information about the effects of *P. vitreus* on wood tissue, e.g. the distribution of degraded pits. Therefore, we use the FGM to discuss the penetration work and the penetration capacity of *P. vitreus*

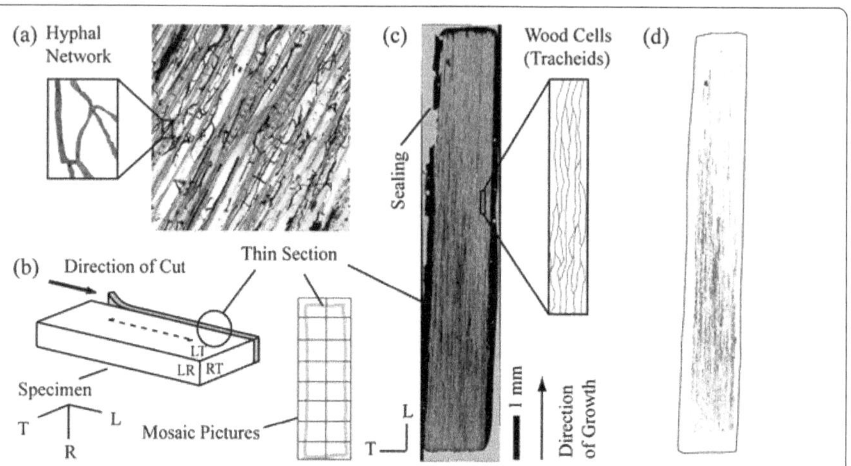

Figure 2 ***Light Microcopy*****: (a) Wood-decay fungi such as the white-rot *Physisporinus vitreus* exploit a wood tissue by forming a root-like network (mycelium) consisting of filaments (hyphae) with a diameter of approximately 2 µm.** This hyphal network is visible under the light microscope. Specimens are usually cut into thin section as shown in (**b**), in order to analyze the distribution of mycelium in the wood by a sequence of light microscopy pictures. **(c)** These single pictures are stitched together and form a mosaic picture covering a whole thin section. **(d)** Using digital image processing one can identify the fungus and analyze mycelium's morphology in a whole sample. The principal directions of wood are usually termed as longitudinal, radial and tangential, denoted by the letters L, R and T.

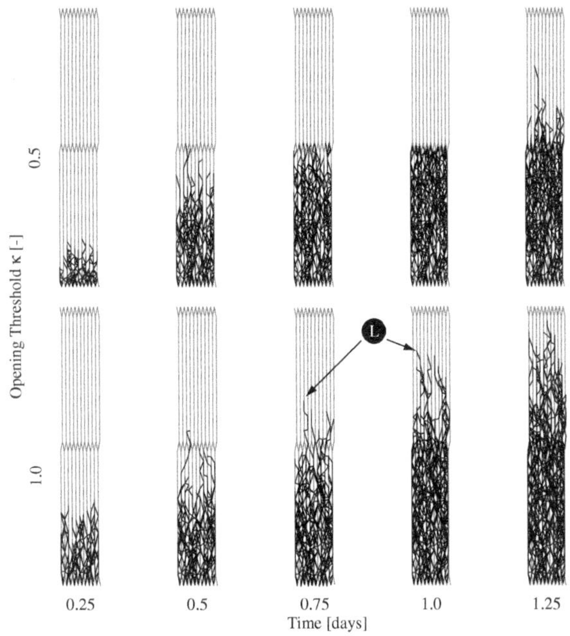

Figure 3 ***Growth pattern.*** The evolution of the mycelium begins from the left longitudinal face by using the model parameters given in Table 2. Fixed boundaries are imposed on all faces. The fungus reaches the right longitudinal face after approximately 32 and 42 hours for $\kappa = 1.0$ and 0.5 (Eq. 1) respectively. *L* denotes leading hyphae.

(Figures 5 and 6). We propose these universal measures for wood-decay fungi, because common network measures such as the hyphal growth unit (Plomley 1958) do not provide information about the effects of a fungus on its resource.

Methods

Specimen material

The white-rot basidiomycete *Physisporinus vitreus* EMPA 642 was cultivated in 9 cm Petri dishes on 2% malt extract agar (MEA) at 23°C and 70% relative humidity (RH). After 2 weeks we placed sterilized (121°C, 20 min and 2 bar vapor pressure) Norway spruce (*Picea abies* (L.) H. Karst.) samples with a size of approximately 10 mm (longitudinal) × 2 mm (radial) and 10 mm (tangential) on the mycelium. The faces of the samples, with the exception of one radial tangential face, were sealed with a topcoat (Nuvovern ACR Emaillack, Walter Mäder AG, Killwangen, Switzerland) by brushing, so that the *P. vitreus* colonized the wood via the unsealed side in a longitudinal direction (Figure 2a).

Light microscopy

After 10 days of incubation at 23°C and 70% RH we cut thin sections of approximately 30 μm thickness from the wood samples using a microtome as shown in Figure 2b. We stained the fungus with lactophenol blue, took mosaic images with a pixel size of approximately 0.65 μm × 0.65 μm using a Zeiss LSM 510-NLO (Figure 2c) and analyzed the images by digital image processing (Figure 2d).

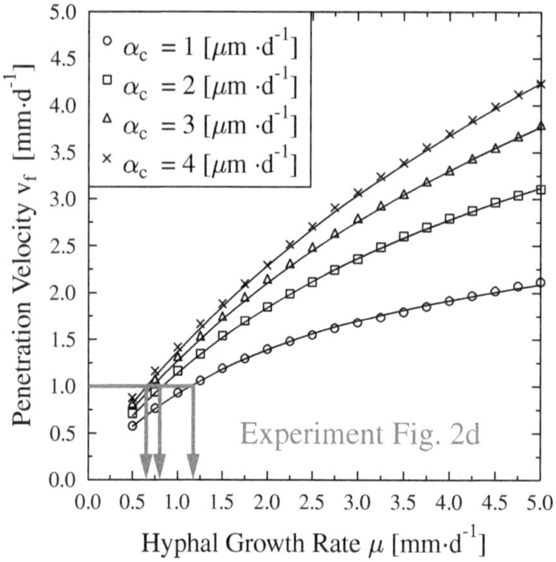

Figure 4 ***Penetration velocity.*** The hyphal growth rate μ plotted versus the growth velocity (Eq. 8). The symbols represent simulations for α_c = 1, 2, 3 and 4 μm·d^{-1} using the parameters of Table 2 and the tracheid framework of Figure 7a. The solid lines are the corresponding analytical models (Eq. 12a) using Eq. 12a – 12d and N_t = 2.65, p = 0.5 and q = 3. Each symbol represents the average over 1000 realizations and the uncertainty of the data points is within the range of the symbols.

Hyphal growth model

To analyze the penetration of the wood-decay fungus *P. vitreus* we use a two-dimensional discrete modeling approach in which both the mycelium and the nutrients are considered discrete structures. The present FGM is a two-dimensional version of the simulation scheme developed by Fuhr et al. (2011a). Typical model parameters used throughout this work are given in Table 2.

Inspired by the approach of Comstock (1970; Siau 1984) we model the Norway spruce tracheids as elongated rectangles with tapered ends. The shape of the tracheids is determined by their length l_L, width l_T and the length of the overlapping zone l_O (Figure 7a) where most of the bordered pits are located (Sirviö & Kärenlampi 1998). The diameter of a bordered pit is described by the parameter D. These pits play an important role in the growth of many wood-decay fungi (Schwarze et al. 2006; Liese & Schmid 1961; Daniel 1994; Green et al. 1995; Green & Highley 1997), because they are the shortest path for a wood-decay fungus in a longitudinal direction as shown by the dotted line in Figure 7a. The nutrient points are randomly uniformly distributed along the cell walls of the tracheids with the density ρ mm^{-1}.

Table 1 ***Pit degradation*****: Pit degradation rates α_c (enzymatic, Eq. 3) and α_p (pressure, Eq. 7) evaluated for the pit penetration times $\tau = \tau_c = \tau_p$ = 0.5, 1, 1.5 and 2 days**

Pit penetration time [d]τ	Pit degradation rate [μm/d]	
	α_c (κ = 0.2)	α_p (β$_1$ = 0.175, β$_2$ = −1.5)
0.5	4	2.77
1	3	0.87
1.5	2	0.44
2	1	0.27

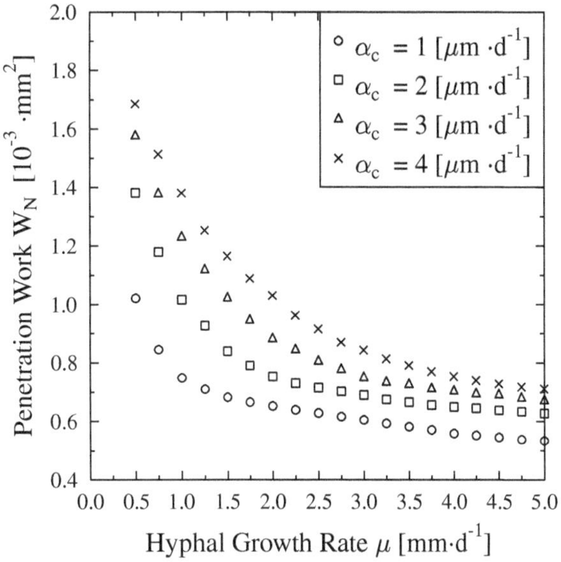

Figure 5 ***Penetration work.*** The hyphal growth rate μ plotted versus the penetration work (Eq.9) of *P. vitreus*. The symbols represent simulations for α_c = 1, 2, 3 and 4 $\mu m\cdot d^{-1}$ using the same simulation setup as in Figure 4. Each symbol represents on average over 1000 realizations and the uncertainty of the data points is within the size of the symbols.

Edges, nodes and tips form the mycelium of the fungus, whereby the position of the nodes is restricted to the position of the nutrient points as shown in Figure 7b. The characteristic growth of the fungus *P. vitreus*, which degrades the initially closed bordered pit membranes in its first stage of wood colonization, is simulated by a pit-to-pit growth of the hyphae. Thereby the mean growth rate of a hypha is given by μ. The evolution of the mycelium is driven by key processes such as the uptake and transport of nutrients, branching and polarization of the hyphae.

At each iteration step m of the algorithm the mycelium may grow by one edge. Thereby the length of the new edge and the angle measured to its predecessor is restricted by the growth cut-off length ξ and the growth cut-off angle θ. The mean edge length is λ. Each extension of the mycelium costs the fungus a specific amount of nutrients depending on the length of the edge. The fungus accumulates these nutrients in the nodes by degrading the pits at each time step initially for $m = m_0$ by α_I and for $m > m_0$ by α_c, where m_0 is the time when a new node is created at a pit. Branching occurs if the amount of nutrients at a node exceeds a certain level β_t (at a tip node) and β_s (at an interior node). A tip node is a node with one edge, whereas an interior node has more than one edge.

The cell walls limit the available set of nutrient points within a tracheid, in contrast to the unrestricted growth of the mycelium in Petri-dishes (Fuhr et al. 2011b). In the present model, an active hypha can only reach all pits within the same tracheid, but according to Fuhr et al. (2011a), we assume that a hypha is able to penetrate through the degraded torus into an adjacent tracheid if

$$F_j^{(m))} < \kappa, \qquad (1)$$

where F_j, the amount of nutrients at point (j) and $\kappa \in [0,\nu]$, denotes a specific pit opening threshold. Details about the model construction are given in (Fuhr et al. 2011a).

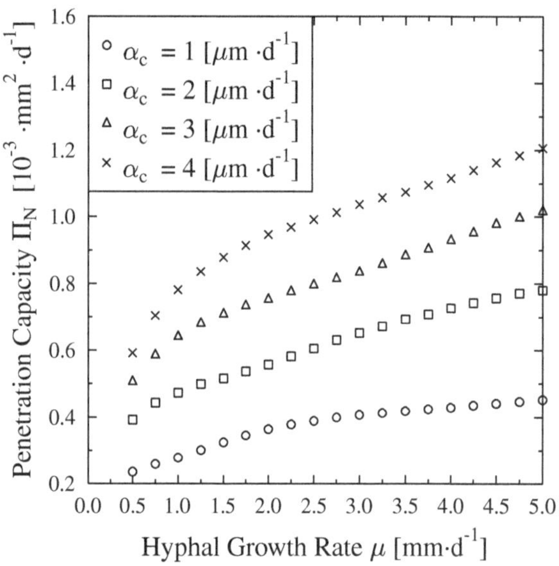

Figure 6 ***Penetration capacity.*** The hyphal growth rate μ plotted versus the penetration capacity (Eq. 10) of *P. vitreus*. The symbols represent simulations for α_c = 1, 2, 3 and 4 μm·d^{-1} using the same simulation setup as in Figure 4. Each symbol represents on average over 1000 realizations and the uncertainty of the data points is within the range of the symbols.

The simulation begins by placing n_k^0 starting nodes, called pellets, with an initial nutrient concentration n_n^0 on the longitudinal face of the wood specimen (Figure 7a). All pits are initially closed (i.e. $F_j > \kappa$) and their initial amount of nutients is ν. The unit of the resource that sustains fungal growth, called pit nutrient, is given in micrometers (μm), because we prefer to express the fungal activity (Eq. 1) with a measure, which is observable in laboratory experiments, e.g. holes in the torus of the bordered pits (directly) or a permeability (indirectly). However, the pit nutrient, given in micrometers, can be converted into other quantities of interest. For example, the mass of degraded lignin may be calculated by multiplying the pit nutrient by the degraded area given in μm^2 times the density of the lignin given in microgram per volume [μg/μm^3].

Analytical growth model

The shortest time T_s of a hypha to grow in a longitudinal direction from a tracheid into an adjacent one, indicated in Figure 7a by the dotted line connecting the points 1 and 3 (by the super indices of T) may be given by

$$T_S^{(1-3)} = \frac{l_S}{2\cdot\mu_{12}} + \frac{l_S}{2\cdot\mu_{23}} + \tau, \tag{2}$$

where μ_{ij} is the growth velocity of a hyphae (i.e. hyphal growth rate) within a tracheid measured between the points *(i)* and *(j)* (see Figure 7a), l_S is the length of the shortest path and τ is the pit penetration time, i.e. the time required for a hypha to grow through a bordered pit. Generally, there are two modes of pit penetration by a hypha.

Table 2 ***Model parameters*****: typical model parameters used throughout this work**

Wood:	Symbol	Value	Unit
Tracheids	$[l_L, l_T, l_O]$	[2.5, 0.08, 0.04]	mm
Density of nutrient point	ρ	40	1*/^d-1
Diameter of bordered pit	D	10	μm
Fungus:			
Mean hyphal growth rate	μ	1	l*d^-1
Mean edge length	λ	2/3*ξ	l
Growth cut-off length	ξ	0.125	l
Growth cut-off angle	θ	90	°
Growth costs	$[a,b]$	[0, 0]	-
Pit initial nutrient	v	D	μm
Pit initial degradation rate	a_I	0	-
Pit degradation rate	a_C	4	μm*d^-1
Pit opening	κ	0.1	-
Apical branching threshold	β_t	0.6*v	-
Lateral branching threshold	β_s	0.35*v	-
Simulation:			
Initial number of pellet	n_k^0	60	-
Initial number of tips	n_s^0	1	-
Initial nutrient concentration	n_n^0	3/2*β_t	-

We use the same notation as in Fuhr et al. (2011a).

In the first case the torus is eroded by enzymes, e.g. *P. vitreus* (Schwarze et al. 2006). Whereas, in the second case a hypha breaks through the torus by mechanical pressure, e.g. blue stain fungi (Liese & Schmid 1961), which is typically indicated by cracks in the lignified torus. We indicate the mode of penetration by lower indices, i.e. enzymatic erosion (c) and pressure (p).

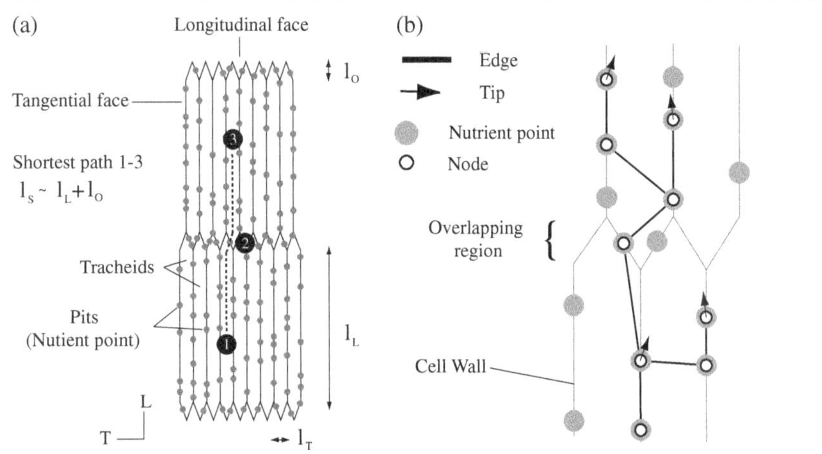

Figure 7 ***Model description.*** **(a)** A longitudinal-tangential section of Norway spruce wood is characterized by the cellular structure of the tracheids. The principal directions of wood are usually termed as longitudinal, radial and tangential, denoted by the letters L, R and T. The pits are the substrate for the fungus (i.e. nutrient points) and distributed with a density ρ on the substratum of the fungus (i.e. tracheid cell walls) by using uniform random numbers. **(b)** The mycelium is modeled as an assemblage of edges, nodes and tips. The evolution of the mycelium is characterized by a pit-to-pit growth of the fungus. Details about the model construction can be found in Fuhr et al. (Fuhr et al. 2011a).

For the first case we may write for the pit penetration time

$$\tau_c = \frac{D \cdot \kappa}{\alpha_c}, \tag{3}$$

where D is the diameter of a bordered pit and $\kappa \in [0, 1]$ is an opening threshold. Thus, $D \cdot \kappa$ is the diameter of a hole, which is prerequisite for a vegetative hypha to penetrate through the torus. α_c is the pit degradation rate, given in $\mu m \cdot d^{-1}$, to dissolve the torus.

For the second case, the pit penetration by pressure, we assume that the fungus bores a hole in the torus with a constant rate α_p, i.e. the diameter d of a hole in the torus after t days is given by

$$d(t) = \alpha_p \cdot t. \tag{4}$$

Bardage and Daniel (1998) studied the penetration of seven fungal strains through two different types of membranes with various pore sizes. They found that the ability of fungi to penetrate micropores depends on time, i.e. smaller pores require longer penetration times. They reported penetration times between 1 and 10 days for micropores of size 0.6 – 0.2 μm. Whereby, no fungus was able to penetrate micropores smaller than 0.2 μm within 15 days. Based on these experiments, we assume that the relation between the diameter of a hole d and the penetration time t is roughly given by the power law

$$\frac{t}{t_{max}} = \beta_1 \left(\frac{d}{d_{max}} \right)^{\beta_2}, \tag{5}$$

using t_{max} = 10 d, d_{max} = 0.6 μm, β_1 = 0.175 and β_2 = −1.5 as shown in Figure 8. Therefore, the diameter of a hole after τ_p days (Eq. 4) is equal to a hole that is prerequisite for penetration after τ_p days (inserting Eq. 4 into Eq. 5), i.e.

$$\frac{\tau_p}{t_{max}} = \beta_1 \left(\frac{\alpha_p \cdot \tau_p}{d_{max}} \right)^{\beta_2}. \tag{6}$$

Solving Eq. (6) for the penetration time τ_p leads to

$$\tau_p = \left[t_{max} \cdot \beta_1 \cdot \left(\frac{\alpha_p}{d_{max}} \right)^{\beta_2} \right]^{\frac{1}{1-\beta_2}} \tag{7}$$

For example, a hypha bores a hole into a cell wall by α_p = 0.25 μm per day. After approximately 2.12 days the diameter of the hole is approximately 0.53 μm (Eq. 4). This hole size corresponds to a penetration time of 2.12 days (inserting d = 0.53 μm into Eq. 5). Thus, the time of fungal activity at the cell wall is equal to the penetration time and therefore the hypha is able to grow through the bored hole. We directly obtain the penetration time τ_p of approximately 2.12 days by inserting α_p = 0.25 into Eq. 7.

Penetration velocity, work and capacity

For biotechnological applications of wood-decay fungi, it is important to understand their capability to penetrate into a specific direction of a wood tissue in a certain amount of time. Therefore, we define the penetration velocity of a fungus (or hypha) as

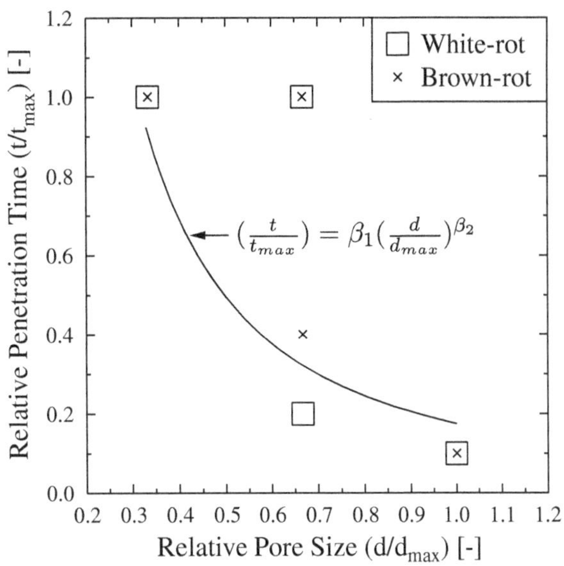

Figure 8 ***Penetration Time*****: Pore size plotted versus penetration time of wood-decay fungi using d_{max} = 0.6 μm and t_{max} = 10 days.** The solid line represents $\beta_1 = 0.175$ and $\beta_2 = -1.5$. Experimental data reproduced after Bardage and Daniel (Bardage & Daniel 1998).

$$v_f = \frac{Penetration\ depth}{Time}. \tag{8}$$

Furthermore, we define the penetration work of a fungus as

$$W_Q = Quantity{\cdot}Penetration\ depth, \tag{9}$$

where the subscript letter Q denotes a quantity of interest, e.g. the total hyphal length of the mycelium, the total number of open pits or the mass loss of the wood. In addition, we propose the penetration capacity of a fungus as

$$\Pi_Q = \frac{Penetration\ Work}{Time}. \tag{10}$$

Obviously, the penetration capacity of a fungus strongly depends on factors such as temperature, water activity and pH, the microclimate, the presence of wood preservatives and other factors, since wood decay-fungi are sensitive to their environment. The scalar Π_Q may figure as a measure for the efficiency of a specific fungus to colonize a certain wood tissue.

For simplicity, throughout this work, we use the terms 'penetration work' and 'penetration capacity' for the specific penetration work and penetration capacity using the total amount of degraded pit nutrients as the quantity of interest.

Results

Growth pattern

After 10 days of incubation, we observe that the fungus starting from the unsealed longitudinal face penetrates approximately 10 mm into the wood sample (Figure 2d). This corresponds to a penetration velocity of approximately 1 mm·d^{-1}. The morphology of the growth front is characterized by hyphae growing at different rates (i.e. leading hyphae). The density of the mycelium at the unsealed end is much higher than on the growth front. We measure that a hypha crosses 3–4 tapered ends from the unsealed to the opposite face.

Figure 3 shows the evolution of the mycelium over a time span of 36 hours for $\kappa = 0.5$ and 1.0 (see Eq. 1) using the model parameters of Table 2. The system consists of 20 tracheids with 1070 pits in total. The fungus begins growing from the left longitudinal face and fixed boundaries are imposed to all faces. For $\kappa = 1.0$ the fungus grows without any resistance by the pit membranes and after approximately 32 hours the first hypha reaches the longitudinal face on the right side, whereas the fungus requires in the second case (i.e. $\kappa = 0.5$) approximately 42 hours for the same distance. The total hyphal length of the mycelium is approximately 130 mm for both conditions, whereas the number of tips is approximately 350 and 500 for $\kappa = 1.0$ and 0.5 respectively. In addition, the model shows that for $\kappa = 0.5$ *P. vitreus* requires approximately 15 per cent more nutrients to reach the right longitudinal face.

Penetration velocity

Figure 4 shows the velocity of the mycelium v_f (Eq. 8) evaluated for hyphal growth rates between 0.5 and 5 mm·d^{-1} and pit degradation rates between 1 and 4 μm·d^{-1} (Eq. 3) using the model parameters of Table 2 and the tracheid framework of Figure 7a. The system consists of 2 (longitudinal) × 10 (tangential) tracheids and fixed boundaries are imposed to all faces. The fungus begins growing from the lower longitudinal face and the velocity is evaluated by measuring the time of the mycelium between points 1 and 3. The numerical results of these experiments are shown in Figure 4 by the symbols ○ (α_c =1 μm·d^{-1}), □ (α_c = 2 μm·d^{-1}), Δ (α_c = 3 μm·d^{-1}) and × (α_c = 4 μm·d^{-1}).

We observe that the penetration velocity v_f increases with increasing values of μ and α_c. For example, at a hyphal growth rate of $\mu = 0.5$ the relative difference between the penetration velocity at $\alpha_c = 1$ and $\alpha_c = 4$ is smaller than 1, whereas at $\mu = 5$ the difference is larger than 2. The influence of the pit degradation rate on the relationship between the hyphal growth rate and the penetration velocity seems to be nonlinear. It would be interesting to know more about this relationship, because the hyphal growth rate depends on environmental factors and may allow us to optimize the process of bioincising. Thus, the next two paragraphs develop such a relationship.

In a first step, we assume that the penetration velocity between the points 1 and 2 differs from the penetration velocity between the points 2 and 3 (see Figure 7a), i.e. before and after the overlapping area. The overlapping area consists of the tapered ends of the tracheids (see Figure 7b), where most of the pit are located. Thus, according to Eq. 2 and Eq. 8 we write for the velocity of the mycelium between the points 1 and 3

$$v_f = \frac{l_S}{T_s^{1-3}} = \frac{l_S}{\frac{l_S}{2}\left(\frac{1}{\mu_{12}} + \frac{1}{\mu_{23}}\right) + \tau}. \qquad (11)$$

Second, we assume that the hyphal growth rate of the leading hypha (Figure 7) before (μ_{12}) and after (μ_{23}) the overlapping area is faster than the rest of the mycelium by a factor μ^* (before) and $q{\cdot}\mu^*$ (after) respectively. In addition, the penetration time depends on the number of nodes per pits N_t, i.e. the more nodes per pit the higher the pit degradation. For the pit penetration time τ, we may write the hyphal growth rate μ_{12} and μ_{23} and the factor μ*

$$\tau = \frac{\tau_c}{N_t} = \frac{D \cdot \kappa}{N_t \cdot \alpha_c}, \tag{12a}$$

$$\mu_{12=\mu+\mu^*}, \tag{12b}$$

$$\mu_{23=\mu+q\cdot\mu^*}, \tag{12c}$$

$$\mu^* = N_t \cdot \alpha_c \cdot \left(\mu \cdot \frac{\tau}{L_S}\right)^p. \tag{12d}$$

The solid lines in Figure 4 show Eq. 11 using Eq. 12a, 12b, 12c, 12d and N_t = 2.65, q=3 and p=0.5. The velocity of 1 mm·d^{-1} measured from the experiment of Figure 2d corresponds approximately to hyphal growth rates μ (α_c) = 0.66 (4), 0.73 (3), 0.84 (2) and 1.14 (1) mm·d^{-1}. The pit degradation rates α_c = 1, 2, 3 and 4 μm·d^{-1} correspond to pit penetration times τ_c = 2, 1, 0.67 and 0.5 days (Eq. 12a) and pit degradation rates by pressure α_p = 2.77, 0.87, 0.44 and 0.27 μm·d^{-1} as shown in Table 1.

Penetration work and penetration capacity

Figure 5 and Figure 6 show the penetration work W_N (Eq. 9) and the penetration capacity Π_N (Eq. 10) of *P. vitreus* measuring as quantity Q the amount of degraded nutrients given in micrometer. Since we measure the penetration depth in millimeters, the unit of the penetration work and the penetration capacity are 10^{-3}·mm^2 and 10^{-3}·mm^2·d^{-1} respectively. A penetration capacity of 1.0 10^{-3}·mm^2·d^{-1} means that the growth front of the mycelium penetrating 1 mm·d^{-1} into the wood degrades 1 micrometer pit membranes in total. We use the same simulation setup as in Figure 4. We observe that the penetration work decreases with increasing hyphal growth rates, whereas the penetration capacity increases with increasing hyphal growth rates. In addition, higher pit degradation rates result in a higher penetration work and penetration capacity.

Discussion

The growth of physisporinus vitreus

The penetration behavior of *P. vitreus* into Norway spruce heartwood is supposed to be characterized by a stepwise capture of wood tissue (Fuhr et al. 2011a), because the aspirated and lignified bordered pits hinder the expansion of the mycelium (Figure 2). Only the degrading of either the bordered pits or the cell wall enables the fungus to grow from one tracheid to another. Thereby the ratio between the velocity of the hypha within the tracheids (i.e. hyphal growth rate) and the pit degradation rate is of interest, because this ratio influences the density of the mycelium, the number of tips in the system and the penetration velocity of the growth front as shown by our simulations (Figures 3 and 4).

Rays are not considered in the model, because they affect fungal growth in radial direction. The present study focuses on the penetration of the mycelium in longitudinal direction as shown in Figure 3.

Penetration velocity

The FGM assumes that the hyphal growth rate (μ) and the pit degradation rate (α_c) are the key factors for the colonization of Norway spruce wood in its first stage of growth. Using the analytical model (Eq. 11), we are able to quantify the influence of both factors on the penetration velocity of *P. vitreus*. The results suggest that a doubling of the hyphal growth rate enables *P. vitreus* to reduce the pit degradation rate by a factor 4 (Figure 4), to reach a penetration velocity of 1 mm·d^{-1} (Figure 2d). The hyphal growth rate is mainly influenced by the water activity, temperature and pH (Schubert et al. 2010), whereas the effect of environmental factors on the ability of *P. vitreus* to penetrate the bordered pits is unknown. Thus, changing the incubation conditions offers an optimization of the bioincising process (see next section 'Optimization of bioincising').

The measured hyphal growth rates between approximately 0.5 and 1.5 mm d^{-1} (Figure 4) are in good agreement with *in vivo* experiment of *P. vitreus* at standard conditions (Stührk 2011) and therefore confirm our model assumption. We are able to estimate for the first time the pit degradation rate of *P. vitreus*. Our simulations in combination with laboratory experiments show that the pit degradation rates (α_c and α_p) are approximately 1 to 4 and 0.3 – 8 μm d^{-1} for a penetration either by enzymes or by pressure (Table 1).

The penetration time τ_p is based on the of the experiments of Bardage and Daniel (1998). However, the influence of environmental effects such as temperature and pH on the penetration behavior of hyphae of white-rot fungi such as *P. vitreus* in wood may not covered by the experiments of Bardage and Daniel (Bardage & Daniel 1998) and may limit the extension of our model besides standard conditions. Thus, more quantitative and qualitative experimental studies about the penetration of hyphae through pores are needed.

Optimization of bioincising

We use a discrete modeling approach to study the biotechnological process of bioincising, because such a model provides information about the effects of *P. vitreus*, e.g. the amount of degraded pits, and therefore enables an optimization of the bioincising process. For example, our simulation shows that an increase of the hyphal growth rate from 1 to 2 μm·d^{-1} results in an increase of the growth velocity of the mycelium from 0.8 to 1.75 μm·d^{-1}, a decrease of the penetration work from 1 to 0.75 10^{-3}·mm^{-2} and an increase of the penetration capacity from 0.5 to 0.6 10^{-3}·mm^{2}·d^{-1} using a pit degradation rate of 2 μm·d^{-1} (Figures 4, 5 and 6). A penetration work of 1 to 0.75 shows that, for the tracheid framework given in Figure 7a, the growth front of the mycelium penetrating 1 mm into the wood degrades approximately 1 to 0.75 micrometer pit membranes in total (see Figure 5). Thereby, a penetration capacity of 0.5 to 0.6 10^{-3}·mm^{2}·d^{-1} indicates that the growth front of the mycelium penetrating 1 mm·d^{-1} into the wood degrades approximately 0.5 to 0.6 micrometer pit membranes in total, which is a measure for the permeability of the wood (Fuhr et al. 2011a). Moreover, the penetration capacity may figure as a measure for the efficiency of wood-decay fungi to colonize wood, since a high pit degradation rate may facilitate the capture of their resource. Thus, it would be interesting to measure and compare various penetration capacities from several wood-decay fungi, e.g. choosing as quantity of interest the biomass, amount of degraded pits or permeability.

Conclusions

The last step would be an optimization of bioincising on a larger scale using the modeling framework as described above. On a macroscopic scale, the influence of the distribution of pellets, which are the inocula on the surface of wood blocks, is of interest because the penetration velocity of *P. vitreus* in radial (i.e. rays) and longitudinal (i.e. tracheids) direction is much higher than in tangential direction. Thus, combining environmental factors with the amount and distribution of the inocula on the wood surface will help assist designing incubation conditions that are required to induce a certain degree of wood permeability by *P. vitreus.*

Abbreviations

EMPA: Swiss federal laboratories for materials and technology; FGM: Fungal growth model; MEA: Malt extract agar; pH: Potential hydrogenium; RH: Relative humidity.

Competing interests

The authors declare no conflict of interest.

Authors' contributions

MF carried out the analytical model and the simulations using the FGM, and participated in the LM experiments. MF and MS drafted the manuscript. CS carried out the LM experiments and helped to draft the manuscript's experimental section. MF, MS, FS and HS conceived of the study, and participated in its design and coordination. HH and FS helped to draft the manuscript. All authors read approved the final manuscript.

Authors' information

There are no relevant information about the authors for the interpretation of the article.

Acknowledgements

The authors gratefully acknowledge J. Hehl (ETH Light Microscopy Center) for helpful discussions on image processing and express their gratitude to the Swiss National Science Foundation (SNF No. 205321-121701) for its financial support.

Author details

[1]ETH Zurich, Institute for Building Materials, Computational Physics for Engineering Materials, Schafmattstrasse 6, HIF E18, Zurich CH-8093, Switzerland. [2]EMPA, Swiss Federal Laboratories for Materials Science and Technology, Wood Laboratory, Group of Wood Protection and Biotechnology, Lerchenfeldstrasse 5, St Gallen CH-9014, Switzerland.

References

Bardage SL, Daniel G: **The ability of fungi to penetrate micropores: implications for wood surface coatings.** *Material und Organismen* 1998, **31**:233–245.
Comstock G: **Directional permeability of softwoods.** *Wood and Fiber Science* 1970, **1**:283–289.
Daniel G: **Use of electron microscopy for aiding our understanding of wood biodegradation.** *FEMS Microbiol Rev* 1994, **13**:199–233.
Fuhr MJ, Schubert M, Schwarze FWMR, Herrmann HJ: **Modelling the hyphal growth of the wood-decay fungus Physisporinus vitreus.** *Fungal Biol* 2011a, **115**:919–932.
Fuhr MJ, Stührk C, Schubert M, Schwarze F, Herrmann HJ: **Modelling the effect of environmental factors on the hyphal growth of the basidiomycete Physisporinus vitreus.** *J Basic Microbiol* 2011b, **52**:523–530.
Fuhr MJ, Stührk C, Münch B, Schwarze F, Schubert M: **Automated quantification of the impact of the wood-decay fungus physisporinus vitreus on the cell wall structure of norway spruce by tomographic microscopy.** *Wood Sci Technol* 2012, **46**:769–779.
Green F, Highley T: **Mechanism of brown-rot decay: paradigm or paradox1.** *Int Biodeterior Biodegradation* 1997, **39**:113–124.
Green F, Tschernitz JL, Kuster TA, Highley TL: **Hydrolysis of bordered pits during colonization of conifers by brown-rot fungi.** In *26th Annual meeting Helsingør, Denmark 11–16 june 1995.* Helsingør: International Research Group on Wood Preservation; 1995.
Lehringer C, Hillebrand K, Richter K, Arnold M, Schwarze F, Militz H: **Anatomy of bioincised Norway spruce wood.** *Int Biodeter Biodegr* 2010, **64**:346–355.
Lehringer C, Koch G, Adusumalli RB, Mook WM, Richter K, Militz H: **Effect of physisporinus vitreus on wood properties of norway spruce. Part 1: aspects of delignification and surface hardness.** *Holzforschung* 2011a, **65**:711–719.
Lehringer C, Saake B, Zivkovic V, Richter K, Militz H: **Effect of physisporinus vitreus on wood properties of Norway spruce. Part 2: aspects of microtensile strength and chemical changes.** *Holzforschung* 2011b, **65**:721–727.
Liese W, Bauch J: **On the closure of bordered pits in conifers.** *Wood Science and Technology* 1967, **1**:1–13.
Liese W, Schmid R: **Licht- und elektronenmikroskopische Untersuchungen über das Wachstum von Bläuepilzen in Kiefer- und Fichtenholz.** *Holz als Roh- und Werkstoff* 1961, **19**:329–337.
Plomley NJB: **Formation of the colony in the fungus Chaetomium.** *Aust J Biol Sci* 1958, **12**:53–64.

Rayner ADM, Boddy L: *Fungal decomposition of wood: Its biology and ecology.* Chichester, UK: John Wiley & Sons; 1988.
Schubert M, Schwarze F: **Response surface analyse (RSM) der wirkung kombinierter parameter auf das wachstum von physisporinus vitreus.** *Holztechnologie* 2009, **50:**16–20.
Schubert M, Mourad S, Schwarze F: **Radial basis function neural networks for modeling growth rates of the basidiomycetes Physisporinus vitreus and Neolentinus lepideus.** *Appl Microbiol Biotechnol* 2010, **85:**703–712.
Schwarze FWMR, Schubert M: **Enhance uptake of wood modification agents in 'bioincised' wood.** *40th Annual IRG Meeting Beijing, China* 2009, **IRG/WP 09–40445**.
Schwarze FWMR, Schubert M: **Physisporinus vitreus: a versatile white rot fungus for engineering value-added wood products.** *Appl Microbiol Biotechnol* 2011, **92:**431–440.
Schwarze FWMR, Landmesser H, Zgraggen B, Heeb M: **Permeability changes in heartwood of Picea abies and Abies alba induced by incubation with Physisporinus vitreus.** *Holzforschung* 2006, **60:**450–454.
Schwarze FWMR, Spycher M, Fink S: **Superior wood for violins wood decay fungi as a substitute for cold climate.** *New Phytol* 2008, **179:**1095–1104.
Siau JF: *Transport processes in wood.* Verlag Berlin Heidelberg New York: Springer; 1984.
Sirviö J, Kärenlampi P: **Pits s natural irregularities in softwood fibers.** *Wood and Fiber Science* 1998, **30:**27–39.
Spycher M, Schwarze F, Steiger R: **Assessment of resonance wood quality by comparing its physical and histological properties.** *Wood Science and Technology* 2008, **42:**325–342.
Stührk C: *Tomographische visualisierung des hyphensystems von physisporinus vitreus in fichtenholz (picea abies) zur optimierung des bioincising prozesses.* Gallen: EMPA St; 2011.
Stührk C, Fuhr M, Schubert M, Schwarze F: **Analyzing hyphal growth of the bioincising fungus Physisporinus vitreus with light-, confocal laser scanning- and, Synchrotron X-ray tomographic microscopy.** *41st Annual Meeting Biarritz, France* 2010, **IRG/WP 10–20438**.

Permissions

All chapters in this book were first published in CASM, by Springer; hereby published with permission under the Creative Commons Attribution License or equivalent. Every chapter published in this book has been scrutinized by our experts. Their significance has been extensively debated. The topics covered herein carry significant findings which will fuel the growth of the discipline. They may even be implemented as practical applications or may be referred to as a beginning point for another development.

The contributors of this book come from diverse backgrounds, making this book a truly international effort. This book will bring forth new frontiers with its revolutionizing research information and detailed analysis of the nascent developments around the world.

We would like to thank all the contributing authors for lending their expertise to make the book truly unique. They have played a crucial role in the development of this book. Without their invaluable contributions this book wouldn't have been possible. They have made vital efforts to compile up to date information on the varied aspects of this subject to make this book a valuable addition to the collection of many professionals and students.

This book was conceptualized with the vision of imparting up-to-date information and advanced data in this field. To ensure the same, a matchless editorial board was set up. Every individual on the board went through rigorous rounds of assessment to prove their worth. After which they invested a large part of their time researching and compiling the most relevant data for our readers.

The editorial board has been involved in producing this book since its inception. They have spent rigorous hours researching and exploring the diverse topics which have resulted in the successful publishing of this book. They have passed on their knowledge of decades through this book. To expedite this challenging task, the publisher supported the team at every step. A small team of assistant editors was also appointed to further simplify the editing procedure and attain best results for the readers.

Apart from the editorial board, the designing team has also invested a significant amount of their time in understanding the subject and creating the most relevant covers. They scrutinized every image to scout for the most suitable representation of the subject and create an appropriate cover for the book.

The publishing team has been an ardent support to the editorial, designing and production team. Their endless efforts to recruit the best for this project, has resulted in the accomplishment of this book. They are a veteran in the field of academics and their pool of knowledge is as vast as their experience in printing. Their expertise and guidance has proved useful at every step. Their uncompromising quality standards have made this book an exceptional effort. Their encouragement from time to time has been an inspiration for everyone.

The publisher and the editorial board hope that this book will prove to be a valuable piece of knowledge for researchers, students, practitioners and scholars across the globe.

List of Contributors

Előd Méhes
Department of Biological Physics, Eötvös University, Budapest, Hungary

Tamás Vicsek
Department of Biological Physics, Eötvös University, Budapest, Hungary
Statistical and Biological Physics Research Group of the Hungarian Academy of Sciences, Budapest, Hungary

Anita Sobe
Institute of Networked and Embedded Systems/Lakeside Labs, Alpen-Adria-Universität Klagenfurt, Klagenfurt, Austria
Computer Science Department, University of Neuchâtel, Neuchâtel, Switzerland

Wilfried Elmenreich
Institute of Networked and Embedded Systems/Lakeside Labs, Alpen-Adria-Universität Klagenfurt, Klagenfurt, Austria
Complex Systems Engineering, Universität Passau, Passau, Germany

Konglin Zhu and Xiaoming Fu
Institute of Computer Science, University of Goettingen, Goettingen, Germany

Wenzhong Li
Institute of Computer Science, University of Goettingen, Goettingen, Germany
State Key Laboratory for Novel Software and Technology, Nanjing University, Nanjing, China

Aditya Kurve and Khashayar Kotobi
EE Department, The Pennsylvania State University, University Park, PA 16802, USA

George Kesidis
CSE and EE Department, The Pennsylvania State University, University Park, PA 16802, USA

Michael J North
Decision and Information Sciences Division, Argonne National Laboratory, 9700 South Cass Avenue, Lemont, IL 60439, USA
University of Chicago, 5801 South Ellis Avenue, Chicago, IL 60637, USA

Marshall Abrams
Department of Philosophy, University of Alabama at Birmingham 900 13th Street South, HB 414A Birmingham, AL 35294-1260, USA

Terry Bossomaier
Centre for Research in Complex Systems, Charles Sturt University, Panorama Ave, Bathurst NSW 2795, Australia

Lionel Barnett
Department of Informatics, University of Sussex, Falmer, Brighton BN1 9QH, UK

Michael Harré
Department of Civil Engineering, University of Sydney, Sydney, NSW, Australia

Fanny Boulaire and Robin Drogemuller
Queensland University of Technology, Gardens Point - P Block level 8, Brisbane Qld 4000, Australia

Mark Utting
University of the Sunshine Coast, Sippy Downs QLD 4556, Australia
Work undertaken at QUT, now at University of the Sunshine Coast, Sippy Downs QLD 4556, Australia

Mark Abdollahian and Zining Yang
School of Politics and Economics, Claremont Graduate University, 170 E. Tenth St. Claremont, CA 91711, USA

Travis Coan
School of Law, Harvard University, 1563 Massachusetts Ave. Cambridge, MA 02138, USA

Birol Yesilada
Hatfield School of Government, Portland State University, 1825 SW Broadway Portland, OR 97201, USA

Soufiane Bouarfa and Richard Curran
Delft University of Technology, Faculty of Aerospace Engineering, Kluyverweg 1, 2629, HS Delft, The Netherlands

Henk AP Blom
Delft University of Technology, Faculty of Aerospace Engineering, Kluyverweg 1, 2629, HS Delft, The Netherlands
National Aerospace Laboratory NLR, Anthony Fokkerweg 2, 1059, CM Amsterdam, The Netherlands

Mariken HC Everdij
National Aerospace Laboratory NLR, Anthony Fokkerweg 2, 1059, CM Amsterdam, The Netherlands

Matthias Jörg Fuhr
ETH Zurich, Institute for Building Materials, Computational Physics for Engineering Materials, Schafmattstrasse 6, HIF E18, Zurich CH-8093, Switzerland
EMPA, Swiss Federal Laboratories for Materials Science and Technology, Wood Laboratory, Group of Wood Protection and Biotechnology, Lerchenfeldstrasse 5, St Gallen CH-9014, Switzerland

Mark Schubert
EMPA, Swiss Federal Laboratories for Materials Science and Technology, Wood Laboratory, Group of Wood Protection and Biotechnology, Lerchenfeldstrasse 5, St Gallen CH-9014, Switzerland

Chris Stührk
ETH Zurich, Institute for Building Materials, Computational Physics for Engineering Materials, Schafmattstrasse 6, HIF E18, Zurich CH-8093, Switzerland
EMPA, Swiss Federal Laboratories for Materials Science and Technology, Wood Laboratory, Group of Wood Protection and Biotechnology, Lerchenfeldstrasse 5, St Gallen CH-9014, Switzerland

Francis WMR Schwarze
EMPA, Swiss Federal Laboratories for Materials Science and Technology, Wood Laboratory, Group of Wood Protection and Biotechnology, Lerchenfeldstrasse 5, St Gallen CH-9014, Switzerland

Hans Jürg Herrmann
ETH Zurich, Institute for Building Materials, Computational Physics for Engineering Materials, Schafmattstrasse 6, HIF E18, Zurich CH-8093, Switzerland